IGCSE
Chemistry

Second
Edition

IGCSE
Chemistry

Second
Edition

**Bryan Earl
and Doug Wilford**

HODDER
EDUCATION
AN HACHETTE UK COMPANY

International hazard warning symbols

You will need to be familiar with these symbols when undertaking practical experiments in the laboratory.

Corrosive
These substances attack or destroy living tissues, including eyes and skin.

Oxidising
These substances provide oxygen which allows other materials to burn more fiercely.

Harmful
These substances are similar to toxic substances but less dangerous.

Toxic
These substances can cause death.

Irritant
These substances are not corrosive but can cause reddening or blistering of the skin.

Highly flammable
These substances can easily catch fire.

Orders: please contact Bookpoint Ltd, 130 Milton Park, Abingdon, Oxon OX14 4SB. Telephone: (44) 01235 827720. Fax: (44) 01235 400454. Lines are open 9.00–5.00, Monday to Saturday, with a 24-hour message answering service. Visit our website at www.hoddereducation.co.uk

© Bryan Earl and Doug Wilford 2002
First published in 2002 by
Hodder Education, an Hachette UK Company
338 Euston Road
London NW1 3BH

This second edition published 2009

Impression number 5 4 3 2
Year   2012 2011 2010 2009

Cover photo © Andrew Brookes/Corbis
Illustrations by Wearset Ltd
Typeset in Bembo 11.5pt by Fakenham Photosetting Ltd
Printed in Dubai

A catalogue record for this title is available from the British Library

ISBN 978 0 340 98188 7

Contents

Preface to the reader xii
IGCSE Chemistry syllabus and textbook matching grid xiii

1 All about matter

1

Solids, liquids and gases 2

The kinetic theory of matter 2
Explaining the states of matter

Changes of state 4
An unusual state of matter
An unusual change of state
Heating and cooling curves

Diffusion – evidence for moving particles 6
Brownian motion

Gas laws 7
Charles' Law
Boyle's Law
Combining the gas laws

Checklist 9

Additional questions 10

2 Elements, compounds and mixtures

11

Elements 12
Atoms – the smallest particles
Molecules

Compounds 15
More about formulae
Balancing chemical equations
Instrumental techniques

Mixtures 17
What is the difference between mixtures and compounds?

Separating mixtures 18
Separating solid/liquid mixtures
Separating liquid/liquid mixtures
Separating solid/solid mixtures
Criteria for purity

Gels, sols, foams and emulsions 25

Mixtures for strength 27
Composite materials

Checklist 29

Additional questions 31

3 Atomic structure and the periodic table

32

Inside atoms   **32**
Proton number and nucleon number
Ions
Isotopes
The mass spectrometer
Relative atomic mass

The arrangement of electrons in atoms   **36**

The periodic table   **37**
Electronic structure and the periodic table
Group 1 – the alkali metals
Group 2 – the alkaline earth metals
Flame colours
Group 7 – the halogens
Group 0 – the noble gases
Transition elements
The position of hydrogen

Checklist   **49**

Additional questions   **50**

4 Bonding and structure

51

Ionic bonding   **51**
Ionic structures
Properties of ionic compounds
Formulae of ionic substances

Covalent bonding   **56**
Other covalent compounds
Covalent structures
Properties of covalent compounds
Allotropy
Allotropes of carbon

Glasses and ceramics   **64**
Glasses
Ceramics

Metallic bonding   **65**
Properties of metals

Checklist   **66**

Additional questions   **67**

5 Chemical calculations

68

Relative atomic mass	68
Reacting masses	68
Calculating moles	70
Moles and elements	
Moles and compounds	
Moles and gases	
Moles and solutions	
Calculating formulae	73
Finding the formula	
Moles and chemical equations	75
Checklist	77
Additional questions	78

6 Electrolysis and its uses

79

Electrolysis of lead(II) bromide	80
Electrolysis of aluminium oxide	81
Anodising	
Extraction of sodium	84
Electrolysis of aqueous solutions	85
Electrolysis of water	
Chlor-alkali industry	
Tests for aqueous cations	
Electrolysis of copper(II) sulfate solution	88
Purification of copper	88
Electrolysis guidelines	90
Electroplating	90
Plating plastics	
Calculations in electrolysis	92
Checklist	94
Additional questions	95

7 Acids, bases and salts

96

Acids and alkalis	96
Theories of acids and bases	
The relative strengths of acids and bases	
Neutralising an acid	
Formation of salts	100
Methods of preparing soluble salts	
Methods of preparing insoluble salts	
More about salts	
Testing for different salts	
Crystal hydrates	105
Calculation of water of crystallisation	
Solubility of salts in water	107
Calculating solubility	
Solubility curves	
Titration	108
Checklist	110
Additional questions	111

8 Inorganic carbon chemistry

113

Limestone — 113
Direct uses of limestone
Indirect uses of limestone

Carbonates and hydrogencarbonates — 118
Properties of carbonates
Sodium carbonate – an important industrial chemical
Hydrogencarbonates
Uses of water
Hardness in water

Carbon dioxide — 121
Uses of carbon dioxide
Laboratory preparation of carbon dioxide gas
Properties of carbon dioxide gas

Checklist — 125

Additional questions — 126

9 Metal extraction and chemical reactivity

128

Metal reactions — 129
With acid
With air/oxygen
With water/steam

Reactivity of metals and their uses — 131
Using the reactivity series
Competition reactions in the solid state
Competition reactions in aqueous solutions

Identifying metal ions — 132
Amphoteric hydroxides

Discovery of metals — 133

Extraction of metals from their ores — 134
Extraction of reactive metals
Extraction of fairly reactive metals
Extraction of unreactive metals

Recycling metals — 138

Rusting of iron — 138
Rust prevention
Corrosion

Alloys — 141
Production of steel
Alloys to order

Checklist — 144

Additional questions — 145

10 Atmosphere and oceans

147

The developing atmosphere	147
The structure of the atmosphere	149
The composition of the atmosphere	
Fractional distillation of liquid air	151
Uses of the gases	151
Oxygen	
Nitrogen	
Noble gases	
Resources in the oceans	153
Bromine	
Magnesium	
Sodium chloride	
The water cycle	155
Pollution	155
Water pollution	
Atmospheric pollution	
Checklist	159
Additional questions	160

11 Rates of reaction

162

Factors that affect the rate of a reaction	163
Surface area	
Concentration	
Temperature	
Light	
Catalysts	
Enzymes	168
Checklist	170
Additional questions	171

12 The petroleum industry

173

Substances from oil	173
Oil refining	
Alkanes	176
Naming the alkanes	
Structural isomerism	
The chemical behaviour of alkanes	
The ozone hole problem	
Methane – another greenhouse gas!	
Other uses of alkanes	
Alkenes	179
Where do we get alkenes from?	
The chemical behaviour of alkenes	
A test for unsaturated compounds	
Checklist	182
Additional questions	183

13 Energy sources 185

Fossil fuels 185

What is a fuel? 186
How are fossil fuels used?

Alternative sources of energy 187
Nuclear power
Hydroelectric power
Geothermal energy
Wave power
Tidal power
Wind power
Solar energy
Biomass and biogas
Hydrogen – the fuel of the future

Chemical energy 191
Combustion
Enthalpy of neutralisation (molar heat of neutralisation)

Change of state 193

Cells and batteries 193

Fuel cells 194

Checklist 195

Additional questions 196

14 The (wider) organic manufacturing industry 198

Plastics and polymers 198
Other addition polymers
Condensation polymers
Thermosoftening and thermosetting plastics
Disposal of plastics

Some biopolymers 203
Hydrolysis of starch

Ethanol (C_2H_5OH) – an alcohol 204

Cholesterol – a steroid 205

Biotechnology 205
Baking
New applications

Ethanoic acid (CH_3COOH) – a carboxylic acid 206

Other carboxylic acids 207
Aspirin
Ascorbic acid (vitamin C)
Citric acid

Soaps and detergents 208
Soapless detergents

Amino acids 209
Protein analysis

DNA 211

Pharmaceuticals 211
Drug abuse

Checklist 213

Additional questions 215

15 Nitrogen

217

Nitrogen – an essential element 217

The Haber process 218
Obtaining nitrogen
Obtaining hydrogen
Making ammonia

Ammonia gas 220
Making ammonia in a laboratory
Physical properties of ammonia
Chemical properties of ammonia

Nitric acid 222
Manufacture of nitric acid
Properties of nitric acid

Artificial fertilisers 223
Manufacture of ammonium nitrate
Problems with fertilisers

Checklist 224

Additional questions 225

16 Sulfur

227

Sulfur – the element 227
The Frasch process
Uses of sulfur
Properties of sulfur
Sulfur dioxide

Acid rain 228

Industrial manufacture of sulfuric acid – the Contact process 229

Properties of sulfuric acid 230
Dilute sulfuric acid
Sulfates
Concentrated sulfuric acid

Manufacture of a soapless detergent 232

Checklist 233

Additional questions 234

236

Experimental chemistry 236

IGCSE exam questions (practical papers) 247

IGCSE exam questions 262

Answers to numerical questions 281

Data tables 283

Acknowledgements 286

Index 289

Preface to the reader

This textbook has been written to help you in your study of chemistry to IGCSE. The different chapters in this book are split up into short topics. At the end of many of these topics are questions to test whether you have understood what you have read. At the end of each chapter there are larger study questions. Try to answer as many of the questions as you can as you come across them because asking and answering questions is at the heart of your study of chemistry.

Some questions selected from the IGCSE examination papers are included at the end of the book. In many cases they are designed to test your ability to apply your chemical knowledge. The questions may provide certain facts and ask you to make an interpretation of them. In such cases, the factual information may not be covered in the text.

To help draw attention to the more important words, scientific terms are printed in **bold** the first time they are used. There are also checklists at the end of each chapter summarising the important points covered.

As you read through the book, you will notice three sorts of shaded area in the text.

Material highlighted in green is for the IGCSE Extended curriculum.

Areas highlighted in yellow contain material that is not part of the IGCSE syllabus. It is extension work and will not be examined.

Questions are highlighted by a blue tint box.

You will see from the box at the foot of this page that the book is divided into four different areas of chemistry. However, although the four chapters on basic principles occur at the start of the book, the other chapters are not all grouped together according to these areas. This is because we felt some topics led naturally on to other topics not in the same area. You can, of course, read and study the chapters in your own preferred order and the colour coding will help you with this.

The matching grid on the following pages will help you identify which parts of this book match with the various parts of the syllabus for IGCSE Chemistry. Core material is described in black text; material described in blue text is for the Extended curriculum and appears on the main pages in a green tint box (see above).

The accompanying **Revision CD-ROM** provides invaluable exam preparation and practice. It has two sections:

- *Revision Support* includes sample examination papers and answers.
- *Test Your Knowledge* includes interactive tests that cover both the Core and Extended curriculum. These are organised by syllabus topic.

Together, the textbook and CD-ROM will provide you with the information you need for the IGCSE syllabus. We hope you enjoy using them.

Bryan Earl and Doug Wilford

We use coloured strips at the edges of pages to define different areas of chemistry:

- ■ 'starter' chapters – basic principles
- ■ physical chemistry
- ■ inorganic chemistry
- ■ organic chemistry and the living world.

IGCSE Chemistry syllabus and textbook matching grid

Key for syllabus levels
IGCSE Core curriculum material
IGCSE Extended curriculum material

Notes:

■ Definitions are given during and at the end of appropriate chapters.

IGCSE Chemistry syllabus topic	Cross-reference to textbook	Cross-reference to CD-ROM
1 The particulate nature of matter		
States of matter and their interconversion in terms of the kinetic particle theory Description and explanation of diffusion Evidence for the movement of particles in gases and liquids Rate of diffusion compared to molecular mass	pp. 1–5 p. 6 p. 6 p. 6	
2 Experimental techniques		Tests 1 and 2
2.1 Measurement	pp. 236–7	
2.2 a Criteria of purity	p. 25	
2.2 b Methods of purification Chromatography applied to colourless substances Interpret simple chromatograms, including the use of R_f values	pp. 18–24 p. 24 p. 24	
3 Atoms, elements and compounds		Tests 3–6
3.1 Atomic structure and relationship to the Periodic Table	pp. 32–9	
3.2 Bonding: the structure of matter Differences between elements, mixtures and compounds Differences between metals and non-metals Alloys	pp. 12–18 p. 13 pp. 141–3	
3.2 a Formation of ions and ionic bonds (Groups 1 and 7) Lattice structure of ionic compounds	pp. 51–2 p. 54	
3.2 b Molecules and covalent bonds Formation of single covalent bonds Electron arrangement in more complex covalent molecules Differences between ionic and covalent compounds	 pp. 56–8 p. 59 pp. 54, 60	
3.2 c Some macromolecules – the giant covalent structures of graphite and diamond and relationship to uses Macromolecular structure of silicon(IV) oxide (silicon dioxide) and its properties	pp. 61–2 p. 60	
3.2 d Metallic bonding and description of electrical conductivity and malleability of metals	p. 65	
4 Stoichiometry		Tests 7–9
Symbols of the elements and the formulae of simple compounds Word equations and simple balanced chemical equations Relative atomic mass, A_r and relative molecular mass, M_r	pp. 13, 14, 38 p. 15 pp. 35, 70	
4.1 The mole concept The mole and the Avogadro constant Use of the molar gas volume Reacting masses and volumes of gases Solutions, solution concentrations Empirical formulae and molecular formulae % yield and % purity	 p. 69 p. 71 p. 76 p. 72 pp. 73, 74 p. 76	

IGCSE Chemistry syllabus topic	Cross-reference to textbook	Cross-reference to CD-ROM
5 Electricity and chemistry		Tests 10 and 11 Tests 12–14
The electrode products in the electrolysis of:		
molten substances	pp. 80–4	
solutions	pp. 85–90	
General principles of electrolysis	p. 79	
Electroplating of metals and the uses of electroplating	pp. 90–1	
Use of copper and (steel-cored) aluminium in cables	p. 88	
Use of plastics and ceramics as insulators	p. 64	
Electrolysis in terms of the ions present and reactions at the electrodes	pp. 79–92	
The products of electrolysis related to the electrolyte and electrodes used with aqueous copper(II) sulfate	pp. 88–90	
The refining of copper	pp. 88–90	
6 Chemical changes		Tests 15 and 16 Tests 17–19
6.1 Energetics of a reaction		
Meaning of exothermic and endothermic reactions	p. 191	
Bond breaking is endothermic and bond forming is exothermic	pp. 191–2	
6.2 Production of energy		
Production of heat energy by burning fuels	pp. 177, 186, 190	
Hydrogen as a fuel	p. 190	
Radioactive isotopes, such as ^{235}U, as a source of energy	pp. 187–8	
Electrical energy from simple cells	p. 193	
Use of batteries as a convenient, portable energy source	p. 193	
Describe the use of hydrogen as a potential fuel reacting with oxygen to generate electricity in a fuel cell	p. 194	
7 Chemical reactions		Tests 15 and 16 Tests 17–19
7.1 Speed of reaction		
The effect on the speeds of reactions of:		
concentration	pp. 164–5	
particle size	pp. 163–4	
catalysts (including enzymes)	pp. 166–9	
temperature	p. 165	
A practical method for investigating the speed of a reaction involving a gas being produced	p. 163	
The application of the above factors to the danger of explosive combustion with fine powders and gases	p. 164	
7.2 Reversible reactions		
Some chemical reactions can be reversed by changing the reaction conditions	pp. 106, 218–19	
Concept of equilibrium	pp. 218–9	
7.3 Redox		
Oxidation and reduction in terms of oxygen loss/gain	p. 15	
Redox in terms of electron transfer	p. 82	
8 Acids, bases and salts		Tests 20–22 Tests 23–25
8.1 The characteristic properties of acids and bases		
The characteristic properties of acids	p. 96	
The characteristic properties of bases	p. 96	
Concept of pH measured using universal indicator paper	pp. 96, 97	
Controlling soil acidity	p. 115	
Acids and bases in terms of proton transfer	p. 98	
Weak and strong acids and bases	p. 99	
8.2 Types of oxides		
Acidic or basic oxides	p. 135	
Classify other oxides as neutral or amphoteric	pp. 133, 135	
8.3 Preparation of salts		
The preparation, separation and purification of soluble salts	pp. 100–03	
The preparation of insoluble salts	p. 103	
8.4 Identification of ions and gases		
Tests to identify aqueous:		
cations	pp. 132–3, 283	
anions	pp. 104–5, 283	
gases	p. 283	

IGCSE Chemistry syllabus topic	Cross-reference to textbook	Cross-reference to CD-ROM
9 The Periodic Table		Tests 26–28
The Periodic Table as a method of classifying elements	pp. 37–9	
9.1 Periodic trends across a period	p. 39	
Group number, number of valency electrons and metallic/non-metallic character	p. 40	
9.2 Group properties		
Properties and trends in Group I	pp. 40–2	
Properties and trends in Group VII	pp. 43–4	
9.3 Transition elements		
Properties of transition elements (metals)	pp. 47–8	
9.4 Noble gases		
Properties and uses of the inert/noble gases	p. 46	
10 Metals		Tests 29–32
		Test 33
10.1 Properties of metals	pp. 13, 129–30	
Alloys and their structure	p. 143	
10.2 Reactivity series		
Order of reactivity	p. 129	
The reduction of metal oxides with carbon	p. 135	
The reactivity series related to the tendency of a metal to form its positive ion	p. 132	
in displacement reactions	pp. 131, 284	
The action of heat on metal compounds	pp. 83, 129	
The apparent unreactivity of aluminium in terms of the surface oxide layer		
10.3 a Extraction of metals		
Obtaining metals from their ores in relation to the reactivity series	pp. 134–7	
The extraction of iron from haematite	p. 135	
The conversion of iron into steel	pp. 141–2	
The extraction of zinc from zinc blende	p. 136	
The main ore of aluminium is bauxite	p. 81	
10.3 b Uses of metal		
The uses of aluminium in relation to its properties	p. 83	
Steel alloys and their uses	p. 143	
Uses of zinc related to properties	p. 136	
Uses of copper related to properties	p. 88	
11 Air and water		Test 34
		Test 35
Chemical test for water	p. 106	
The purification of the water supply	p. 156	
The uses of water	p. 119	
The composition of clean air	p. 149	
The common pollutants in the air	pp. 157–8	
The adverse effect of common pollutants	pp. 157–8	
The uses of oxygen	p. 151	
Methods of rust prevention	pp. 138–40	
Fertilisers	p. 223	
The displacement of ammonia from its salts	p. 220	
The sources of carbon dioxide and methane, state that they are greenhouse gases and may contribute to climate change	pp. 79, 122, 179	
The formation of carbon dioxide	pp. 122, 123, 186	
Catalytic removal of oxides of nitrogen from car exhausts	p. 167	
Fractional distillation of liquid air	p. 151	
Sacrificial protection as a method of rust prevention	p. 140	
The manufacture of ammonia by the Haber process	p. 219	
Describe the carbon cycle	pp. 121–2	
12 Sulfur		
Sources of sulfur	p. 227	
The uses of sulfur dioxide	p. 228	
The manufacture of sulfuric acid by the Contact process	pp. 229–30	
The properties of dilute sulfuric acid	pp. 230–1	

IGCSE Chemistry syllabus topic	Cross-reference to textbook	Cross-reference to CD-ROM
13 Carbonates		Tests 29–32
The manufacture of lime	p. 116	
The uses of lime and slaked lime	p. 116	
The uses of calcium carbonate	pp. 114–17	
14 Organic Chemistry		Tests 36–39 Tests 40 and 41
14.1 Names of compounds Name, and draw the structures of methane, ethane, ethanol, ethanoic acid and the products of their reactions	Chapters 12 and 14	
14.2 Fuels Fractions of oil The type of compound given a chemical name, ending in: –ane –ene –ol –oic acid The fossil fuels Crude oil, its separation by fractional distillation the uses of the fractions the structures and naming of unbranched hydrocarbons alcohols acids	p. 175 pp. 176–7 pp. 179–81 p. 204 pp. 206–8 pp. 185–6 pp. 174–5 p. 175 pp. 176, 179 p. 204 p. 206	
14.3 Homologous series The concept of homologous series Characteristics of an homologous series: structural isomerism	p. 176 p. 177	
14.4 Alkanes The properties of alkanes The bonding in alkanes Substitution reactions of alkanes with chlorine	p. 177 pp. 176–7 p. 178	
14.5 Alkenes The manufacture of alkenes Testing for saturated versus unsaturated hydrocarbons The formation of poly(ethene) The properties of alkenes	p. 180 p. 181 p. 198 p. 180	
14.6 Alcohols The formation of ethanol by: fermentation the catalytic addition of steam to ethene The properties of ethanol The uses of ethanol	 p. 205 p. 180 p. 204 p. 204	
14.7 Acids The formation of ethanoic acid by the oxidation of ethanol Ethanoic acid as a typical weak acid Formation of the ester ethyl ethanoate	p. 204 p. 99 p. 207	
14.8 Macromolecules Synthetic polymers Uses of plastics and man-made fibres The pollution problems caused by non-biodegradable plastics The structure of the polymer product from a given alkene and vice versa The formation and structure of: nylon (a polyamide) terylene (a polyester) Natural macromolecules proteins, fats and carbohydrates structure of proteins the hydrolysis of proteins soap as a product of hydrolysis of fats the acid hydrolysis of complex carbohydrates the fermentation of simple sugars the usefulness of chromatography in separating and identifying the product of hydrolysis of proteins	 p. 200 p. 202 p. 199 p. 200 p. 201 pp. 203, 208, 210 pp. 209–10 p. 210 p. 208 p. 203 p. 205 p. 210	

1 *All about matter*

Solids, liquids and gases

The kinetic theory of matter
Explaining the states of matter

Changes of state
An unusual state of matter
An unusual change of state
Heating and cooling curves

Diffusion – evidence for moving particles
Brownian motion

Gas laws
Charles' law
Boyle's law
Combining the gas laws

Checklist

Additional questions

Chemistry is about what **matter** is like and how it behaves, and our explanations and predictions of its behaviour. What is matter? This word is used to cover all the substances and materials from which the physical universe is composed. There are many millions of different substances known, and all of them can be categorised as solids, liquids or gases (Figure 1.1). These are what we call the **three states of matter**.

b liquid

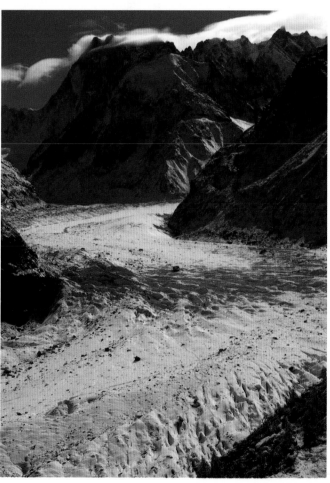

a solid

Figure 1.1 Water in three different states.

c gas

Solids, liquids and gases

A **solid**, at a given temperature, has a definite volume and shape which may be affected by changes in temperature. Solids usually increase slightly in size when heated (**expansion**) (Figure 1.2) and usually decrease in size if cooled (**contraction**).

A **liquid**, at a given temperature, has a fixed volume and will take up the shape of any container into which it is poured. Like a solid, a liquid's volume is slightly affected by changes in temperature.

A **gas**, at a given temperature, has neither a definite shape nor a definite volume. It will take up the shape of any container into which it is placed and will spread out evenly within it. Unlike those of solids and liquids, the volumes of gases are affected quite markedly by changes in temperature.

Liquids and gases, unlike solids, are relatively **compressible**. This means that their volume can be reduced by the application of pressure. Gases are much more compressible than liquids.

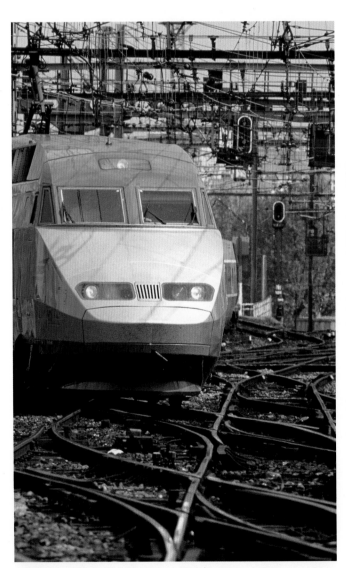

Figure 1.2 Without expansion gaps between the rails, the track would buckle in hot weather.

The kinetic theory of matter

The **kinetic theory** helps to explain the way in which matter behaves. The evidence is consistent with the idea that all matter is made up of tiny **particles**. This theory explains the physical properties of matter in terms of the movement of its constituent particles.

The main points of the theory are:

* all matter is made up of tiny, moving particles, invisible to the naked eye. Different substances have different types of particles (atoms, molecules or ions) which have different sizes
* the particles move all the time. The higher the temperature, the faster they move on average
* heavier particles move more slowly than lighter ones at a given temperature.

The kinetic theory can be used as a scientific model to explain how the arrangement of particles relates to the properties of the three states of matter.

Explaining the states of matter

In a solid the particles attract one another. There are attractive forces between the particles which hold them close together. The particles have little freedom of movement and can only vibrate about a fixed position. They are arranged in a regular manner, which explains why many solids form crystals.

a A model of a chrome alum crystal.

b An actual chrome alum crystal.

Figure 1.3

It is possible to model such crystals by using spheres to represent the particles (Figure 1.3a). If the spheres are built up in a regular way then the shape compares very closely with that of a part of a chrome alum crystal (Figure 1.3b).

Figure 1.4 A modern X-ray crystallography instrument, used for studying crystal structure.

Studies using X-ray crystallography (Figure 1.4) have confirmed how the particles are arranged in crystal structures. When crystals of a pure substance form under a given set of conditions, the particles present are always packed in the same way. However, the particles may be packed in different ways in crystals of different substances. For example, common salt (sodium chloride) has its particles arranged to give cubic crystals as shown in Figure 1.5.

Figure 1.5 Sodium chloride crystals.

In a liquid the particles are still close together but they move around in a random way and often collide with one another. The forces of attraction between the particles in a liquid are weaker than those in a solid. Particles in the liquid form of a substance have more energy on average than the particles in the solid form of the same substance.

In a gas the particles are relatively far apart. They are free to move anywhere within the container in which they are held. They move randomly at very high velocities, much more rapidly than those in a liquid. They collide with each other, but less often than in a liquid, and they also collide with the walls of the container. They exert virtually no forces of attraction on each other because they are relatively far apart. Such forces, however, are very significant. If they did not exist we could not have solids or liquids (see Changes of state, p. 4).

The arrangement of particles in solids, liquids and gases is shown in Figure 1.6.

solid
Particles only vibrate about fixed positions. Regular structure.

liquid
Particles have some freedom and can move around each other. Collide often.

gas
Particles move freely and at random in all the space available. Collide less often than in liquid.

Figure 1.6 The arrangement of particles in solids, liquids and gases.

Questions

1 When a metal such as copper is heated it expands. Explain what happens to the metal particles as the solid metal expands.

2 Use your research skills on the Internet to find out about the technique of X-ray crystallography and how this technique can be used to determine the crystalline structure of solid substances such as sodium chloride.

Changes of state

The kinetic theory model can be used to explain how a substance changes from one state to another. If a solid is heated the particles vibrate faster as they gain energy. This makes them 'push' their neighbouring particles further away from themselves. This causes an increase in the volume of the solid and the solid expands. Expansion has taken place.

Eventually, the heat energy causes the forces of attraction to weaken. The regular pattern of the structure breaks down. The particles can now move around each other. The solid has melted. The temperature at which this takes place is called the **melting point** of the substance. The temperature of a pure melting solid will not rise until it has all melted. When the substance has become a liquid there are still very significant forces of attraction between the particles, which is why it is a liquid and not a gas.

Solids which have high melting points have stronger forces of attraction between their particles than those which have low melting points. A list of some substances with their corresponding melting and boiling points is shown in Table 1.1.

Table 1.1

Substance	Melting point/°C	Boiling point/°C
Aluminium	661	2467
Ethanol	−117	79
Magnesium oxide	−827	3627
Mercury	−30	357
Methane	−182	−164
Oxygen	−218	−183
Sodium chloride	801	1413
Sulfur	113	445
Water	0	100

If the liquid is heated the particles will move around even faster as their average energy increases. Some particles at the surface of the liquid have enough energy to overcome the forces of attraction between themselves and the other particles in the liquid and they escape to form a gas. The liquid begins to **evaporate** as a gas is formed.

Eventually, a temperature is reached at which the particles are trying to escape from the liquid so quickly that bubbles of gas actually start to form inside the bulk of the liquid. This temperature is called the **boiling point** of the substance. At the boiling point the pressure of the gas created above the liquid equals that in the air — **atmospheric pressure**.

Liquids with high boiling points have stronger forces between their particles than liquids with low boiling points.

When a gas is cooled the average energy of the particles decreases and the particles move closer together. The forces of attraction between the particles now become significant and cause the gas to **condense** into a liquid. When a liquid is cooled it **freezes** to form a solid. In each of these changes energy is given out.

Changes of state are examples of **physical changes**. Whenever a physical change of state occurs, the temperature remains constant during the change (see Heating and cooling curves, opposite). During a physical change no new substance is formed.

An unusual state of matter

Liquid crystals are an unusual state of matter (Figure 1.7). These substances look like liquids, flow like liquids but have some order in the arrangement of the particles, and so in some ways they behave like crystals.

Liquid crystals are now part of our everyday life. They are widely used in displays for digital watches, calculators and lap-top computer displays (Figure 1.8), and in portable televisions. They are also useful in thermometers because liquid crystals change colour as the temperature rises and falls.

Figure 1.7 A polarised light micrograph of liquid crystals.

Figure 1.8 Liquid crystals are used in this TV screen.

An unusual change of state

There are a few substances that when they are heated change directly from a solid to a gas without ever becoming a liquid. This rapid spreading out of the particles is called **sublimation**. Cooling causes a change from a gas directly back to a solid. Examples of substances that behave in this way are carbon dioxide (Figure 1.9) and iodine.

Figure 1.9 Dry ice (solid carbon dioxide) sublimes on heating and can be used to create special effects on stage.

Carbon dioxide is a white solid called dry ice at temperatures below −78 °C. When heated to just above −78 °C it changes into carbon dioxide gas. The changes of state are summarised in Figure 1.10.

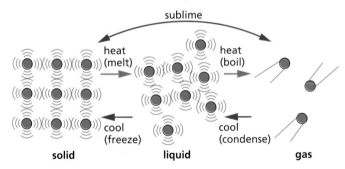

Figure 1.10 Summary of the changes of state.

Heating and cooling curves

The graph shown in Figure 1.11 was drawn by plotting the temperature of water as it was heated steadily from −15 °C to 110 °C. You can see from the curve that changes of state have taken place. When the temperature was first measured only ice was present. After a short time the curve flattens, showing that even though heat energy is being put in, the temperature remains constant.

In ice the particles of water are close together and are attracted to one another. For ice to melt the particles must obtain sufficient energy to overcome the forces of attraction between the water particles to allow relative movement to take place. This is where the heat energy is going.

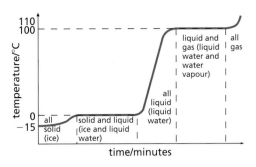

Figure 1.11 Graph of temperature against time for the change from ice at −15 °C to water to steam.

The temperature will begin to rise again only after all the ice has melted. Generally, the heating curve for a pure solid always stops rising at its melting point and gives rise to a sharp melting point. A sharp melting point indicates a pure sample. The addition or presence of impurities lowers the melting point. You can try to find the melting point of a substance using the apparatus shown in Figure 1.12.

Figure 1.12 Apparatus shown here if heated slowly can be used to find the melting point of a substance such as the solid in the melting point tube.

In the same way, if you want to boil a liquid such as water you have to give it some extra energy. This can be seen on the graph (Figure 1.11) where the curve levels out at 100 °C − the boiling point of water.

Solids and liquids can be identified from their characteristic melting and boiling points.

The reverse processes of condensing and freezing occur on cooling. This time, however, energy is given out when the gas condenses to the liquid and the liquid freezes to give the solid.

Questions

1 Write down as many uses as you can for liquid crystals.

2 Why do gases expand more than solids for the same increase in temperature?

3 Ice on a car windscreen will disappear as you drive along, even without the heater on. Explain why this happens.

4 When salt is placed on ice the ice melts. Explain why.

5 Draw and label the graph you would expect to produce if water at 100 °C was allowed to cool to −5 °C.

Diffusion – evidence for moving particles

When you walk past a cosmetics counter in a department store you can usually smell the perfumes. For this to happen gas particles must be leaving open perfume bottles and be spreading out through the air in the store. This spreading out of a gas is called **diffusion** and it takes place in a haphazard and random way.

All gases diffuse to fill the space available to them. As you can see from Figure 1.13, after a day the brown–red fumes of gaseous bromine have spread evenly throughout both gas jars from the liquid present in the lower gas jar.

Figure 1.13 After 24 hours the bromine fumes have diffused throughout both gas jars.

Gases diffuse at different rates. If one piece of cotton wool is soaked in concentrated ammonia solution and another is soaked in concentrated hydrochloric acid and these are put at opposite ends of a dry glass tube, then after a few minutes a white cloud of ammonium chloride appears (Figure 1.14). This shows the position at which the two gases meet and react. The white cloud forms in the position shown because the ammonia particles are lighter than the hydrogen chloride particles (released from the hydrochloric acid) and so move faster.

Diffusion also takes place in liquids (Figure 1.15) but it is a much slower process than in gases. This is because the particles of a liquid move much more slowly.

When diffusion takes place between a liquid and a gas it is known as **intimate mixing**. The kinetic theory can be used to explain this process. It states that collisions are taking place between particles in a liquid or a gas and that there is sufficient space between the particles of one substance for the particles of the other substance to move into.

Figure 1.14 Hydrochloric acid (left) and ammonia (right) diffuse at different rates.

a   b

Figure 1.15 Diffusion within nickel(II) sulfate solution can take days to reach the stage shown on the right.

Brownian motion

Evidence for the movement of particles in liquids came to light in 1827 when a botanist, Robert Brown, observed that fine pollen grains on the surface of water were not stationary. Through his microscope he noticed that the grains were moving about in a random way. It was 96 years later, in 1923, that another scientist called Norbert Wiener explained what Brown had observed. He said that the pollen grains were moving because the much smaller and faster-moving water particles were constantly colliding with them (Figure 1.16a).

This random motion of visible particles (pollen grains) caused by much smaller, invisible ones (water particles) is called **Brownian motion** (Figure 1.16b), after the scientist who first observed this phenomenon.

Figure 1.16a Pollen particle being bombarded by water molecules.

Figure 1.16b Brownian motion causes the random motion of the visible particle.

Questions

1 When a jar of coffee is opened, people in all parts of the room soon notice the smell. Use the kinetic theory to explain how this happens.

2 Describe, with the aid of diagrams, the diffusion of nickel(II) sulfate solution.

3 Explain why diffusion is faster in gases than in liquids.

■ *Gas laws*

What do you think has caused the difference between the balloons in Figure 1.17?

Figure 1.17 The balloon is partly inflated at room temperature (left) and becomes fully expanded at a higher temperature (right).

The **pressure** inside a balloon is caused by the gas particles striking the inside surface of the balloon (Figure 1.18). There is an increased pressure inside the balloon at a higher temperature due to the gas particles having more energy, moving around faster and producing a higher collision rate with the inside surface of the balloon. Each collision with the surface takes place with a greater force as the gas particles have more energy. This in turn leads to an increase in pressure. Since the balloon is an elastic envelope, the increased pressure causes the skin to stretch and the volume to increase. This increase in volume with increased temperature is a property of all gases. In 1781, from this sort of observation, a French scientist called Jacques Alexandre César Charles concluded that when the temperature of a gas increased the volume also increased, at a constant pressure. This law is known as Charles' Law.

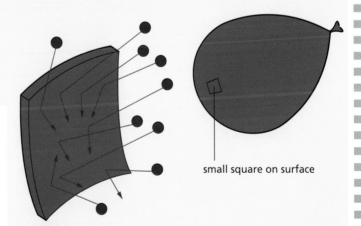

small square on surface

enlargement of surface showing bombardment by air particles

Figure 1.18 The gas particles striking the surface create the pressure.

Charles' Law

Charles' Law states that the volume, V, of a fixed mass of gas is directly proportional to its absolute temperature, T, if the pressure is kept constant. The absolute temperature (in kelvin) is found by adding 273 to the °C value. Mathematically,

$$V \propto T$$

or

$$V = \text{constant} \times T$$

or

$$\frac{V}{T} = \text{constant}$$

Boyle's Law

In 1662 a scientist called Robert Boyle (Figure 1.19) deduced from experiments he carried out on various gases that when the pressure was increased, the volume of the gas was reduced.
Mathematically,

$$V \propto \frac{1}{p}$$

or

$$V = \frac{\text{constant}}{p}$$

or

$$pV = \text{constant}$$

Figure 1.20 Scuba diving can be dangerous even at shallow depths, as nitrogen in the compressed air can dissolve in body tissues and make the diver feel 'drunk'.

Why is it important to learn about how the volume of a gas changes with pressure? One popular use of this knowledge is in scuba diving (scuba stands for self-contained underwater breathing apparatus). To breathe under water, divers carry tanks of compressed gas (Figure 1.20). As they dive deeper, the water exerts more and more pressure on their bodies, and on the gas in their tanks. So that they can still breathe, the 'air' in the tanks has to be regulated and the pressure reduced so that it is about the same as that in the surrounding water.

Divers can descend fairly quickly as the volume of air in their bodies decreases as the pressure increases. When they want to return to the surface, though, they have to take it slowly because from 10 m below the surface to the surface itself the pressure doubles. This means that if divers come up too quickly or while holding their breath they can burst their lungs.

Combining the gas laws

By combining Boyle's and Charles' Laws it is possible to show the relationship between the pressure, p, the volume, V, and the temperature, T, of a fixed mass of gas as:

$$\frac{pV}{T} = \text{constant}$$

or

$$\frac{p_1 V_1}{T_1} = \frac{p_2 V_2}{T_2}$$

where p_1, V_1 and T_1 and p_2, V_2 and T_2 are the pressure (in pascals), volume and temperature (in kelvin) in two different situations.

Figure 1.19 Robert Boyle (1627–1691).

An example of the use of this relationship is given next. In the decomposition of hydrogen peroxide, if the volume of oxygen gas collected at 40 °C and a pressure of 1×10^5 Pa was 100 cm^3, what would be the volume of the gas at a temperature of 10 °C and a pressure of 2×10^5 Pa?

Using

$$\frac{p_1 V_1}{T_1} = \frac{p_2 V_2}{T_2}$$

$p_1 = 1 \times 10^5$ Pa $\qquad p_2 = 2 \times 10^5$ Pa

$V_1 = 100$ cm^3 $\qquad V_2 = ?$

$T_1 = (40 + 273)$ $\qquad T_2 = (10 + 273)$

$\quad = 313$ K $\qquad\qquad = 283$ K

$$\frac{(1 \times 10^5) \times 100}{313} = \frac{(2 \times 10^5) \times V_2}{283}$$

$$V_2 = \frac{(1 \times 10^5) \times 100 \times 283}{313 \times (2 \times 10^5)}$$

$$V_2 = 45.21 \text{ cm}^3$$

Questions

1 When a gas is heated the particles move more quickly. Explain what will happen to the volume of the heated gas if the pressure is kept constant.

2 A bubble of methane gas rises from the bottom of the North Sea.
   a What will happen to the size of the bubble as it rises to the surface?
   b Explain your answer to **a**.

3 A gas syringe contains 50 cm^3 of oxygen gas at 20 °C. If the temperature was increased to 45 °C, what would be the volume occupied by this gas, assuming constant pressure throughout?

4 A bicycle pump contains 50 cm^3 of air at a pressure of 1×10^5 pa. What would be the volume of the air if the pressure was increased to 2.1×10^5 Pa at constant temperature?

5 If the volume of a gas collected at 60 °C and 1×10^5 Pa pressure was 70 cm^3, what would be the volume at a temperature of 0 °C and a pressure of 4×10^5 Pa?

■ *Checklist*

After studying Chapter 1 you should know and understand the following terms.

■ **Absolute temperature** A temperature measured with respect to absolute zero on the Kelvin scale. Absolute zero is the lowest possible temperature for all substances. The Kelvin scale is usually denoted by T.

$$T \text{ K} = °C + 273$$

■ **Atmospheric pressure** The pressure exerted by the atmosphere on the surface of the Earth due to the weight of the air.

■ **Boiling point** The temperature at which the pressure of the gas created above a liquid equals atmospheric pressure.

■ **Boyle's Law** At a constant temperature the volume of a given mass of gas is inversely proportional to the pressure.

$$V \propto \frac{1}{p}$$

■ **Charles' Law** At constant pressure the volume of a given mass of gas is directly proportional to the absolute temperature.

$$V \propto T$$

■ **Condensation** The change of a vapour or a gas into a liquid. This process is accompanied by the evolution of heat.

■ **Diffusion** The process by which different substances mix as a result of the random motions of their particles.

■ **Evaporation** A process occurring at the surface of a liquid involving the change of state of a liquid into a vapour at a temperature below the boiling point.

■ **Kinetic theory** A theory which accounts for the bulk properties of matter in terms of the constituent particles.

■ **Matter** Anything which occupies space and has a mass.

■ **Melting point** The temperature at which a solid begins to liquefy. Pure substances have a sharp melting point.

■ **Solids, liquids and gases** The three states of matter to which all substances belong.

■ **Sublimation** The direct change of state from solid to gas and the reverse process.

All about matter
Additional questions

1 a Draw diagrams to show the arrangement of particles in:
  (i) solid lead
  (ii) molten lead
  (iii) gaseous lead.
  b Explain how the particles move in these three states of matter.
  c Explain, using the kinetic theory, what happens to the particles in oxygen as it is cooled down.

2 Explain the meaning of each of the following terms. In your answer include an example to help with your explanation.
  a Expansion.
  b Contraction.
  c Physical change.
  d Sublimation.
  e Diffusion.
  f Random motion.

3 a Why do solids not diffuse?
  b Give two examples of diffusion of gases and liquids found in the house.

4 Use the kinetic theory to explain the following:
  a When you take a block of butter out of the fridge, it is quite hard. However, after 15 minutes it is soft enough to spread.
  b When you come home from school and open the door you can smell your tea being cooked.
  c A football is blown up until it is hard on a hot summer's day. In the evening the football feels softer.
  d When a person wearing perfume enters a room it takes several minutes for the smell to reach the back of the room.
  e A windy day is a good drying day.

5 The apparatus shown below was set up. Give explanations for the following observations.

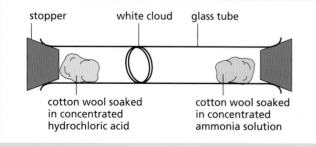

  a The formation of a white cloud.
  b It took a few minutes before the white cloud formed.
  c The white cloud formed further from the cotton wool soaked in ammonia.
  d Cooling the concentrated ammonia and hydrochloric acid before carrying out the experiment increased the time taken for the white cloud to form.

6 The following diagram shows the three states of matter and how they can be interchanged.

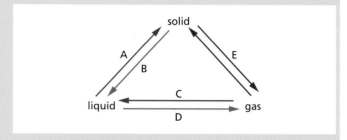

  a Name the changes **A** to **E**.
  b Name a substance which will undergo change **E**.
  c Name a substance which will undergo changes from solid to liquid to gas between 0 °C and 100 °C.
  d Describe what happens to the particles of the solid during change **E**.
  e Which of the changes **A** to **E** will involve:
    (i) an input of heat energy?
    (ii) an output of heat energy?

7 Some nickel(II) sulfate solution was carefully placed in the bottom of a beaker of water. The beaker was then covered and left for several days.

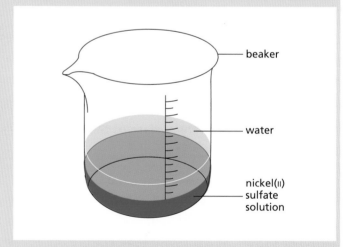

  a Describe what you would see after:
    (i) a few hours
    (ii) several days.
  b Explain your answer to **a** using your ideas of the kinetic theory of particles.
  c What is the name of the physical process that takes place in this experiment?

8 An electric light bulb has a volume of 200 cm³. It contains argon gas at a pressure of 1.1×10^5 Pa and a temperature of 25 °C. When the light is switched on, the pressure increases to a steady 1.8×10^5 Pa. What is the temperature of the argon in the electric light bulb that creates this new steady pressure?

2 Elements, compounds and mixtures

Elements
Atoms – the smallest particles
Molecules

Compounds
More about formulae
Balancing chemical equations
Instrumental techniques

Mixtures
What is the difference between mixtures and compounds?

Separating mixtures
Separating solid/liquid mixtures
Separating liquid/liquid mixtures
Separating solid/solid mixtures
Criteria for purity

Gels, sols, foams and emulsions

Mixtures for strength
Composite materials

Checklist

Additional questions

The universe is made up of a very large number of substances (Figure 2.1), and our own world is no exception. If this vast array of substances is examined more closely, it is found that they are made up of some basic substances which were given the name **elements** in 1661 by Robert Boyle, who we met in Chapter 1.

In 1803, John Dalton (Figure 2.2) suggested that each element was composed of its own kind of particles, which he called **atoms**. Atoms are much too small to be seen. We now know that about 20×10^6 of them would stretch over a length of only 1 cm.

Figure 2.2 John Dalton (1766–1844).

Figure 2.1 The planets in the universe are made of millions of substances. These are made up mainly from just 91 elements which occur naturally on the Earth.

Elements

Robert Boyle used the name element for any substance that cannot be broken down further, into a simpler substance. This definition can be extended to include the fact that each element is made up of only one kind of atom. The word atom comes from the Greek word *atomos* meaning 'unsplittable'.

For example, aluminium is an element which is made up of only aluminium atoms. It is not possible to obtain a simpler substance chemically from the aluminium atoms. You can only make more complicated substances from it, such as aluminium oxide, aluminium nitrate or aluminium sulphate.

There are 117 elements which have now been identified. Twenty-four of these do not occur in nature and have been made artificially by scientists. They include elements such as plutonium, curium and unnilpentium. Ninety-one of the elements occur naturally and range from some very reactive gases, such as fluorine and chlorine, to gold and platinum, which are unreactive elements.

All elements can be classified according to their various properties. A simple way to do this is to classify them as **metals** or **non-metals** (Figures 2.3 and 2.4). Table 2.1 shows the physical data for some common metallic and non-metallic elements.

You will notice that many metals have high densities, high melting points and high boiling points, and that most non-metals have low densities, low melting points and low boiling points. Table 2.2 summarises the different properties of metals and non-metals.

A discussion of the chemical properties of metals is given in Chapters 3, 4 and 9. The chemical properties of certain non-metals are discussed in Chapters 3, 4, 15 and 16.

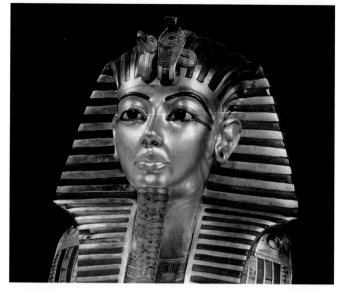

a Gold is very decorative.

b Aluminium has many uses in the aerospace industry.

c These coins contain nickel.

Figure 2.3 Some metals.

Table 2.1 Physical data for some metallic and non-metallic elements at room temperature and pressure.

Element	Metal or non-metal	Density/g cm^{-3}	Melting point/°C	Boiling point/°C
Aluminium	Metal	2.70	660	2580
Copper	Metal	8.92	1083	2567
Gold	Metal	19.29	1065	2807
Iron	Metal	7.87	1535	2750
Lead	Metal	11.34	328	1740
Magnesium	Metal	1.74	649	1107
Nickel	Metal	8.90	1453	2732
Silver	Metal	10.50	962	2212
Zinc	Metal	7.14	420	907
Carbon	Non-metal	2.25	2652	Sublimes
Hydrogen	Non-metal	0.07[a]	−259	−253
Nitrogen	Non-metal	0.88[b]	−210	−196
Oxygen	Non-metal	1.15[c]	−218	−183
Sulfur	Non-metal	2.07	113	445

Source: Earl B., Wilford L.D.R. Chemistry data book. Nelson Blackie, 1991 [a] At −254 °C [b] At −197 °C [c] At −184 °C.

Table 2.2 How the properties of metals and non-metals compare.

Property	Metal	Non-metal
Physical state at room temperature	Usually solid (occasionally liquid)	Solid, liquid or gas
Malleability	Good	No – usually soft or brittle when solid
Ductility	Good	
Appearance (solids)	Shiny (lustrous)	Dull
Melting point	Usually high	Usually low
Boiling point	Usually high	Usually low
Density	Usually high	Usually low
Conductivity (thermal and electrical)	Good	Very poor

a A premature baby needs oxygen.

b Artists often use charcoal (carbon) to produce an initial sketch.

c Neon is used in advertising signs.

Figure 2.4 Some non-metals.

Atoms – the smallest particles

Everything is made up of billions of atoms. The atoms of all elements are extremely small; in fact they are too small to be seen. The smallest atom known is hydrogen, with each atom being represented as a sphere having a diameter of 0.000 000 07 mm (or 7×10^{-8} mm) (Table 2.3). Atoms of different elements have different diameters as well as different masses. How many atoms of hydrogen would have to be placed side by side along the edge of your ruler to fill just one of the 1 mm divisions?

Table 2.3 Sizes of atoms.

Atom	Diameter of atom/mm
Hydrogen	7×10^{-8}
Oxygen	12×10^{-8}
Sulphur	20.8×10^{-8}

Chemists use shorthand symbols to label the elements and their atoms. The symbol consists of one, two or three letters, the first of which must be a capital. Where several elements have the same initial letter, a second letter of the name or subsequent letter is added. For example, **C** is used for **carbon**, **Ca** for **calcium** and **Cl** for **chlorine**. Some symbols seem to have no relationship to the name of the element, for example **Na** for **sodium** and **Pb** for **lead**. These symbols come from their Latin names, natrium for sodium and plumbum for lead. A list of some common elements and their symbols is given in Table 2.4 (overleaf).

Table 2.4 Some common elements and their symbols.

Element	Symbol	Physical state at room temperature and pressure
Aluminium	Al	Solid
Argon	Ar	Gas
Barium	Ba	Solid
Boron	B	Solid
Bromine	Br	Liquid
Calcium	Ca	Solid
Carbon	C	Solid
Chlorine	Cl	Gas
Chromium	Cr	Solid
Copper (Cuprum)	Cu	Solid
Fluorine	F	Gas
Germanium	Ge	Solid
Gold (Aurum)	Au	Solid
Helium	He	Gas
Hydrogen	H	Gas
Iodine	I	Solid
Iron (Ferrum)	Fe	Solid
Lead (Plumbum)	Pb	Solid
Magnesium	Mg	Solid
Mercury (Hydragyrum)	Hg	Liquid
Neon	Ne	Gas
Nitrogen	N	Gas
Oxygen	O	Gas
Phosphorus	P	Solid
Potassium (Kalium)	K	Solid
Silicon	Si	Solid
Silver (Argentum)	Ag	Solid
Sodium (Natrium)	Na	Solid
Sulfur	S	Solid
Tin (Stannum)	Sn	Solid
Zinc	Zn	Solid

The complete list of the elements with their corresponding symbols is shown in the periodic table on p. 285.

Molecules

The atoms of some elements are joined together in small groups. These small groups of atoms are called **molecules**. For example, the atoms of the elements hydrogen, oxygen, nitrogen, fluorine, chlorine, bromine and iodine are each joined in pairs and they are known as **diatomic** molecules. In the case of phosphorus and sulfur the atoms are joined in larger numbers, four and eight respectively (P_4, S_8). In chemical shorthand the molecule of chlorine shown in Figure 2.5 is written as Cl_2.

$$Cl — Cl$$

a As a letter-and-stick model.

b As a space-filling model.

Figure 2.5 A chlorine molecule.

The gaseous elements helium, neon, argon, krypton, xenon and radon are composed of separate and individual atoms. When an element exists as separate atoms, then the molecules are said to be **monatomic**. In chemical shorthand these monatomic molecules are written as He, Ne, Ar, Kr, Xe and Rn respectively.

Molecules are not always formed by atoms of the same type joining together. For example, water exists as molecules containing oxygen and hydrogen atoms.

Questions

1 How would you use a similar chemical shorthand to write a representation of the molecules of iodine and fluorine?

2 Using the periodic table on p. 285 write down the symbols for the following elements and give their physical states at room temperature:
   a chromium
   b krypton
   c osmium.

Compounds

Compounds are pure substances which are formed when two or more elements chemically combine together. Water is a simple compound formed from the elements hydrogen and oxygen (Figure 2.6). This combining of the elements can be represented by a word equation:

hydrogen + oxygen → water

| Hydrogen a pure element | Oxygen a pure element | Hydrogen and oxygen mixed together | Water a pure compound formed from hydrogen burning in oxygen |

Figure 2.6 The element hydrogen reacts with the element oxygen to produce the compound water.

Water molecules contain two atoms of hydrogen and one atom of oxygen, and hence water has the **chemical formula** H_2O. Elements other than hydrogen will also react with oxygen to form compounds called oxides. For example, magnesium reacts violently with oxygen gas to form the white powder magnesium oxide (Figure 2.7). This reaction is accompanied by a release of energy as new chemical bonds are formed.

When a new substance is formed during a chemical reaction, a **chemical change** has taken place.

magnesium + oxygen → magnesium oxide

When substances such as hydrogen and magnesium combine with oxygen in this way they are said to have been **oxidised**. The process is known as **oxidation**.

Reduction is the opposite of oxidation. In this process oxygen is removed instead of being added. For example, the oxygen has to be removed in the extraction of iron from iron(III) oxide. This can be done in a blast furnace with carbon monoxide. The iron(III) oxide loses oxygen to the carbon monoxide and is reduced to iron while carbon monoxide is oxidised to carbon dioxide. You will deal in more detail with this extraction process in Chapter 9.

iron(III) + carbon → iron + carbon
oxide   monoxide   dioxide

Both **reduction** and **oxidation** have taken place in this chemical process, and so this is known as a **redox** reaction. A further discussion of oxidation and reduction takes place in Chapter 6.

Figure 2.7 Magnesium burns brightly in oxygen to produce magnesium oxide.

More about formulae

The formula of a compound is made up from the symbols of the elements present and numbers to show the ratio in which the different atoms are present. Carbon dioxide has the formula CO_2. This tells you that it contains one carbon atom for every two oxygen atoms. The 2 in the formula tells you that there are two oxygen atoms present in each molecule of carbon dioxide. For further discussion see p. 55.

Table 2.5 (overleaf) shows the names and formulae of some common compounds which you will meet in your study of chemistry.

The ratio of atoms within a chemical compound is usually constant. Compounds are made up of fixed proportions of elements: they have a fixed composition. Chemists call this the **Law of constant composition**.

Table 2.5 Names and formulae of some common compounds.

Compound	Formula
Ammonia	NH_3
Calcium hydroxide	$Ca(OH)_2$
Carbon dioxide	CO_2
Copper sulfate	$CuSO_4$
Ethanol (alcohol)	C_2H_5OH
Glucose	$C_6H_{12}O_6$
Hydrochloric acid	HCl
Nitric acid	HNO_3
Sodium carbonate	Na_2CO_3
Sodium hydroxide	$NaOH$
Sulfuric acid	H_2SO_4

Balancing chemical equations

Word equations are a useful way of representing chemical reactions but a better and more useful method is to produce a **balanced chemical equation**. This type of equation gives the formulae of the reactants and the products as well as showing the relative numbers of each particle involved. Balanced equations often include the physical state symbols:

(s) = solid, (l) = liquid, (g) = gas, (aq) = aqueous solution

The word equation to represent the reaction between iron and sulfur is:

$$\text{iron} + \text{sulfur} \xrightarrow{\text{heat}} \text{iron(II) sulfide}$$

When we replace the words with symbols for the reactants and the products and include their physical state symbols, we obtain:

$$Fe(s) + S(s) \xrightarrow{\text{heat}} FeS(s)$$

Since there is the same number of each type of atom on both sides of the equation this is a **balanced** chemical equation.

In the case of magnesium reacting with oxygen, the word equation was:

$$\text{magnesium} + \text{oxygen} \xrightarrow{\text{heat}} \text{magnesium oxide}$$

When we replace the words with symbols for the reactants and the products and include their physical state symbols, it is important to remember that oxygen is a diatomic molecule:

$$Mg(s) + O_2(g) \xrightarrow{\text{heat}} MgO(s)$$

In the equation there are two oxygen atoms on the left-hand side (O_2) but only one on the right (MgO). We cannot change the formula of magnesium oxide, so

to produce the necessary two oxygen atoms on the right-hand side we will need 2MgO – this means $2 \times MgO$. The equation now becomes:

$$Mg(s) + O_2(g) \xrightarrow{\text{heat}} 2MgO(s)$$

There are now two atoms of magnesium on the right-hand side and only one on the left. By placing a 2 in front of the magnesium, we obtain the following balanced chemical equation:

$$2Mg(s) + O_2(g) \xrightarrow{\text{heat}} 2MgO(s)$$

This balanced chemical equation now shows us that two atoms of magnesium react with one molecule of oxygen gas when heated to produce two units of magnesium oxide.

Instrumental techniques

Elements and compounds can be detected and identified by a variety of instrumental methods. Scientists have developed instrumental techniques that allow us to probe and discover which elements are present in the substance as well as how the atoms are arranged within the substance.

Many of the instrumental methods that have been developed are quite sophisticated. Some methods are suited to identifying elements. For example, atomic absorption spectroscopy allows the element to be identified and also allows the quantity of the element that is present to be found (Figure 2.8).

Figure 2.8 This instrument allows the quantity of a particular element to be found. It is used extensively throughout industry for this purpose. It will allow even tiny amounts of a particular element to be found.

Some methods are particularly suited to the identification of compounds. For example, infrared spectroscopy is used to identify compounds by showing the presence of particular groupings of atoms (Figure 2.9).

Figure 2.9 This is a modern infrared spectrometer. It is used in analysis to obtain a so-called fingerprint spectrum of a substance that will allow the substance to be identified.

Infrared spectroscopy is used by the pharmaceutical industry to identify and discriminate between drugs that are similar in structure, for example penicillin-type drugs. Used both with organic and inorganic molecules, this method assumes that each compound has a unique infrared spectrum. Samples can be solid, liquid or gas and are tiny. However, Ne, He, O_2, N_2 or H_2 cannot be used.

This method is also used to monitor environmental pollution, and has biological uses in monitoring tissue physiology including oxygenation, respiratory status and blood flow damage.

Forensic scientists make use of both these techniques because they are very accurate but they only require tiny amounts of sample – often only small amounts of sample are found at crime scenes. Other techniques utilised are nuclear magnetic resonance spectroscopy and ultraviolet/visible spectroscopy.

Questions

1 Write the word and balanced chemical equations for the reactions which take place between:
a calcium and oxygen
b copper and oxygen.

2 Write down the ratio of the atoms present in the formula for each of the compounds shown in Table 2.5.

3 Iron is extracted from iron(III) oxide in a blast furnace by a redox reaction. What does the term 'redox reaction' mean?

Mixtures

Many everyday things are not pure substances, they are mixtures. A mixture contains more than one substance (elements and/or compounds). An example of a common mixture is sea water (Figure 2.10).

Figure 2.10 Sea water is a common mixture.

What is the difference between mixtures and compounds?

There are differences between compounds and mixtures. This can be shown by considering the reaction between iron filings and sulfur. A mixture of iron filings and sulfur looks different from the individual elements (Figure 2.11). This mixture has the properties of both iron and sulfur; for example, a magnet can be used to separate the iron filings from the sulfur (Figure 2.12).

Figure 2.11 The elements (top left to right) iron and sulfur, and below (left) a mixture of iron and (right) sulfur and black iron(II) sulfide.

Figure 2.12 A magnet will separate the iron from the mixture. ■ **17**

Substances in a mixture have not undergone a chemical reaction and it is possible to separate them provided that there is a suitable difference in their physical properties. If the mixture of iron and sulfur is heated a chemical reaction occurs and a new substance is formed called iron(II) sulfide (Figure 2.11). The word equation for this reaction is:

$$\text{iron} + \text{sulfur} \xrightarrow{\text{heat}} \text{iron(II) sulfide}$$

During the reaction heat energy is given out as new chemical bonds are formed. This is called an **exothermic** reaction and accompanies a chemical change (Chapter 13, p. 191). The iron(II) sulfide formed has totally different properties to the mixture of iron and sulfur (Table 2.6). Iron(II) sulfide, for example, would not be attracted towards a magnet.

In iron(II) sulfide, FeS, one atom of iron has combined with one atom of sulfur. No such ratio exists in a mixture of iron and sulfur, because the atoms have not chemically combined. Table 2.7 summarises how mixtures and compounds compare.

Separating mixtures

Many mixtures contain useful substances mixed with unwanted material. In order to obtain these useful substances, chemists often have to separate them from the impurities. Chemists have developed many different methods of separation, particularly for separating compounds from complex mixtures. Which separation method they use depends on what is in the mixture and the properties of the substances present. It also depends on whether the substances to be separated are solids, liquids or gases.

Separating solid/liquid mixtures

If a solid substance is added to a liquid it may **dissolve** to form a **solution**. In this case the solid is said to be **soluble** and is called the **solute**. The liquid it has dissolved in is called the **solvent**. An example of this type of process is when sugar is added to tea or coffee. What other examples can you think of where this type of process takes place?

Sometimes the solid does not dissolve in the liquid. This solid is said to be **insoluble**. For example, tea leaves themselves do not dissolve in boiling water when tea is made from them, although the soluble materials from which tea is made are seen to dissolve from them.

Table 2.6 Different properties of iron, sulphur, an iron/sulphur mixture and iron(II) sulphide.

Substance	Appearance	Effect of a magnet	Effect of dilute hydrochloric acid
Iron	Dark grey powder	Attracted to it	Very little action when cold. When warm, a gas is produced with a lot of bubbling (effervescence)
Sulfur	Yellow powder	None	No effect when hot or cold
Iron/sulfur mixture	Dirty yellow powder	Iron powder attracted to it	Iron powder reacts as above
Iron(II) sulfide	Dark grey solid	No effect	A foul-smelling gas is produced with some effervescence

Table 2.7 The major differences between mixtures and compounds.

Mixture	Compound
It contains two or more substances	It is a single substance
The composition can vary	The composition is always the same
No chemical change takes place when a mixture is formed	When the new substance is formed it involves chemical change
The properties are those of the individual elements	The properties are very different to those of the component elements
The components may be separated quite easily by physical means	The components can only be separated by one or more chemical reactions

Question

1 Make a list of some other common mixtures, stating what they are mixtures of.

Filtration

When a cup of tea is poured through a tea strainer you are carrying out a **filtering** process. **Filtration** is a common separation technique used in chemistry laboratories throughout the world. It is used when a solid needs to be separated from a liquid. For example, sand can be separated from a mixture with water by filtering through filter paper as shown in Figure 2.13.

The filter paper contains holes that, although too small to be seen, are large enough to allow the molecules of water through but not the sand particles. It acts like a sieve. The sand gets trapped in the filter paper and the water passes through it. The sand is called the **residue** and the water is called the **filtrate**.

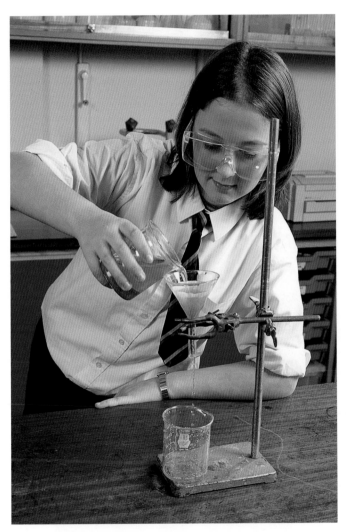

Figure 2.13 It is important when filtering not to overfill the filter paper.

Decanting

Vegetables do not dissolve in water. When you have boiled some vegetables it is easy to separate them from the water by pouring it off. This process is called **decanting**. This technique is used quite often to separate an insoluble solid, which has settled at the bottom of a flask, from a liquid.

Centrifuging

Another way to separate a solid from a liquid is to use a **centrifuge**. This technique is sometimes used instead of filtration. It is usually used when the solid particles are so small that they spread out (disperse) throughout the liquid and remain in **suspension**. They do not settle to the bottom of the container, as heavier particles would do, under the force of gravity. The technique of **centrifuging** or **centrifugation** involves the suspension being spun round very fast in a centrifuge so that the solid gets flung to the bottom of the tube (Figure 2.14a and b).

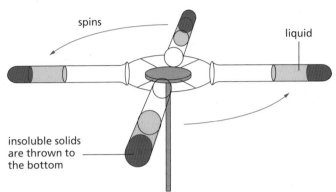

a The sample is spun round very fast and the solid is flung to the bottom of the tube.

b An open centrifuge.

Figure 2.14

The pure liquid can be decanted after the solid has been forced to the bottom of the tube. This method of separation is used extensively to separate blood cells from blood plasma (Figure 2.15). In this case, the solid particles (the blood cells) are flung to the bottom of the tube, allowing the liquid plasma to be decanted.

Figure 2.15 Whole blood (top) is separated by centrifuging into blood cells and plasma (bottom).

Evaporation

If the solid has dissolved in the liquid it cannot be separated by filtering or centrifuging. Instead, the solution can be heated so that the liquid evaporates completely and leaves the solid behind. The simplest way to obtain salt from its solution is by slow evaporation as shown in Figure 2.16.

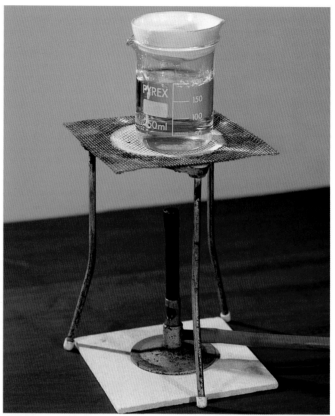

Figure 2.16 Apparatus used to slowly evaporate a solvent.

Crystallisation

In many parts of the world salt is obtained from sea water on a vast scale. This is done by using the heat of the sun to evaporate the water to leave a saturated solution of salt known as brine. A **saturated solution** is defined as one that contains as much solute as can be dissolved at a particular temperature. When the solution is saturated the salt begins to **crystallise**, and it is removed using large scoops (Figure 2.17).

Figure 2.17 Salt is obtained in Rio de Janeiro, Brazil, by evaporation of sea water.

Simple distillation

If we want to obtain the solvent from a solution, then the process of **distillation** can be carried out. The apparatus used in this process is shown in Figure 2.18.

Figure 2.19 This plant produces large quantities of drinking water in Saudi Arabia.

Separating liquid/liquid mixtures

In recent years there have been many oil tanker disasters, just like the one shown in Figure 2.20. These have resulted in millions of litres of oil being washed into the sea. Oil and water do not mix easily. They are said to be **immiscible**. When cleaning up disasters of this type, a range of chemicals can be added to the oil to make it more soluble. This results in the oil and water mixing with each other. They are now said to be **miscible**. The following techniques can be used to separate mixtures of liquids.

flask. In hot and arid countries such as Saudi Arabia this sort of technique is used on a much larger scale to obtain pure water for drinking (Figure 2.19). This process is carried out in a desalination plant.

Figure 2.18 Water can be obtained from salt water by distillation.

Water can be obtained from salt water using this method. The solution is heated in the flask until it boils. The steam rises into the Liebig condenser, where it condenses back into water. The salt is left behind in the

Figure 2.20 Millions of litres of oil are spilt in tanker disasters and cleaning-up is a slow and costly process.

Liquids which are immiscible

If two liquids are immiscible they can be separated using a **separating funnel**. The mixture is poured into the funnel and the layers allowed to separate. The lower layer can then be run off by opening the tap as shown in Figure 2.21.

Figure 2.21 The pink liquid is more dense than the clear oil and so sinks to the bottom of the separating funnel. When the tap is opened the pink liquid can be run off.

Liquids which are miscible

If miscible liquids are to be separated, then this can be done by **fractional distillation**. The apparatus used for this process is shown in the photo and diagram in Figure 2.22, and could be used to separate a mixture of ethanol and water.

Fractional distillation relies upon the liquids having different boiling points. When an ethanol and water mixture is heated the vapours of ethanol and water boil off at different temperatures and can be condensed and collected separately.

Ethanol boils at 78 °C whereas water boils at 100 °C. When the mixture is heated the vapour produced is mainly ethanol with some steam. Because water has the higher boiling point of the two, it condenses out from the mixture with ethanol. This is what takes place in the fractionating column. The water condenses and drips back into the flask while the ethanol vapour moves up the column and into the condenser, where it condenses into liquid ethanol and is collected in the receiving flask. When all the ethanol has distilled over, the temperature reading on the thermometer rises steadily to 100 °C, showing that the steam is now entering the condenser. At this point the receiver can be changed and the condensing water can now be collected.

Figure 2.22 Typical fractional distillation apparatus.

Fractional distillation is used to separate miscible liquids such as those in crude oil (see Figure 2.23a and p. 175), and the technique can also separate individual gases, such as nitrogen, from the mixture we call air (see Figure 2.23b and p. 151).

a Fractional distillation unit for crude oil.

b Gases from the air are extracted in this fractional distillation plant.

Figure 2.23

Separating solid/solid mixtures

You saw earlier in this chapter (p. 17) that it was possible to separate iron from sulfur using a magnet. In that case we were using one of the physical properties of iron, that is, the fact that it is magnetic. In a similar way, it is possible to separate scrap iron from other metals by using a large electromagnet like the one shown in Figure 2.24.

It is essential that when separating solid/solid mixtures you pay particular attention to the individual physical properties of the components. If, for example, you wish to separate two solids, one of which sublimes, then this property should dictate the method you employ.

Figure 2.24 Magnetic separation of iron-containing materials.

In the case of an iodine/salt mixture the iodine sublimes but salt does not. Iodine can be separated by heating the mixture in a fume cupboard as shown in Figure 2.25. The iodine sublimes and re-forms on the cool inverted funnel.

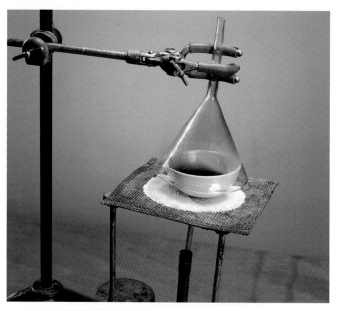

Figure 2.25 Apparatus used to separate an iodine/salt mixture. The iodine sublimes on heating.

Chromatography

What happens if you have to separate two or more solids that are soluble? This type of problem is encountered when you have mixtures of coloured materials such as inks and dyes. A technique called **chromatography** is widely used to separate these materials so that they can be identified.

There are several types of chromatography; however, they all follow the same basic principles. The simplest kind is paper chromatography. To separate the different-coloured dyes in a sample of black ink, a spot of the ink is put on to a piece of chromatography paper. This paper is then set in a suitable solvent as shown in Figure 2.26 on the next page.

As the solvent moves up the paper, the dyes are carried with it and begin to separate. They separate because the substances have different solubilities in the solvent and are absorbed to different degrees by the chromatography paper. As a result, they are separated gradually as the solvent moves up the paper. The **chromatogram** in Figure 2.26b shows how the ink contains three dyes, P, Q and R.

Numerical measurements (retardation factors) known as R_f **values** can be obtained from chromatograms. An R_f value is defined as the ratio of the distance travelled by the solute (for example P, Q or R) to the distance travelled by the solvent.

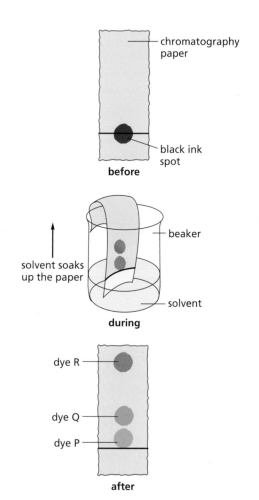

a Chromatographic separation of black ink.

chromatography paper

black ink spot

before

beaker

solvent soaks up the paper

solvent

during

dye R

dye Q

dye P

after

b The black ink separates into three dyes: P, Q and R.

Figure 2.26

Chromatography and electrophoresis (separation according to charge) are used extensively in medical research and forensic science laboratories to separate a variety of mixtures (Figure 2.27).

Figure 2.27 Protein samples are separated by electrophoresis in medical research laboratories.

The substances to be separated do not have to be coloured. Colourless substances can be made visible by spraying the chromatogram with a **locating agent**. The locating agent will react with the colourless substances to form a coloured product. In other situations the position of the substances on the chromatogram may be located using ultraviolet light.

Solvent extraction

Sugar can be obtained from crushed sugar cane by adding water. The water dissolves the sugar from the sugar cane (Figure 2.28). This is an example of **solvent extraction**. In a similar way some of the green substances can be removed from ground-up grass using ethanol. The substances are extracted from a mixture by using a solvent which dissolves only those substances required.

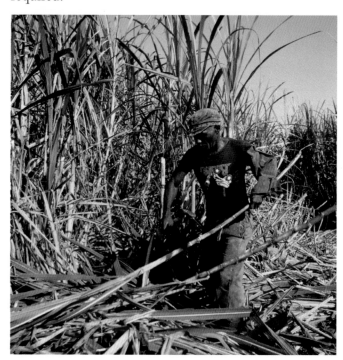

Figure 2.28 Cutting sugar cane, from which sugar can be extracted by using a suitable solvent.

Criteria for purity

Drugs are manufactured to a very high degree of purity (Figure 2.29). To ensure that the highest possible purity is obtained, the drugs are dissolved in a suitable solvent and subjected to fractional crystallisation.

Figure 2.29 Drugs are manufactured to a high degree of purity by fractional crystallisation.

It is illegal to put anything harmful into food. Also, government legislation requires that a lot of testing takes place before a new pharmaceutical is marketed.

Throughout the chemical, pharmaceutical and food industries it is essential that the substances used are pure. The purity of a substance can be gauged by:

- its melting point – if it is a pure solid it will have a sharp melting point. If an impurity is present then melting takes place over a range of temperatures.
- its boiling point – if it is a pure liquid the temperature will remain steady at its boiling point. If the substance is impure then the mixture will boil over a temperature range.
- chromatography – if it is a pure substance it will produce only one well-defined spot on a chromatogram. If impurities are present then several spots will be seen on the chromatogram (see Figure 2.26, p. 24).

Figure 2.30 These pharmaceuticals must have been through a lot of testing before they can be sold in a chemist's shop.

Questions

1 Use your research techniques (including the Internet) to obtain as many examples as you can in which a centrifuge is used.

2 What is the difference between simple distillation and fractional distillation?

3 Describe how you would use chromatography to show whether blue ink contains a single pure dye or a mixture of dyes.

4 Explain the following terms, with the aid of examples:
   a miscible
   b immiscible
   c evaporation
   d condensation
   e solvent extraction.

5 Devise a method for obtaining salt (sodium chloride) from sea water in the school laboratory.

6 What criteria can be used to test the purity of a substance?

■ *Gels, sols, foams and emulsions*

Gels, sols, foams and emulsions are all examples of mixtures which are formed by mixing two substances (or phases) which cannot mix. These mixtures are often referred to as **colloids**. Colloids are formed if the suspended particles are between 1 nm and 1000 nm in size (1 nm = 1×10^{-9} m).

Generally colloids cannot be separated by filtration since the size of the dispersed particles is smaller than that of the pores found in the filter paper. Look closely at the substances shown in Figure 2.31 to see examples of these mixtures.

a These jelly-like mixtures of solid and liquid in fruit jelly and cold custard are examples of 'gels'.

b Emulsion paint is an example of a 'sol'.

c These foams have been formed by trapping bubbles of gas in liquids or solids.

d Emulsions are formed by mixing immiscible liquids.

Figure 2.31

25

When you mix a solid with a liquid you sometimes get a gel. A gel is a semi-solid which can move around but not as freely as a liquid. Within a gel the solid makes a kind of network which traps the liquid and makes it unable to flow freely (Figure 2.32).

network of
gelatine molecules

water molecules trapped
in a network of gelatine

Figure 2.32 The network within a gel.

A gelatine gel is made with warm water. Gelatine is a protein. Proteins are natural polymers (Chapter 14, p. 209) and the molecules of protein are very large. The large molecules disperse in water to form a gel. As the gelatine-in-water mixture cools, the gelatine molecules are attracted to each other and form a continuous network. In this way, the jelly you eat as a pudding is formed. The kind of gel which you put into your hair is made from water and an oil (Figure 2.33).

Figure 2.33 Hair gel is a mixture of water and an oil plus a perfume.

A **sol** is similar to a gel; however, the mixture will flow, for example emulsion paint, or PVA glue.

When you pour out a glass of fizzy drink, the frothy part at the top of the drink is a gas/liquid mixture called a **foam**. The gas, carbon dioxide, has formed tiny bubbles in the liquid but has not dissolved in it. If left to stand, foams like this one collapse as the tiny bubbles join together to form bigger bubbles which then escape. It is possible to form solid foams where the gases are trapped in a solid structure. This happens in foam rubber and bread (Figure 2.34).

Emulsions are mixtures of liquids which are immiscible. Earlier in this chapter you found out that when two liquids are immiscible they do not mix but form two different layers. Oil and water are like this but if you shake the mixture it becomes cloudy.

Figure 2.34 Examples of solid foams.

The apparent mixing that you see is due to the fact that one of the liquids has been broken into tiny droplets which float suspended in the other liquid. If the mixture of oil and water is now left to stand the two layers will re-form. To make emulsions, such as mayonnaise, an **emulsifier** is used to stop the droplets joining back together again to form a separate layer. The emulsifier used when making mayonnaise is egg yolk. In many countries of the world, if you examine the ingredients on the side of many packets found in kitchen cupboards you will find that emulsifiers have 'E-numbers' in the range E322 to E494. For example, ammonium phosphatide E442 is used as the emulsifier in cocoa and chocolate. Other food additives such as colourings and preservatives are also given E-numbers but in different ranges to that of the emulsifiers.

It is worth noting that gels, foams and emulsions are all examples of different kinds of solutions. In true solutions the two phases completely mix together but in these systems the two phases are separate.

To produce a stable colloid, the particles dispersed must not only be of the right size (1 –1000 nm) but also be prevented from joining back together (coagulating). One way of doing this is to ensure that all the particles possess the same electrical charge. This causes the particles to repel one another.

A colloidal suspension can be destroyed by bringing the dispersed particles together. This process is known as **flocculation**. A method of doing this involves adding ionic substances such as aluminium chloride or aluminium sulphate to the particular colloid. The dispersed particles interact with the added highly charged ions and form particles which are large enough either to settle out under the force of gravity or simply be filtered out. During the treatment of water, aluminium sulphate is added to water prior to filtering to remove suspended solids (Figure 2.35).

Figure 2.35 Water is treated to remove suspended solids by the addition of aluminium sulphate.

Questions

1 Explain the following terms:
 a colloid
 b emulsifier
 c foam
 d 'E' number
 e sol.

2 Use your research skills (including the Internet) to obtain information about as many common gels, sols, foams and emulsions as you can, other than those given in the text.

Mixtures for strength
Composite materials

Composite materials are those that combine the properties of two constituents in order to get the exact properties needed for a particular job.

Glass-reinforced plastic (GRP) is an example of a composite material combining the properties of two different materials. It is made by embedding short fibres of glass in a matrix of plastic. The glass fibres give the plastic extra strength so that it does not break when it is bent or moulded into shape. The finished product has the lightness of plastic as well as the strength and flexibility of the glass fibres (Figure 2.36).

Figure 2.36 GRP consists of glass fibres (rod shapes) embedded in plastic, in this case polyester.

Figure 2.37 The glass-reinforced plastic used to make boats like this is a composite material.

Figure 2.38 Bone is a composite material.

Figure 2.39 The combination of cellulose fibres and lignin makes the cell walls hard, thick and very strong. These properties reinforce the cells against collapse.

With a little investigation you will find that many composite materials are found in the natural world. Our bones, for example, are a composite material formed from strands of the protein collagen and the mineral calcium phosphate (Figure 2.38). The calcium phosphate is hard and therefore gives strength to the bone. Another example is wood. Wood consists of cellulose fibres mixed with lignin (Figure 2.39), which is largely responsible for the strength of the wood.

Questions

1 Why are composite materials often used instead of single materials?

2 Using the information in the text and any other information available to you, give a use other than those already mentioned for each of the following composite materials:
   a reinforced concrete
   b glass-reinforced plastic
   c laminate
   d glass fibre.

Checklist

After studying Chapter 2 you should know and understand the following terms.

■ **Atom** The smallest part of an element that can exist as a stable entity.

■ **Centrifuging** The separation of the components of a mixture by rapid spinning. The denser particles are flung to the bottom of the containing tubes. The liquid can then be decanted off.

■ **Chemical change** A permanent change in which a new substance is formed.

■ **Chemical formula** A shorthand method of representing chemical elements and compounds.

■ **Chromatography** A technique employed for the separation of mixtures of dissolved substances.

■ **Colloid** Systems in which there are two or more phases, with one (the dispersed phase) distributed in the other (the continuous phase). One of the phases has particles in the range 1 to 1000 nm (1 nm = 1×10^{-9} m).

■ **Composite materials** Materials which combine the properties of two substances in order to get the exact properties required for a particular job.

■ **Compound** A substance formed by the combination of two or more elements in fixed proportions.

■ **Crystallisation** The process of forming crystals from a liquid.

■ **Decanting** The process of removing a liquid from a solid which has settled or from an immiscible heavier liquid by pouring.

■ **Diatomic molecule** A molecule containing two atoms, for example hydrogen, H_2, and oxygen, O_2.

■ **Distillation** The process of boiling a liquid and then condensing the vapour produced back into a liquid. It is used to purify liquids and to separate mixtures of liquids.

■ **Element** A substance which cannot be further divided into simpler substances by chemical methods.

■ **Emulsion** The apparent mixing of two immiscible liquids by the use of an emulsifier which breaks down one of the liquids into tiny droplets. The droplets of this liquid float suspended in the other liquid so that they do not separate out into different layers.

■ **Exothermic reaction** A chemical reaction in which heat energy is produced.

■ **Filtrate** The liquid which passes through the filter paper during filtration.

■ **Filtration** The process of separating a solid from a liquid, using a fine filter paper which does not allow the solid to pass through.

■ **Flocculation** The destruction of a colloidal suspension by bringing the dispersed particles together.

■ **Foam** A mixture formed between a gas and a liquid. The gas forms tiny bubbles in the liquid but has not dissolved in it.

■ **Gel** A mixture formed between a solid and a liquid in which the solid forms a network which traps the liquid so that it cannot flow freely.

■ **Immiscible** When two liquids form two layers when mixed together, they are said to be immiscible.

■ **Insoluble** If the solute does not dissolve in the solvent it is said to be insoluble.

■ **Instrumental techniques** Instrumental methods of analysis that are particularly useful when the amount of sample is very small. Examples are atomic absorption spectroscopy and infrared spectroscopy.

■ **Law of constant composition** Compounds always have the same elements joined together in the same proportions.

■ **Metals** A class of chemical elements which have a characteristic lustrous appearance and which are good conductors of heat and electricity.

■ **Miscible** When two liquids form a homogeneous layer when mixed together, they are said to be miscible.

■ **Mixture** A system of two or more substances that can be separated by physical means.

■ **Molecule** A group of atoms chemically bonded together.

continued

■ **Monatomic molecule**   A molecule which consists of only one atom, for example neon and argon.

■ **Non-metals**   A class of chemical elements that are typically poor conductors of heat and electricity.

■ **Oxidation**   The process of combining with oxygen.

■ **Redox reaction**   A reaction which involves the two processes of reduction and oxidation.

■ **Reduction**   The process of removing oxygen.

■ **Residue**   The solid left behind in the filter paper after filtration has taken place.

■ R_f **value**   This is the ratio of the distance travelled by the solute to the distance travelled by the solvent in chromatography.

■ **Saturated solution**   This is a solution which contains as much dissolved solute as it can at a particular temperature.

■ **Sol**   A mixture formed between a solid and a liquid, which then forms a network that can flow.

■ **Soluble**   If the solute dissolves in the solvent it is said to be soluble.

■ **Solution**   This is formed when a substance (solute) disappears (dissolves) into another substance (solvent).

$$\text{solute} + \text{solvent} \xrightarrow{\text{dissolves}} \text{solution}$$

Elements, compounds and mixtures
Additional questions

1 Define the following terms using specific examples to help with your explanation:
  a element
  b metal
  c non-metal
  d compound
  e molecule
  f mixture
  g flocculation
  h gel
  i foam
  j emulsion
  k sol.

2 Which of the substances listed below are:
  a metallic elements?
  b non-metallic elements?
  c compounds?
  d mixtures?
  Silicon, sea water, calcium, argon, water, air, carbon monoxide, iron, sodium chloride, diamond, brass, copper, dilute sulfuric acid, sulfur, oil, nitrogen, ammonia.

3 At room temperature and pressure (rtp), which of the substances listed below is:
  a a solid element?
  b a liquid element?
  c a gaseous mixture?
  d a solid mixture?
  e a liquid compound?
  f a solid compound?
  Bromine, carbon dioxide, helium, steel, air, oil, marble, copper, water, sand, tin, bronze, mercury, salt.

4 A student heated a mixture of iron filings and sulfur strongly. He saw a red glow spread through the mixture as the reaction continued. At the end of the experiment a black solid had been formed.
  a Explain what the red glow indicates.
  b Give the chemical name of the black solid.
  c Write a word equation and a balanced chemical equation to represent the reaction which has taken place.
  d The black solid is a compound. Explain the difference between the mixture of sulfur and iron and the compound formed by the chemical reaction between them.

5 Name the method which is most suitable for separating the following:
  a oxygen from liquid air
  b red blood cells from plasma
  c petrol and kerosene from crude oil
  d coffee grains from coffee solution
  e pieces of steel from engine oil
  f amino acids from fruit juice solution
  g ethanol and water.

6 The table below shows the melting points, boiling points and densities of substances A to D.

Substance	Melting point/ °C	Boiling point/ °C	Density/ g cm^{-3}
A	1110	2606	9.1
B	−266	−252	0.07
C	40	94	1.6
D	−14	60	0.9

  a Which substance is a gas at room temperature?
  b Which substance is a liquid at room temperature?
  c Which substances are solids at room temperature?
  d Which substance is most likely to be a metal?
  e Which substance will be a liquid at −260 °C?
  f What is the melting point of the least dense non-metal?
  g Which substance is a gas at 72 °C?

7 a How many atoms of the different elements are there in the formulae of the compounds given below?
    (i) nitric acid, HNO_3
    (ii) methane, CH_4
    (iii) copper nitrate, $Cu(NO_3)_2$
    (iv) ethanoic acid, CH_3COOH
    (v) sugar, $C_{12}H_{22}O_{11}$
    (vi) phenol, C_6H_5OH
    (vii) ammonium sulfate, $(NH_4)_2SO_4$
  b Balance the following equations:
    (i) $Zn(s) + O_2(g) \rightarrow ZnO(s)$
    (ii) $Fe(s) + Cl_2(g) \rightarrow FeCl_3(s)$
    (iii) $Li(s) + O_2(g) \rightarrow Li_2O(s)$
    (iv) $H_2(g) + O_2(g) \rightarrow H_2O(g)$
    (v) $Mg(s) + CO_2(g) \rightarrow MgO(s) + C(s)$

8 Carbon-fibre-reinforced plastic (CRP) is used in the manufacture of golf clubs and tennis rackets.
  a What are composite materials?
  b Which two substances are used to manufacture this composite material?

Consider the data below.

Material	Strength/ GPa	Stiffness/ GPa	Density/ g cm^{-3}	Relative cost
Aluminium	0.2	75	2.7	low
Steel	1.1	200	7.8	low
CRP	1.8	195	1.6	high

  c Discuss the advantages and disadvantages of using the three materials above in the manufacture of golf clubs.

31

3 Atomic structure and the periodic table

Inside atoms
Proton number and nucleon number
Ions
Isotopes
The mass spectrometer
Relative atomic mass

The arrangement of electrons in atoms

The periodic table
Electronic structure and the periodic table
Group 1 – the alkalli metals
Group 2 – the alkaline earth metals
Flame colours
Group 7 – the halogens
Group 0 – the noble gases
Transition elements
The position of hydrogen

Checklist
Additional questions

We have already seen in Chapter 2 that everything you see around you is made out of tiny particles called atoms (Figure 3.1). When John Dalton developed his atomic theory, about 200 years ago (1807/1808), he stated that the atoms of any one element were identical and that each atom was 'indivisible'. Scientists in those days believed that atoms were solid particles like marbles.

However, in the last hundred years or so it has been proved by great scientists, such as Niels Bohr, Albert Einstein, Henry Moseley, Joseph Thomson, Ernest Rutherford and James Chadwick, that atoms are in fact made up of even smaller 'sub-atomic' particles. The most important of these are **electrons**, **protons** and **neutrons**, although 70 sub-atomic particles have now been discovered.

Inside atoms

The three sub-atomic particles are found in distinct and separate regions. The protons and neutrons are found in the centre of the atom, which is called the **nucleus**. The neutrons have no charge and protons are positively charged. The nucleus occupies only a very small volume of the atom but is very dense.

The rest of the atom surrounding the nucleus is where electrons are most likely to be found. The electrons are negatively charged and move around very quickly in **electron shells** or **energy levels**. The electrons are held within the atom by an **electrostatic force of attraction** between themselves and the positive charge of protons in the nucleus (Figure 3.2).

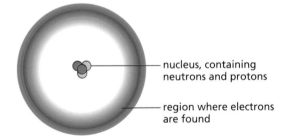

nucleus, containing neutrons and protons

region where electrons are found

Figure 3.2 Diagram of an atom.

About 1837 electrons are equal in mass to the mass of one proton or one neutron. A summary of each type of particle, its mass and relative charge is shown in Table 3.1. You will notice that the masses of all these particles are measured in **atomic mass units (amu)**. This is because they are so light that their masses cannot be measured usefully in grams.

Table 3.1 Characteristics of a proton, a neutron and an electron.

Particle	Symbol	Relative mass/amu	Relative charge
Proton	p	1	+1
Neutron	n	1	0
Electron	e	1/1837	−1

Figure 3.1 Atoms. A field ion micrograph of atoms of the metal element iridium. The tiny dots are the locations of individual atoms, and the ring-like patterns are facets of a single crystal.

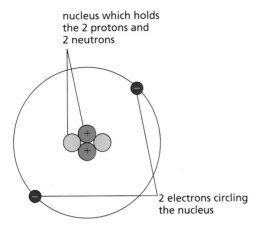

nucleus which holds
the 2 protons and
2 neutrons

2 electrons circling
the nucleus

Figure 3.3 An atom of helium has two protons, two electrons and two neutrons.

Although atoms contain electrically charged particles, the atoms themselves are electrically neutral (they have no overall electric charge). This is because atoms contain equal numbers of electrons and protons. For example, the diagram in Figure 3.3 represents the atom of the non-metallic element helium. The atom of helium possesses two protons, two neutrons and two electrons. The electrical charge of the protons in the nucleus is, therefore, balanced by the opposite charge of the two electrons.

Proton number and nucleon number

The number of protons in the nucleus of an atom is called the **proton number** (or atomic number) and is given the symbol **Z**. Hence in the diagram shown in Figure 3.3, the helium atom has a proton number of 2, since it has two protons in its nucleus. Each element has its own proton number and no two different elements have the same proton number. For example, a different element, lithium, has a proton number of 3, since it has three protons in its nucleus.

Neutrons and protons have a similar mass. Electrons possess very little mass. So the mass of any atom depends on the number of neutrons and protons in its nucleus. The total number of protons and neutrons found in the nucleus of an atom is called the **nucleon number** (or mass number) and is given the symbol **A**.

$$\underset{(A)}{\text{nucleon number}} = \underset{(Z)}{\text{proton number}} + \underset{\text{neutrons}}{\text{number of}}$$

Hence, in the example shown in Figure 3.3 the helium atom has a nucleon number of 4, since it has two protons and two neutrons in its nucleus. If we consider the metallic element lithium, it has three protons and four neutrons in its nucleus. It therefore has a nucleon number of 7.

The proton number and nucleon number of an element are usually written in the following shorthand way:

nucleon number (A)

$_{2}^{4}\text{He}\leftarrow$symbol of the element

proton number (Z)

The number of neutrons present can be calculated by rearranging the relationship between the proton number, nucleon number and number of neutrons to give:

$$\underset{\text{neutrons}}{\text{number of}} = \underset{(A)}{\text{nucleon number}} - \underset{(Z)}{\text{proton number}}$$

For example, the number of neutrons in one atom of $_{12}^{24}\text{Mg}$ is:

$$\underset{(A)\ (Z)}{24 - 12 = 12}$$

and the number of neutrons in one atom of $_{82}^{207}\text{Pb}$ is:

$$\underset{(A)\ (Z)}{207 - 82 = 125}$$

Table 3.2 shows the number of protons, neutrons and electrons in the atoms of some common elements.

Table 3.2 Number of protons, neutrons and electrons in some elements.

Element	Symbol	Proton number	Number of electrons	Number of protons	Number of neutrons	Nucleon number
Hydrogen	H	1	1	1	0	1
Helium	He	2	2	2	2	4
Carbon	C	6	6	6	6	12
Nitrogen	N	7	7	7	7	14
Oxygen	O	8	8	8	8	16
Fluorine	F	9	9	9	10	19
Neon	Ne	10	10	10	10	20
Sodium	Na	11	11	11	12	23
Magnesium	Mg	12	12	12	12	24
Sulfur	S	16	16	16	16	32
Potassium	K	19	19	19	20	39
Calcium	Ca	20	20	20	20	40
Iron	Fe	26	26	26	30	56
Zinc	Zn	30	30	30	35	65

Ions

An ion is an electrically charged particle. When an atom loses one or more electrons it becomes a positively charged ion. For example, during the chemical reactions of potassium, each atom loses an electron to form a positive ion, K^+.

$$_{19}K^+ \quad \begin{array}{ll} 19 \text{ protons} & = 19+ \\ 18 \text{ electrons} & = 18- \\ \hline \text{Overall charge} & = 1+ \end{array}$$

When an atom gains one or more electrons it becomes a negatively charged ion. For example, during some of the chemical reactions of oxygen it gains two electrons to form a negative ion, O^{2-}.

$$_{8}O^{2-} \quad \begin{array}{ll} 8 \text{ protons} & = 8+ \\ 10 \text{ electrons} & = 10- \\ \hline \text{Overall charge} & = 2- \end{array}$$

Table 3.3 shows some common ions. You will notice from Table 3.3 that:

- some ions contain more than one type of atom, for example NO_3^-
- an ion may possess more than one unit of charge (either negative or positive), for example Al^{3+}, O^{2-} or SO_4^{2-}.

Table 3.3 Some common ions.

Name	Formula
Lithium ion	Li^+
Sodium ion	Na^+
Potassium ion	K^+
Magnesium ion	Mg^{2+}
Calcium ion	Ca^{2+}
Aluminium ion	Al^{3+}
Zinc ion	Zn^{2+}
Ammonium ion	NH_4^+
Fluoride ion	F^-
Chloride ion	Cl^-
Bromide ion	Br^-
Hydroxide ion	OH^-
Oxide ion	O^{2-}
Sulfide ion	S^{2-}
Carbonate ion	CO_3^{2-}
Nitrate ion	NO_3^-
Sulphate ion	SO_4^{2-}

Isotopes

Not all of the atoms in a sample of chlorine, for example, will be identical. Some atoms of the same element can contain different numbers of neutrons. Atoms of the same element which have different numbers of neutrons are called **isotopes**.

The two isotopes of chlorine are shown in Figure 3.4. Generally, isotopes behave in the same way during chemical reactions. The only effect of the extra neutrons is to alter the mass of the atom and properties which depend on it, such as density. Some other examples of atoms with isotopes are shown in Table 3.4.

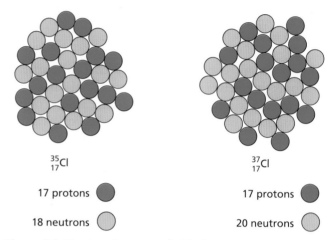

$$_{17}^{35}Cl \qquad\qquad _{17}^{37}Cl$$

17 protons ●   17 protons ●
18 neutrons ○   20 neutrons ○

Figure 3.4 The two isotopes of chlorine.

There are two types of isotopes: those which are stable and those which are unstable. The isotopes which are unstable, as a result of the extra neutrons in their nuclei, are **radioactive** and are called **radioisotopes**. For example, uranium-235 which is used as a source of power in nuclear reactors (p. 187) and cobalt-60 which is used in radiotherapy treatment are both radioisotopes.

Table 3.4 Some atoms and their isotopes.

Element	Symbol	Particles present
Hydrogen	$_1^1H$	1 e, 1 p, 0 n
(Deuterium)	$_1^2H$	1 e, 1 p, 1 n
(Tritium)	$_1^3H$	1 e, 1 p, 2 n
Carbon	$_6^{12}C$	6 e, 6 p, 6 n
	$_6^{13}C$	6 e, 6 p, 7 n
	$_6^{14}C$	6 e, 6 p, 8 n
Oxygen	$_8^{16}O$	8 e, 8 p, 8 n
	$_8^{17}O$	8 e, 8 p, 9 n
	$_8^{18}O$	8 e, 8 p, 10 n
Strontium	$_{38}^{86}Sr$	38 e, 38 p, 48 n
	$_{38}^{88}Sr$	38 e, 38 p, 50 n
	$_{38}^{90}Sr$	38 e, 38 p, 52 n
Uranium	$_{92}^{235}U$	92 e, 92 p, 143 n
	$_{92}^{238}U$	92 e, 92 p, 146 n

The mass spectrometer

Figure 3.5 A diagram of a mass spectrometer.

How do we know isotopes exist? They were first discovered by scientists using apparatus called a mass spectrometer (Figure 3.5). The first mass spectrometer was built by the British scientist Francis Aston in 1919 and enabled scientists to compare the relative masses of atoms accurately for the first time.

A vacuum exists inside a mass spectrometer. A sample of the vapour of the element is injected into a chamber where it is bombarded by electrons. The subsequent collisions that take place cause the atoms in the vapour to lose one of their electrons and so form positive ions. The beam of positive ions is accelerated by an electric field and then deflected by a magnetic field. The amount of deflection depends on the different masses of the positive ions. The lighter ions, which were formed from the lighter isotopes, are deflected more than the heavier ones. In this way particles with different masses can be separated and identified. A detector counts the number of each of the ions that fall upon it and so a measure of the percentage abundance of each isotope is obtained. A typical mass spectrum for chlorine is shown in Figure 3.6.

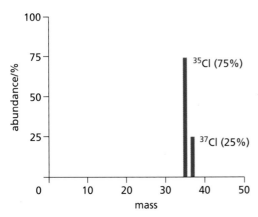

Figure 3.6 The mass spectrum for chlorine.

Relative atomic mass

The average mass of a large number of atoms of an element is called its **relative atomic mass** (symbol A_r). This quantity takes into account the percentage abundance of all the isotopes of an element which exist.

In 1961 the International Union of Pure and Applied Chemistry (IUPAC) recommended that the standard used for the A_r scale was carbon-12. An atom of carbon-12 was taken to have a mass of 12 amu. The A_r of an element is now defined as the average mass of its isotopes compared with one-twelfth the mass of one atom of carbon-12:

$$A_r = \frac{\text{average mass of isotopes of the element}}{\frac{1}{12} \times \text{mass of 1 atom of carbon-12}}$$

Note: $\frac{1}{12}$ of the mass of one carbon-12 atom = 1 amu.

For example, chlorine has two isotopes:

	$^{35}_{17}Cl$	$^{37}_{17}Cl$
% Abundance	75	25

Hence the 'average mass' of a chlorine atom is:

$$\frac{(75 \times 35) + (25 \times 37)}{100} = 35.5$$

$$A_r = \frac{35.5}{1}$$

$$= 35.5 \text{ amu}$$

Questions

1 Calculate the number of neutrons in the following atoms:

    a $^{27}_{13}Al$ **b** $^{31}_{15}P$ **c** $^{262}_{107}Uns$ **d** $^{190}_{76}Os$

2 Given that the percentage abundance of $^{20}_{10}Ne$ is 90% and that of $^{22}_{10}Ne$ is 10%, calculate the A_r of neon.

The arrangement of electrons in atoms

The nucleus of an atom contains the heavier sub-atomic particles – the protons and the neutrons. The electrons, the lightest of the sub-atomic particles, move around the nucleus at great distances from the nucleus relative to their size. They move very fast in electron energy levels very much as the planets orbit the Sun.

It is not possible to give the exact position of an electron in an energy level. However, we can state that electrons can only occupy certain, definite energy levels and that they cannot exist between them. Each of the electron energy levels can hold only a certain number of electrons.

- First energy level holds up to 2 electrons.
- Second energy level holds up to 8 electrons.
- Third energy level holds up to 18 electrons.

There are further energy levels which contain increasing numbers of electrons.

The third energy level can be occupied by a maximum of 18 electrons. However, when eight electrons have occupied this level a certain stability is given to the atom and the next two electrons go into the fourth energy level, and then the remaining ten electrons complete the third energy level.

The electrons fill the energy levels starting from the energy level nearest to the nucleus, which has the lowest energy. When this is full (with two electrons) the next electron goes into the second energy level. When this energy level is full with eight electrons, then the electrons begin to fill the third and fourth energy levels as stated above.

For example, a $^{16}_{8}O$ atom has a proton number of 8 and therefore has eight electrons. Two of the eight electrons enter the first energy level, leaving six to occupy the second energy level, as shown in Figure 3.7. The electron configuration for oxygen can be written in a shorthand way as 2,6.

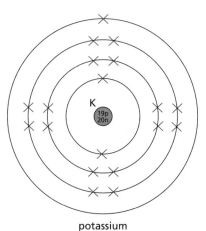

Figure 3.8 Electron arrangements of hydrogen, lithium, sodium, argon and potassium.

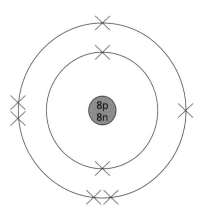

Figure 3.7 Arrangement of electrons in an oxygen atom.

There are 117 elements, and Table 3.5 shows the way in which the electrons are arranged in the first 20 of these elements. The way in which the electrons are distributed is called the **electronic structure** (or electron configuration). Figure 3.8 shows the electronic structure of a selection of atoms.

Table 3.5 Electronic structure of the first 20 elements.

Element	Symbol	Proton number	Number of electrons	Electron structure
Hydrogen	H	1	1	1
Helium	He	2	2	2
Lithium	Li	3	3	2,1
Beryllium	Be	4	4	2,2
Boron	B	5	5	2,3
Carbon	C	6	6	2,4
Nitrogen	N	7	7	2,5
Oxygen	O	8	8	2,6
Fluorine	F	9	9	2,7
Neon	Ne	10	10	2,8
Sodium	Na	11	11	2,8,1
Magnesium	Mg	12	12	2,8,2
Aluminium	Al	13	13	2,8,3
Silicon	Si	14	14	2,8,4
Phosphorus	P	15	15	2,8,5
Sulfur	S	16	16	2,8,6
Chlorine	Cl	17	17	2,8,7
Argon	Ar	18	18	2,8,8
Potassium	K	19	19	2,8,8,1
Calcium	Ca	20	20	2,8,8,2

Questions

1 How many electrons may be accommodated in the first three energy levels?

2 What is the same about the electron structures of:
   a lithium, sodium and potassium?
   b beryllium, magnesium and calcium?

The periodic table

The periodic table was devised in 1869 by the Russian Dmitri Mendeleev, who was the Professor of Chemistry at St Petersburg University (Figure 3.9). His periodic table was based on the chemical and physical properties of the 63 elements that had been discovered at that time.

Figure 3.9 Dmitri Mendeleev (1834–1907).

However, other scientists had also attempted to categorise the known elements. In 1817, Johann Döbereiner noticed that the atomic weight (now called atomic mass) of strontium fell midway between the weights of calcium and barium. These were elements which possessed similar chemical properties. They formed a **triad** of elements. Other triads were also discovered, composed of:

chlorine, bromine, iodine

lithium, sodium, potassium

He called this the '**Law of Triads**'. This encouraged other scientists to search for patterns.

In 1865, John Newlands, an English chemist, arranged the 56 known elements in order of increasing atomic weight. He realised when he did this that every eighth element in the series was similar.

H **Li** Be B C N O F **Na** Mg Al Si P S Cl **K**

He likened this to music and called it the '**Law of Octaves**'. It fell down, however, because some of the weights were inaccurate and there were elements that had not been discovered then.

Period	Group							
	1	2	3	4	5	6	7	8
1	H							
2	Li	Be	B	C	N	O	F	
3	Na	Mg	Al	Si	P	S	Cl	
4	K	Ca	*	Ti	V	Cr	Mn	Fe Co Ni
	Cu	Zn	*	†	As	Se	Br	

Figure 3.10 Mendeleev's periodic table. He left gaps for undiscovered elements.

Mendeleev's classification proved to be the most successful. Mendeleev arranged all the 63 known elements in order of increasing atomic weight but in such a way that elements with similar properties were in the same vertical column. He called the vertical columns **groups** and the horizontal rows **periods** (Figure 3.10). If necessary he left gaps in the table.

As a scientific idea, Mendeleev's periodic table was tested by making predictions about elements that were unknown at that time but could possibly fill the gaps.

Three of these gaps are shown by the symbols ★ and † in Figure 3.10. As new elements were discovered, they were found to fit easily into the classification. For example, Mendeleev predicted the properties of the missing element 'eka-silicon' (†). He predicted the colour, density and melting point as well as its atomic weight.

In 1886 the element we now know as germanium was discovered in Germany by Clemens Winkler; its properties were almost exactly those Mendeleev had predicted. In all, Mendeleev predicted the atomic weight of ten new elements, of which seven were eventually discovered – the other three, atomic weights 45, 146 and 175 do not exist!

The success of Mendeleev's predictions showed that his ideas were probably correct. His periodic table was quickly accepted by scientists as an important summary of the properties of the elements.

Mendeleev's periodic table has been modified in the light of work carried out by Rutherford and Moseley. Discoveries about sub-atomic particles led them to realise that the elements should be arranged by proton number. In the modern periodic table the 117 known elements are arranged in order of increasing proton number (Figure 3.11).

Key
- reactive metals
- transition metals
- poor metals
- metalloids
- non-metals
- noble gases

Figure 3.11 The modern periodic table.

Figure 3.12 Transition elements have a wide range of uses, both as elements and as alloys.

Those elements with similar chemical properties are found in the same columns or **groups**. There are eight groups of elements. The first column is called group 1; the second group 2; and so on up to group 7. The final column in the periodic table is called group 0. Some of the groups have been given names.

Group 1:	The alkali metals
Group 2:	The alkaline earth metals
Group 7:	The halogens
Group 0:	Inert gases or noble gases

The horizontal rows are called **periods** and these are numbered 1–7 going down the periodic table.

Between groups 2 and 3 is the block of elements known as the transition elements (Figure 3.12).

The periodic table can be divided into two as shown by the bold line that starts beneath boron, in Figure 3.11. The elements to the left of this line are metals (fewer than three-quarters) and those on the right are non-metals (fewer then one-quarter). The elements which lie on this dividing line are known as metalloids (Figure 3.13). These elements behave in some ways as metals and in others as non-metals.

If you look at the properties of the elements across a period of the periodic table you will notice certain trends. For example, there is:

- a gradual change from metal to non-metal
- an increase in the number of electrons in the outer energy level of the element
- a change in the structure of the element, from giant metallic in the case of metals (e.g. magnesium, p. 65, Figure 4.33), through giant covalent (e.g. diamond, p. 62, Figure 4.23), to simple molecular (e.g. chlorine, p. 56, Figure 4.10).

Figure 3.13 The metalloid silicon is used to make silicon 'chips'.

Electronic structure and the periodic table

Now that the number of electrons in the outer energy level has been established, it can be seen that it corresponds with the number of the group in the periodic table in which the element is found. For example, the elements shown in Table 3.6 have one electron in their outer energy level and they are all found in group 1. The elements in group 0, however, are an exception to this rule, as they have two or eight electrons in their outer energy level. The outer electrons are mainly responsible for the chemical properties of any element, and, therefore, elements in the same group have similar chemical properties (Tables 3.7 and 3.8).

Table 3.6 Electronic structure of the first three elements of group 1.

Element	Symbol	Proton number	Electronic structure
Lithium	Li	3	2,1
Sodium	Na	11	2,8,1
Potassium	K	19	2,8,8,1

Table 3.7 Electronic structure of the first three elements of group 2.

Element	Symbol	Proton number	Electronic structure
Beryllium	Be	4	2,2
Magnesium	Mg	12	2,8,2
Calcium	Ca	20	2,8,8,2

Table 3.8 Electronic structure of the first three elements in group 7.

Element	Symbol	Proton number	Electronic structure
Fluorine	F	9	2,7
Chlorine	Cl	17	2,8,7
Bromine	Br	35	2,8,18,7

Group 1 – the alkali metals

Group 1 consists of the five metals lithium, sodium, potassium, rubidium and caesium, and the radioactive element francium. Lithium, sodium and potassium are commonly available for use in school. They are all very reactive metals and they are stored under oil to prevent them coming into contact with water or air. These three metals have the following properties.

- They are good conductors of electricity and heat.
- They are soft metals. Lithium is the hardest and potassium the softest.

- They are metals with low densities. For example, lithium has a density of $0.53\ \mathrm{g\ cm^{-3}}$ and potassium has a density of $0.86\ \mathrm{g\ cm^{-3}}$.
- They have shiny surfaces when freshly cut with a knife (Figure 3.14).
- They have low melting points. For example, lithium has a melting point of 181 °C and potassium has a melting point of 64 °C.

Figure 3.14 Freshly cut sodium.

- They burn in oxygen or air, with characteristic flame colours, to form white solid oxides. For example, lithium reacts with oxygen in air to form white lithium oxide, according to the following equation:

$$\text{lithium} + \text{oxygen} \rightarrow \text{lithium oxide}$$
$$4\mathrm{Li(s)} + \mathrm{O_2(g)} \rightarrow 2\mathrm{Li_2O(s)}$$

These group 1 oxides all dissolve in water to form alkaline solutions of the metal hydroxide.

$$\text{lithium oxide} + \text{water} \rightarrow \text{lithium hydroxide}$$
$$\mathrm{Li_2O(s)} + \mathrm{H_2O(l)} \rightarrow 2\mathrm{LiOH(aq)}$$

- They react vigorously with water to give an alkaline solution of the metal hydroxide as well as producing hydrogen gas. For example:

$$\text{potassium} + \text{water} \rightarrow \text{potassium hydroxide} + \text{hydrogen gas}$$
$$2\mathrm{K(s)} + 2\mathrm{H_2O(l)} \rightarrow 2\mathrm{KOH(aq)} + \mathrm{H_2(g)}$$

Of these three metals, potassium is the most reactive towards water (Figure 3.15), followed by sodium and then lithium. Such gradual changes we call **trends**. Trends are useful to chemists as they allow predictions to be made about elements we have not observed in action.

- They react vigorously with halogens, such as chlorine, to form metal halides, for example sodium chloride (Figure 3.16).

$$\text{sodium} + \text{chorine} \rightarrow \text{sodium chloride}$$
$$2\mathrm{Na(s)} + \mathrm{Cl_2(g)} \rightarrow 2\mathrm{NaCl(s)}$$

Considering the group as a whole, the further down the group you go the more reactive the metals become. Francium is, therefore, the most reactive group 1 metal.

a Potassium reacts very vigorously with cold water.

b An alkaline solution is produced when potassium reacts with water.

Figure 3.15

Figure 3.16 A very vigorous reaction takes place when sodium burns in chlorine gas. Sodium chloride is produced.

Table 3.6 shows the electronic structure of the first three elements of group 1. You will notice in each case that the outer energy level contains only one electron. When these elements react they lose this outer electron, and in doing so become more stable, because they obtain the electron configuration of a noble gas. You will learn more about the stable nature of these gases later in this chapter (p. 46).

When, for example, the element sodium reacts it loses its outer electron. This requires energy to overcome the electrostatic attractive forces between the outer electron and the positive nucleus (Figure 3.17).

sodium atom
(Na)

sodium ion
(Na$^+$)

$$Na \longrightarrow Na^+ + e^-$$

Figure 3.17 This sodium atom loses an electron to become a sodium ion.

Look at Figure 3.18. Why do you think potassium is more reactive than lithium or sodium?

Potassium is more reactive because less energy is required to remove the outer electron from its atom than for lithium or sodium. This is because as you go down the group the size of the atoms increases and the outer electron gets further away from the nucleus and becomes easier to remove.

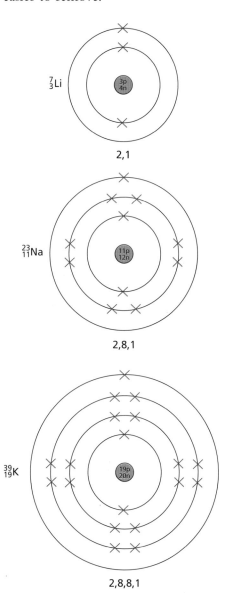

Figure 3.18 Electronic structures of lithium, sodium and potassium.

Questions

1 Write word and balanced chemical equations for the reactions between:
   a sodium and oxygen
   b sodium and water.

2 Write word and balanced chemical equations for the reactions between:
   a magnesium and water
   b calcium and oxygen.

3 Account for the fact that calcium is more reactive than magnesium.

Group 2 – the alkaline earth metals

Group 2 consists of the five metals beryllium, magnesium, calcium, strontium and barium, and the radioactive element radium. Magnesium and calcium are generally available for use in school. These metals have the following properties.

- They are harder than those in group 1.
- They are silvery-grey in colour when pure and clean. They tarnish quickly, however, when left in air due to the formation of a metal oxide on their surfaces (Figure 3.19).
- They are good conductors of heat and electricity.
- They burn in oxygen or air with characteristic flame colours to form solid white oxides. For example:

$$\text{magnesium} + \text{oxygen} \rightarrow \text{magnesium oxide}$$
$$2Mg(s) + O_2(g) \rightarrow 2MgO(s)$$

- They react with water, but they do so much less vigorously than the elements in group 1. For example:

$$\text{calcium} + \text{water} \rightarrow \text{calcium hydroxide} + \text{hydrogen gas}$$
$$Ca(s) + 2H_2O(l) \rightarrow Ca(OH)_2(aq) + H_2(g)$$

Considering the group as a whole, the further down the group you go, the more reactive the elements become.

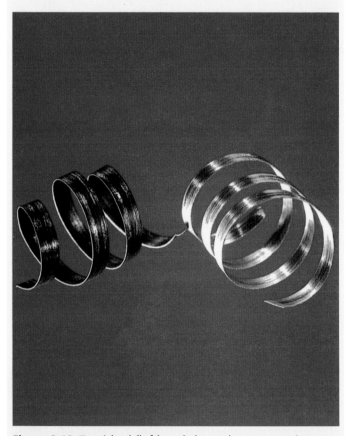

Figure 3.19 Tarnished (left) and cleaned-up magnesium.

Flame colours

If a clean nichrome wire is dipped into a metal compound and then held in the hot part of a Bunsen flame, the flame can become coloured (Figure 3.20). Certain metal ions may be detected in their compounds by observing their flame colours (Table 3.9).

Table 3.9 Characteristic flame colours of some metal ions.

	Metal	Flame colour
Group 1	Lithium	Crimson
	Sodium	Golden yellow
	Potassium	Lilac
	Rubidium	Red
	Caesium	Blue
Group 2	Calcium	Brick red
	Strontium	Crimson
	Barium	Apple green
Others	Lead	Blue–white
	Copper	Green

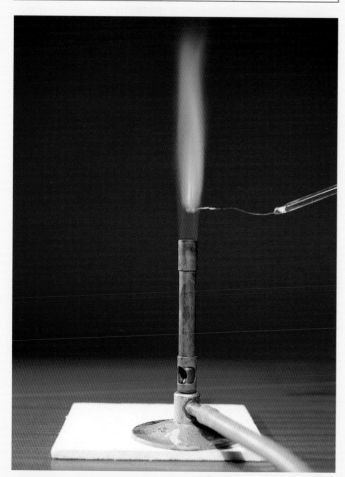

Figure 3.20 The green colour is characteristic of copper.

A flame colour is obtained as a result of the electrons in the particular ions being excited when they absorb energy from the flame which is then emitted as visible light. The different electronic structures of the different ions, therefore, give rise to the different colours (Figure 3.21).

Figure 3.21 Different metal ions are used to produce the colours seen in fireworks.

Group 7 – the halogens

Group 7 consists of the four elements fluorine, chlorine, bromine and iodine, and the radioactive element astatine. Of these five elements, chlorine, bromine and iodine are generally available for use in school.

- These elements are coloured and darken going down the group (Table 3.10).
- They exist as diatomic molecules, for example Cl_2, Br_2 and I_2.
- They show a gradual change from a gas (Cl_2), through a liquid (Br_2), to a solid (I_2) (Figure 3.22, overleaf).
- They form molecular compounds with other non-metallic elements, for example HCl.
- They react with hydrogen to produce the hydrogen halides, which dissolve in water to form acidic solutions.

$$\text{hydrogen} + \text{chlorine} \rightarrow \text{hydrogen chloride}$$
$$H_2(g) + Cl_2(g) \rightarrow 2HCl(g)$$

$$\text{hydrogen chloride} + \text{water} \rightarrow \text{hydrochloric acid}$$
$$HCl(g) + H_2O \rightarrow HCl(aq) \rightarrow H^+(aq) + Cl^-(aq)$$

- They react with metals to produce ionic metal halides, for example chlorine and iron produce iron(III) chloride.

$$\text{iron} + \text{chlorine} \rightarrow \text{iron(III) chloride}$$
$$2Fe(s) + 3Cl_2(g) \rightarrow 2FeCl_3(s)$$

To find out about the extraction of bromine, see Chapter 10, p. 153.

Table 3.10 Colours of some halogens.

Halogen	Colour
Chlorine	Pale green
Bromine	Red–brown
Iodine	Purple–black

a Chlorine, bromine and iodine.

b Chlorine gas bleaches moist indicator paper.

Figure 3.22

Displacement reactions

If chlorine is bubbled into a solution of potassium iodide, the less reactive halogen, iodine, is **displaced** by the more reactive halogen, chlorine, as you can see from Figure 3.23:

potassium + chlorine → potassium + iodine
iodide   chloride
$2KI_{(aq)}$ + $Cl_2{}_{(g)}$ → $2KCl_{(aq)}$ + $I_2{}_{(aq)}$

Figure 3.23 Iodine being displaced from potassium iodide solution as chlorine is bubbled through.

The observed order of reactivity of the halogens, confirmed by similar displacement reactions, is:

Decreasing reactivity

→

chlorine   bromine   iodine

You will notice that, in contrast to the elements of groups 1 and 2, the order of reactivity decreases on going down the group.

Table 3.11 shows the electronic structure for chlorine and bromine. In each case the outer energy level contains seven electrons. When these elements react they gain one electron per atom to gain the stable electron configuration of a noble gas. You will learn more about the stable nature of these gases in the next section. For example, when chlorine reacts it gains a single electron and forms a negative ion (Figure 3.24).

Table 3.11 Electronic structure of chlorine and bromine.

Element	Symbol	Proton number	Electronic structure
Chlorine	Cl	17	2,8,7
Bromine	Br	35	2,8,18,7

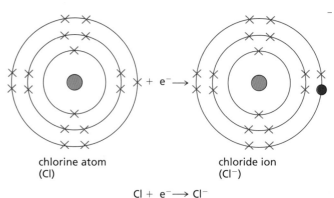

chlorine atom
(Cl)

chloride ion
(Cl⁻)

$Cl + e^- \longrightarrow Cl^-$

Figure 3.24 A chlorine atom gains an electron to form a chloride ion.

Chlorine is more reactive than bromine because the incoming electron is being more strongly attracted into the outer energy level of the smaller atom. The attractive force on it will be greater than in the case of bromine, since the outer energy level of chlorine is closer to the nucleus. As you go down the group this outermost extra electron is further from the nucleus. It will, therefore, be held less securely, and the resulting reactivity of the elements in group 7 will decrease down the group.

The halogens and halogenic compounds are used in a multitude of different ways (Figure 3.25).

In the reaction of chlorine with potassium iodide, both Cl atoms in Cl_2 gain an electron from an iodide ion, I^-, thus forming two chloride ions, Cl^-. The iodine atoms so formed, by the loss of an electron, combine to give an iodine molecule, I_2.

$$2I^-(aq) + Cl_2(g) \rightarrow 2Cl^-(aq) + I_2(aq)$$

Figure 3.25 The halogens have many varied uses – flouride in toothpaste to help reduce dental decay, chlorine in household bleach to kill bacteria, bromine as a fire retardant, and iodine in photographic reproduction.

- Fluorine is used in the form of fluorides in drinking water and toothpaste because it reduces tooth decay by hardening the enamel on teeth.
- Chlorine is used to make PVC plastic as well as household bleaches. It is also used to kill bacteria and viruses in drinking water (Chapter 10, p. 156).
- Bromine (Chapter 10, p. 153) is used to make disinfectants, medicines and fire retardants.
- Iodine is used in medicines and disinfectants and also as a photographic chemical.

Group 0 – the noble gases

Helium, neon, argon, krypton, xenon and the radioactive element radon make up a most unusual group of non-metals, called the noble gases. They were all discovered after Mendeleev had published his periodic table. They were discovered between 1894 and 1900, mainly through the work of the British scientists Sir William Ramsay (Figure 3.26a) and Lord John William Strutt Rayleigh (Figure 3.26b).

- They are colourless gases.
- They exist as individual atoms, for example He, Ne and Ar.
- They are very unreactive.

No compounds of helium, neon or argon have ever been found. However, more recently a number of compounds of xenon and krypton with fluorine and oxygen have been produced, for example XeF_6.

These gases are chemically unreactive because they have electronic structures which are stable and very difficult to change (Table 3.12). They are so stable that other elements attempt to attain these electron configurations during chemical reactions (Chapter 4, p. 52). You have probably seen this in your study of the elements of groups 1, 2 and 7.

Although unreactive, they have many uses. Argon, for example, is the gas used to fill light bulbs to prevent the tungsten filament reacting with air. Neon is used extensively in advertising signs and in lasers. Further uses of these gases are discussed in Chapter 10, p. 152.

Helium is separated from natural gas by the liquefaction of the other gases. The other noble gases are obtained in large quantities by the fractional distillation of liquid air (Chapter 10, p. 151).

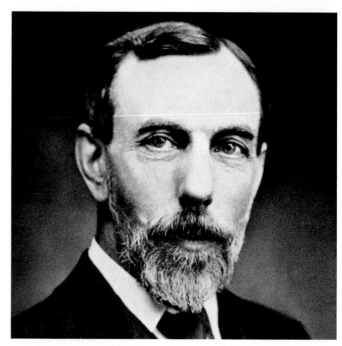

a Sir William Ramsay (1852–1916).

b Lord Rayleigh (1842–1919).

Figure 3.26 Both helped to discover the noble gases and won the Nobel Prize in Chemistry in 1904 for their work.

Table 3.12 Electronic structure of helium, neon and argon.

Element	Symbol	Proton number	Electronic structure
Helium	He	2	2
Neon	Ne	10	2,8
Argon	Ar	18	2,8,8

Questions

1 Write word and balanced chemical equations for the reactions between:
   a bromine and potassium iodide solution
   b bromine and potassium chloride solution.
   If no reaction will take place, write 'no reaction' and explain why.

2 Write down the names and symbols for the noble gases not given in Table 3.12 and use your research skills to find a use for each.

Transition elements

This block of metals includes many you will be familiar with, for example copper, iron, nickel, zinc and chromium (Figure 3.27).

- They are harder and stronger than the metals in groups 1 and 2.
- They have much higher densities than the metals in groups 1 and 2.
- They have high melting points (except for mercury, which is a liquid at room temperature).
- They are less reactive metals.
- They form a range of brightly coloured compounds (Figure 3.28, overleaf).
- They are good conductors of heat and electricity.
- They show catalytic activity (Chapter 11, p. 166) as elements and compounds. For example, iron is used in the industrial production of ammonia gas (Haber process, Chapter 15, p. 218).
- They do not react (corrode) so quickly with oxygen and/or water.
- They form simple ions with more than one oxidation state. For example, copper forms Cu^+ ($Cu(I)$) and Cu^{2+} ($Cu(II)$), in compounds such as Cu_2O and $CuSO_4$, and iron forms Fe^{2+} ($Fe(II)$) and Fe^{3+} ($Fe(III)$), in compounds such as $FeSO_4$ and $FeCl_3$.
- They form more complicated ions with high oxidation states. For example, chromium forms the dichromate(VI) ion, $Cr_2O_7^{2-}$, which contains chromium with a +6 oxidation state ($Cr(VI)$) and manganese forms the manganate(VII) ion, MnO_4^-, which contains manganese with a +7 oxidation state ($Mn(VII)$).

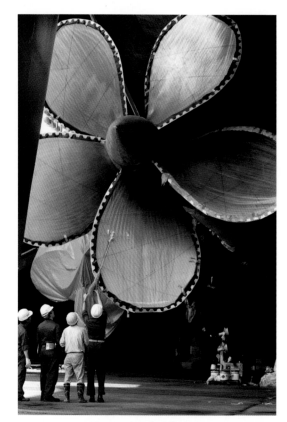

c Monel is an alloy of nickel and copper. It is extremely resistant to corrosion even from that caused by sea water.

d This bucket has been coated with zinc to prevent the steel of the bucket corroding.

a Copper is used in many situations, which involve good heat and electrical conduction. It is also used in medallions and bracelets.

b These gates are made of iron. Iron can easily be moulded into different shapes.

e The alloy stainless steel contains a high proportion of chromium, which makes it corrosion resistant.

Figure 3.27 Everyday uses of transition elements and their compounds. They are often known as the 'everyday metals'.

a Some solutions of coloured transition element compounds.

b The coloured compounds of transition elements can be seen in these pottery glazes.

Figure 3.28

1 Look at the photographs in Figure 3.27 (p. 47) and decide which properties are important when considering the particular use the metal is being put to.

2 Which groups in the periodic table contain:
   a only metals?
   b only non-metals?
   c both metals and non-metals?

The position of hydrogen

Hydrogen is often placed by itself in the periodic table. This is because the properties of hydrogen are unique. However, profitable comparisons can be made with the other elements. It is often shown at the top of either group 1 or group 7, but it cannot fit easily into the trends shown by either group, see Table 3.13.

Table 3.13 Comparison of hydrogen with lithium and fluorine.

Lithium	Hydrogen	Fluorine
Solid	Gas	Gas
Forms a positive ion	Forms positive or negative ions	Forms a negative ion
1 electron in outer energy level	1 electron in outer energy level	1 electron short of a full outer energy level
Loses 1 electron to form a noble gas configuration	Needs 1 electron to form a noble gas configuration	Needs 1 electron to form a noble gas configuration

■ *Checklist*

After studying Chapter 3 you should know and understand the following terms.

■ **Atomic mass unit**   Exactly $\frac{1}{12}$ of the mass of one atom of the most abundant isotope of carbon-12.

■ **Displacement reaction**   A reaction in which a more reactive element displaces a less reactive element from solution.

■ **Electron**   A fundamental sub-atomic particle with a negative charge present in all atoms within energy levels around the nucleus.

■ **Electron energy levels**   The allowed energies of electrons in atoms.

■ **Electronic structure (configuration)**   A shorthand method of describing the arrangement of electrons within the energy levels of an atom.

■ **Electrostatic force of attraction**   A strong force of attraction between opposite charges.

■ **Group**   A vertical column of the periodic table containing elements with similar properties with the same number of electrons in their outer energy levels. They have an increasing number of inner energy levels as you descend the group.

■ **Ion**   An atom or group of atoms which has either lost one or more electrons, making it positively charged, or gained one or more electrons, making it negatively charged.

■ **Isotopes**   Atoms of the same element which possess different numbers of neutrons. They differ in nucleon number (mass number).

■ **Mass spectrometer**   A device in which atoms or molecules are ionised and then accelerated. Ions are separated according to their mass.

■ **Metalloid (semi-metal)**   Any of the class of chemical elements intermediate in properties between metals and non-metals, for example boron and silicon.

■ **Neutron**   A fundamental, uncharged sub-atomic particle present in the nuclei of atoms.

■ **Nucleon number (mass number)**   Symbol A. The total number of protons and neutrons found in the nucleus of an atom.

■ **Periodic table**   A table of elements arranged in order of increasing proton number to show the similarities of the chemical elements with related electronic structures.

■ **Periods**   Horizontal rows of the periodic table. Within a period the atoms of all the elements have the same number of occupied energy levels but have an increasing number of electrons in the outer energy level.

■ **Proton**   A fundamental sub-atomic particle which has a positive charge equal in magnitude to that of an electron. Protons occur in all nuclei.

■ **Proton number (atomic number)**   Symbol Z. The number of protons in the nucleus of an atom. The number of electrons present in an atom. The order of the element within the periodic table.

■ **Radioactive**   A property of unstable isotopes. They disintegrate spontaneously to give off one or more types of radiation.

■ **Radioisotope**   A radioactive isotope.

■ **Relative atomic mass**   Symbol A_r.

$$A_r = \frac{\text{average mass of isotopes of the element}}{\frac{1}{12} \times \text{mass of a carbon-12 atom}}$$

Atomic structure and the periodic table
Additional questions

1 An atom **X** has a proton number of 19 and relative atomic mass of 39.
  a How many electrons, protons and neutrons are there in an atom of **X**?
  b How many electrons will there be in the outer energy level (shell) of an atom of **X**?
  c Write down the symbol for the ion **X** will form.
  d Which group of the periodic table would **X** be in?
  e (i) How would you expect **X** to react with water?
  (ii) Write a word and balanced chemical equation for this reaction.

2 The proton number of barium (Ba) is 56. It is in group 2 of the periodic table.
  a How many electrons would you expect a barium atom to contain in its outer energy level?
  b How would you expect barium to react with chlorine? Write a word and balanced chemical equation for this reaction.
  c How would you expect barium to react with water? Write a word and balanced chemical equation for this reaction.
  d Write down the formulae of the bromide and sulfate of barium.

3 Find the element germanium (Ge) in the periodic table.
  a Which group of the periodic table is this element in?
  b How many electrons will it have in its outer energy level (shell)?
  c Is germanium a metal or a non-metal?
  d What is the formula of the chloride of germanium?
  e Name and give the symbols of the other elements in this group.

4 Three members of the halogens are: $^{35.5}_{17}Cl$, $^{80}_{35}Br$ and $^{127}_{53}I$.
  a (i) Write down the electronic structure of an atom of chlorine.
  (ii) Why is the relative atomic mass of chlorine not a whole number?
  (iii) How many protons are there in an atom of bromine?
  (iv) How many neutrons are there in an atom of iodine?
  (v) State and account for the order of reactivity of these elements.
  b When potassium is allowed to burn in a gas jar of chlorine, in a fume cupboard, clouds of white smoke are produced.
  (i) Why is this reaction carried out in a fume cupboard?
  (ii) What does the white smoke consist of?
  (iii) Write a word and balanced chemical equation for this reaction.
  (iv) Describe what you would expect to see when potassium is allowed to burn safely in a gas jar of bromine vapour. Write a sword and balanced chemical equation for this reaction.

5 'By using displacement reactions it is possible to deduce the order of reactivity of the halogens.' Discuss this statement with reference to the elements bromine, iodine and chlorine only.

6 Use the information given in the table below to answer the questions below concerning the elements **Q, R, S, T** and **X**.

Element	Proton number	Nucleon number	Electronic structure
Q	3	7	2,1
R	20	40	2,8,8,2
S	18	40	2,8,8
T	8	18	2,6
X	19	39	2,8,8,1

  a Which element has 22 neutrons in each atom?
  b Which element is a noble gas?
  c Which two elements form ions with the same electronic structure as neon?
  d Which two elements are in the same group of the periodic table and which group is this?
  e Place the elements in the table above into the periods in which they belong.
  f Which is the most reactive metal element shown in the table?
  g (i) Which of the above elements is calcium?
  (ii) What colour flame would compounds containing calcium produce?

7 a $^{69}_{31}Ga$ and $^{71}_{31}Ga$ are isotopes of gallium. With reference to this example, explain what you understand by the term isotope.
  b A sample of gallium contains 60% of atoms of $^{69}_{31}Ga$ and 40% of atoms of $^{71}_{31}Ga$. Calculate the relative atomic mass of this sample of gallium.

8 Copy and complete the following table with reference to the periodic table on p. 38.

Element name	Symbol	Proton number	Nucleon number	Number of neutrons	$^A_Z X$
		5	11		
			40	22	
		14	28		
			20	10	
		26		30	
			84	48	
		52		76	

4 Bonding and structure

Ionic bonding
Ionic structures
Properties of ionic compounds
Formulae of ionic substances

Covalent bonding
Other covalent compounds
Covalent structures
Properties of covalent compounds
Allotropy
Allotropes of carbon

Glasses and ceramics
Glasses
Ceramics

Metallic bonding
Properties of metals

Checklist

Additional questions

In Chapter 3 we have seen that the noble gases are stable or unreactive because they have full electron energy levels (or shells) (Figure 4.1). It would seem that when elements react to form compounds they do so to achieve full electron energy levels. This idea forms the basis of the electronic theory of chemical bonding.

Ionic bonding

Ionic bonds are usually found in compounds that contain metals combined with non-metals. When this type of bond is formed, electrons are transferred from the metal atoms to the non-metal atoms during the chemical reaction. In doing this, the atoms become more stable by getting full outer energy levels. For example, consider what happens when sodium and chlorine combine to make sodium chloride (Figure 4.2).

sodium + chlorine → sodium chloride

Figure 4.1 The unreactive noble gas neon is used in lasers, as seen in this display in Germany.

Figure 4.2 The properties of salt are very different from those of the sodium and chlorine it was made from. To get your salt you would not eat sodium or inhale chlorine.

Sodium has just one electron in its outer energy level ($_{11}$Na 2,8,1). Chlorine has seven electrons in its outer energy level ($_{17}$Cl 2,8,7). When these two elements react, the outer electron of each sodium atom is transferred to the outer energy level of a chlorine atom (Figure 4.3). In this way both the atoms obtain full outer energy levels and become 'like' the nearest noble gas. The sodium atom has become a sodium ion with an electron configuration like neon, while the chlorine atom has become a chloride ion with an electron configuration like argon. Only the outer electrons are important in bonding, so we can simplify the diagrams by missing out the inner energy levels (Figure 4.4).

The charges on the sodium and chloride ions are equal but opposite. They balance each other and the resulting formula for sodium chloride is NaCl. These oppositely charged ions attract each other and are pulled, or **bonded**, to one another by strong electrostatic forces. This type of bonding is called **ionic bonding**. The alternative name, **electrovalent bonding**, is derived from the fact that there are electrical charges on the atoms involved in the bonding.

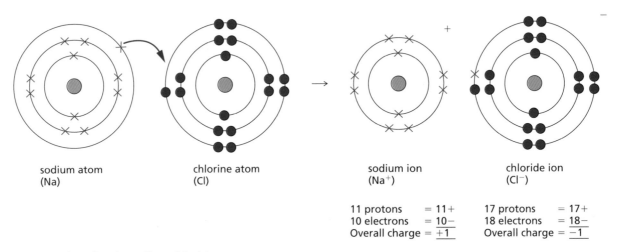

sodium atom (Na)   chlorine atom (Cl)   sodium ion (Na$^+$)   chloride ion (Cl$^-$)

11 protons = 11+	17 protons = 17+
10 electrons = 10−	18 electrons = 18−
Overall charge = +1	Overall charge = −1

Figure 4.3 Ionic bonding in sodium chloride.

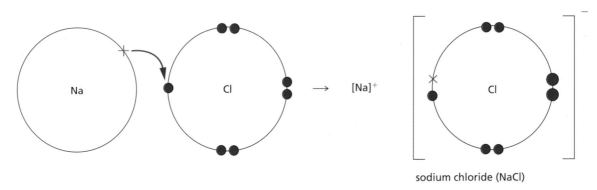

sodium chloride (NaCl)

Figure 4.4 Simplified diagram of ionic bonding in sodium chloride.

magnesium oxide (MgO)

Figure 4.5 Simplified diagram of ionic bonding in magnesium oxide.

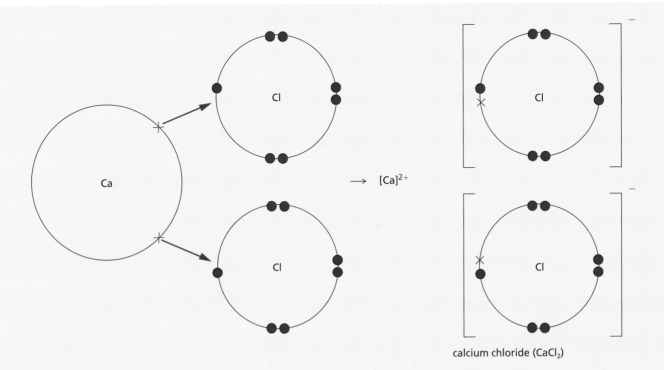

calcium chloride (CaCl₂)

Figure 4.6 The transfer of electrons that takes place during the formation of calcium chloride.

Figure 4.5 shows the electron transfers that take place between a magnesium ion and an oxide ion during the formation of magnesium oxide.

Magnesium obtains a full outer energy level by losing two electrons. These are transferred to the oxygen atom. In magnesium oxide, the Mg^{2+} and O^{2-} are oppositely charged and are attracted to one another. The formula for magnesium oxide is MgO.

Figure 4.6 shows the electron transfers that take place during the formation of calcium chloride. When these two elements react, the calcium atom gives each of the two chlorine atoms one electron. In this case, a compound is formed containing two chloride ions (Cl^-) for each calcium ion (Ca^{2+}). The chemical formula is $CaCl_2$.

Within the lattice, oppositely charged ions attract one another strongly (see Figure 4.7d, p. 54).

Question

1 Draw diagrams to represent the bonding in each of the following ionic compounds:
   a magnesium fluoride (MgF_2)
   b potassium fluoride (KF)
   c lithium chloride (LiCl)
   d calcium oxide (CaO).

Ionic structures

Ionic structures are solids at room temperature and have high melting and boiling points. The ions are packed together in a regular arrangement called a **lattice**. Within the lattice, oppositely charged ions attract one another strongly (see Figure 4.7d, p. 54).

Scientists, using **X-ray diffraction** (Figure 4.7a, overleaf), have obtained photographs that indicate the way in which the ions are arranged (Figure 4.7b). The electron density map of sodium chloride is shown in Figure 4.7c.

Figure 4.7d shows the structure of sodium chloride as determined by the X-ray diffraction technique. The study of crystals using X-ray diffraction was pioneered by Sir William Bragg and his son Sir Lawrence Bragg in 1912. X-rays are a form of electromagnetic radiation. They have a much shorter wavelength than light therefore it is possible to use them to investigate extremely small structures.

When X-rays are passed through a crystal of sodium chloride, for example, you get a pattern of spots called a diffraction pattern (Figure 4.7b). This pattern can be recorded on photographic film and used to work out how the ions or atoms are arranged in the crystal. Crystals give particular diffraction patterns depending on their structure, and this makes X-ray diffraction a particularly powerful technique in the investigation of crystal structures.

Figure 4.7d shows only a tiny part of a small crystal of sodium chloride. Many millions of sodium ions and chloride ions would be arranged in this way in a crystal of sodium chloride to make up the giant ionic structure. Each sodium ion in the lattice is surrounded by six chloride ions, and each chloride ion is surrounded by six sodium ions.

a X-ray technique.

b X-ray diffraction photograph of sodium chloride.

c Electron density map of sodium chloride.

Na$^+$ surrounded by 6 Cl$^-$ ions

Cl$^-$ surrounded by 6 Na$^+$ ions

d The structure of sodium chloride.

Figure 4.7

Not all ionic substances form the same structures. Caesium chloride (CsCl), for example, forms a different structure due to the larger size of the caesium ion compared with that of the sodium ion. This gives rise to the structure shown in Figure 4.8, which is called a body-centred cubic structure. Each caesium ion is surrounded by eight chloride ions and, in turn, each chloride ion is surrounded by eight caesium ions.

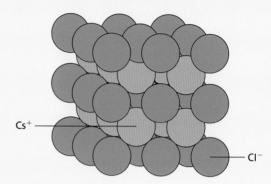

Cs$^+$

Cl$^-$

a The arrangement of ions.

Cs$^+$ surrounded by 8 Cl$^-$ ions
In the same way, each Cl$^-$
is surrounded by 8 Cs$^+$ ions

b The unit cell.

Figure 4.8 Structure of caesium chloride.

Properties of ionic compounds

Ionic compounds have the following properties.

- They are usually solids at room temperature, with high melting points. This is due to the strong electrostatic forces holding the crystal lattice together. A lot of energy is therefore needed to separate the ions and melt the substance.
- They are usually hard substances.
- They usually cannot conduct electricity when solid, because the ions are not free to move.
- They mainly dissolve in water. This is because water molecules are able to bond with both the positive and the negative ions, which breaks up the lattice and keeps the ions apart.
- They usually conduct electricity when in the molten state or in aqueous solution. The forces of attraction between the ions are weakened and the ions are free to move to the appropriate electrode. This allows an electric current to be passed through the molten compound (see Chapter 6, p. 84).

Formulae of ionic substances

On p. 52 we saw that ionic compounds contain positive and negative ions, whose charges balance. For example, sodium chloride contains one Na^+ ion for every Cl^- ion, giving rise to the formula $NaCl$. This method can be used to write down formulae which show the ratio of the number of ions present in any ionic compound.

The formula of magnesium chloride is $MgCl_2$. This formula is arrived at by each Mg^{2+} ion combining with two Cl^- ions, and once again the charges balance. The size of the charge on an ion is a measure of its **valency** or **combining power**. Na^+ has a valency of 1, but Mg^{2+} has a valency of 2. Na^+ can bond (combine) with only one Cl^- ion, whereas Mg^{2+} can bond with two Cl^- ions.

Some elements, such as copper and iron, possess two ions with different valencies. Copper can form the Cu^+ ion and the Cu^{2+} ion, and therefore it can form two different compounds with chlorine, $CuCl$ (copper(I) chloride) and $CuCl_2$ (copper(II) chloride). Iron forms the Fe^{2+} and Fe^{3+} ions and so can also form two different compounds with, for example, chlorine $FeCl_2$ (iron(II) chloride) and $FeCl_3$ (iron(III) chloride).

Table 4.1 shows the valencies of a series of ions you will normally meet in your study of chemistry.

You will notice that Table 4.1 includes groups of atoms which have net charges. For example, the nitrate ion is a single unit composed of one nitrogen atom and three oxygen atoms and has one single negative charge. The formula, therefore, of magnesium nitrate would be $Mg(NO_3)_2$. You will notice that the NO_3 has been placed in brackets with a $_2$ outside the bracket. This indicates that there are two nitrate ions present for every magnesium ion. The ratio of the atoms present is therefore:

$$Mg\ (N\ O_3)_2$$
$$1Mg : 2N : 6O$$

Note that the charge on the element ion is often referred to as its **oxidation state**.

Questions

1 Using the information in Table 4.1, write the formulae for:
   a copper(I) oxide
   b zinc phosphate
   c iron(III) chloride
   d lead bromide.

2 Using the formulae in your answer to question **1**, write down the ratio of atoms present for each of the compounds.

Table 4.1 Valencies of some common substances.

		Valency				
		1		**2**		**3**
Metals	Lithium	(Li$^+$)	Magnesium	(Mg^{2+})	Aluminium	(Al^{3+})
	Sodium	(Na$^+$)	Calcium	(Ca^{2+})	Iron(III)	(Fe^{3+})
	Potassium	(K$^+$)	Copper(II)	(Cu^{2+})		
	Silver	(Ag$^+$)	Zinc	(Zn^{2+})		
	Copper(I)	(Cu$^+$)	Iron(II)	(Fe^{2+})		
			Lead	(Pb^{2+})		
			Barium	(Ba^{2+})		
Non-metals	Fluoride	(F$^-$)	Oxide	(O^{2-})		
	Chloride	(Cl$^-$)	Sulfide	(S^{2-})		
	Bromide	(Br$^-$)				
	Hydrogen	(H$^+$)				
Groups of atoms	Hydroxide	(OH$^-$)	Carbonate	(CO$_3^{2-}$)	Phosphate (PO$_4^{3-}$)	
	Nitrate	(NO$_3^-$)	Sulfate	(SO$_4^{2-}$)		
	Ammonium	(NH$_4^+$)	Dichromate (VI)	(Cr$_2$O$_7^{2-}$)		
	Hydrogencarbonate	(HCO$_3^-$)				
	Manganate (VII)	(MnO$_4^-$)				

■ *Covalent bonding*

Another way in which atoms can gain the stability of the noble gas electron configuration is by sharing the electrons in their outer energy levels. This occurs between non-metal atoms, and the bond formed is called a **covalent bond**. The simplest example of this type of bonding can be seen by considering the hydrogen molecule, H_2.

Each hydrogen atom in the molecule has one electron. In order to obtain a full outer energy level and gain the electron configuration of the noble gas helium each of the hydrogen atoms must have two electrons. To do this, the two hydrogen atoms allow their outer energy levels to overlap (Figure 4.9a). A molecule of hydrogen is formed, with two hydrogen atoms sharing a pair of electrons (Figure 4.9a, b, c). This shared pair of electrons is known as a single covalent bond and is represented by a single line as in hydrogen:

H—H

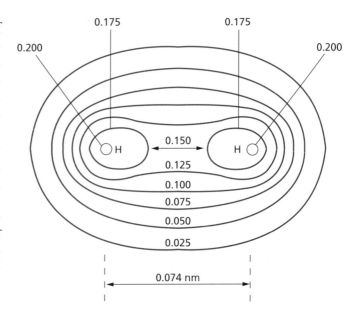

electron densities in electrons per cubic atomic unit of length

Figure 4.9c The electron density map of a hydrogen molecule.

2 hydrogen atoms   hydrogen molecule (H_2)

Figure 4.9a Electron sharing to form the single covalent bond in H_2 molecules.

A similar example exists in the diatomic halogen molecule, chlorine, Cl_2 (Figure 4.10).

2 chlorine atoms

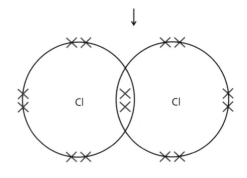

chlorine molecule (Cl_2)

Figure 4.10 The electron sharing to form single covalent bond in Cl_2 molecules (sharing outer electron shells only).

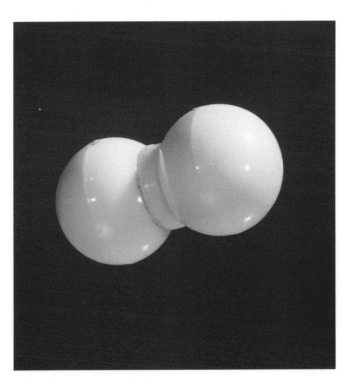

Figure 4.9b Model of hydrogen molecule.

Other covalent compounds

Methane (natural gas) is a gas whose molecules contain atoms of carbon and hydrogen. The electron structures are:

$$_6C\ 2,4 \qquad _1H\ 1$$

The carbon atom needs four more electrons to attain the electron configuration of the noble gas neon. Each hydrogen atom needs only one electron to form the electron configuration of helium. Figure 4.11 shows how the atoms gain these electron configurations by the sharing of electrons. You will note that only the outer electron energy levels are shown. Figure 4.12 shows the shape of the methane molecule.

Ammonia is a gas containing the elements nitrogen and hydrogen. It is used in large amounts to make fertilisers. The electron configurations of the two elements are:

$$_7N\ 2,5 \qquad _1H\ 1$$

The nitrogen atom needs three more electrons to obtain the noble gas structure of neon. Each hydrogen requires only one electron to form the noble gas structure of helium. The nitrogen and hydrogen atoms share electrons, forming three single covalent bonds (Figure 4.13, overleaf). Unlike methane the shape of an ammonia molecule is pyramidal (Figure 4.14, overleaf).

Water is a liquid containing the elements hydrogen and oxygen. The electronic structures of the two elements are:

$$_8O\ 2,6 \qquad _1H\ 1$$

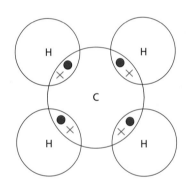

carbon atom   4 hydrogen atoms

methane molecule (CH₄)

Figure 4.11 Formation of methane.

the methane molecule is tetrahedral

a Methane molecule.

Figure 4.12

b Model of the methane molecule. The tetrahedral shape is caused by the repulsion of the C−H bonding pairs of the electrons.

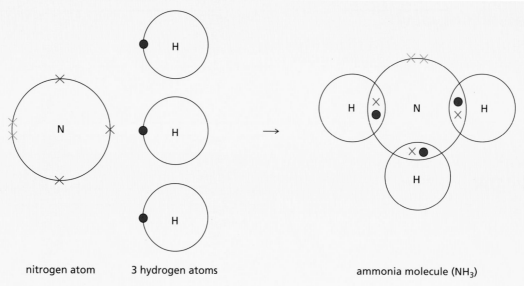

Figure 4.13 The bonding in ammonia.

nitrogen atom   3 hydrogen atoms   ammonia molecule (NH₃)

a Ammonia molecule.

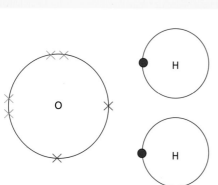

b Model of the ammonia molecule. The pyramidal shape is caused by the repulsion between the bonding pairs of electrons as well as the lone pair (or non-bonding pair) of electrons.

Figure 4.14

The oxygen atom needs two electrons to gain the electron configuration of neon. Each hydrogen requires one more electron to gain the electron configuration of helium. Again, the oxygen and hydrogen atoms share electrons, forming a water molecule with two single covalent bonds as shown in Figure 4.15. A water molecule is V–shaped (Figure 4.16).

a Water molecule.

b Model of a water molecule. This is a V-shaped molecule.

Figure 4.16

oxygen atom   2 hydrogen atoms   water molecule (H₂O)

Figure 4.15 Formation of water.

2 oxygen atoms   carbon atom   carbon dioxide molecule (CO_2)

Figure 4.17 Formation of carbon dioxide.

Carbon dioxide is a gas containing the elements carbon and oxygen. The electronic structures of the two elements are:

$$_6C\ 2,4 \qquad _8O\ 2,6$$

In this case each carbon atom needs to share four electrons to gain the electron configuration of neon. Each oxygen needs to share two electrons to gain the electron configuration of neon. This is achieved by forming two **double covalent bonds** in which two pairs of electrons are shared in each case, as shown in Figure 4.17. Carbon dioxide is a linear molecule (Figure 4.18).

$$O {=\!=} C {=\!=} O$$

a Carbon dioxide molecule. Note the double covalent bond is represented by a double line.

b Model of the linear carbon dioxide molecule.

Figure 4.18

Questions

1 Draw diagrams to represent the bonding in each of the following covalent compounds:
   a tetrachloromethane (CCl_4)
   b oxygen gas (O_2)
   c hydrogen sulphide (H_2S)
   d hydrogen chloride (HCl)
   e ethene (C_2H_4)
   f methanol (CH_3OH)
   g nitrogen (N_2).

2 Explain why the water molecule in Figure 4.16 is V-shaped.

Covalent structures

Compounds containing covalent bonds have molecules whose structures can be classified as either **simple molecular** or **giant molecular**.

Simple molecular structures are simple, formed from only a few atoms. They have strong covalent bonds between the atoms within a molecule (**intramolecular bonds**) but have weak bonds between the molecules (**intermolecular bonds**) (Figure 4.19). One type of weak bond between molecules is known as the **van der Waals' bond** (or force), and these forces increase steadily with the increasing size of the molecule. Examples of simple molecules are iodine, methane, water and ethanol.

Figure 4.19 Strong covalent and weak intermolecular forces in iodine.

Giant **molecular** or **macromolecular** structures contain many hundreds of thousands of atoms joined by strong covalent bonds. Examples of substances showing this type of structure are diamond, graphite, silicon(IV) oxide (Figure 4.20) and plastics such as polythene (polymers, Chapter 14, p. 198). Plastics are a tangled mass of very long molecules in which the atoms are joined together by strong covalent bonds to form long chains. Molten plastics can be made into fibres by being forced through hundreds of tiny holes in a 'spinneret' (Figure 4.21). This process aligns the long chains of atoms along the length of the fibre. For a further discussion of plastics, see Chapter 14, pp. 198–9.

a The silicon(IV) oxide structure in quartz.

b Quartz is a hard solid at room temperature. It has a melting point of 1610 °C and a boiling point of 2230 °C.

Figure 4.20

Figure 4.21 Nylon ribbon is formed by forcing molten plastic through hundreds of tiny holes.

Properties of covalent compounds

Covalent compounds have the following properties.

- As simple molecular substances, they are usually gases, liquids or solids with low melting and boiling points. The melting points are low because of the weak intermolecular forces of attraction which exist between simple molecules. Giant molecular substances have higher melting points, because the whole structure is held together by strong covalent bonds.
- Generally, they do not conduct electricity when molten or dissolved in water. This is because they do not contain ions. However, some molecules actually react with water to form ions. For example, hydrogen chloride gas produces aqueous hydrogen ions and chloride ions when it dissolves in water:

$$HCl(g) \xrightarrow{\text{water}} H^+(aq) + Cl^-(aq)$$

- Generally, they do not dissolve in water. However, water is an excellent solvent and can interact with and dissolve some covalent molecules better than others.

Allotropy

When an element can exist in more than one physical form in the same state it is said to exhibit **allotropy** (or polymorphism). Each of the different physical forms is called an **allotrope**. Allotropy is actually quite a common feature of the elements in the periodic table. Some examples of elements which show allotropy are sulfur, tin, iron and carbon.

Allotropes of carbon

Carbon is a non-metallic element which exists in more than one solid structural form. Its allotropes are called **graphite** and **diamond**. Each of the allotropes has a different structure (Figures 4.22, opposite, and 4.23 on p. 62) and so the allotropes exhibit different physical properties (Table 4.2). The different physical properties that they exhibit lead to the allotropes being used in different ways (Table 4.3, opposite, and Figure 4.24 on p. 62).

Table 4.2 Physical properties of graphite and diamond.

Property	Graphite	Diamond
Appearance	A dark grey, shiny solid	A colourless transparent crystal which sparkles in light
Electrical conductivity	Conducts electricity	Does not conduct electricity
Hardness	A soft material with a slippery feel	A very hard substance
Density/g cm^{-3}	2.25	3.51

Table 4.3 Uses of graphite and diamond.

Graphite	Diamond
Pencils	Jewellery
Electrodes	Glass cutters
Lubricant	Diamond-studded saws
	Drill bits
	Polishers

Graphite

Figure 4.22a shows the structure of graphite. This is a layer structure. Within each layer each carbon atom is bonded to three others by strong covalent bonds. Each layer is therefore like a giant molecule. Between these layers there are weak forces of attraction (van der Waals' forces) and so the layers will pass over each other easily.

With only three covalent bonds formed between carbon atoms within the layers, an unbonded electron is present on each carbon atom. These 'spare' (or **delocalised**) electrons form electron clouds between the layers and it is because of these spare electrons that graphite conducts electricity.

In recent years a set of interesting compounds known as **graphitic compounds** have been developed. In these compounds different atoms have been fitted in between the layers of carbon atoms to produce a substance with a greater electrical conductivity than pure graphite. Graphite is also used as a component in certain sports equipment, such as tennis and squash rackets.

one layer

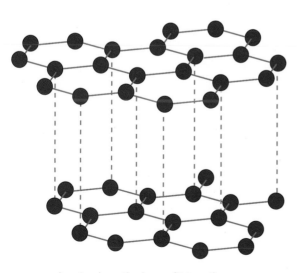

showing how the layers fit together

a A portion of the graphite structure.

Figure 4.22

b A piece of graphite as imaged through a scanning tunnelling microscope.

Diamond

Figure 4.23a shows the diamond structure. Each of the carbon atoms in the giant structure is covalently bonded to four others. They form a tetrahedral arrangement similar to that found in silicon(IV) oxide (p. 60). This bonding scheme gives rise to a very rigid, three-dimensional structure and accounts for the extreme hardness of the substances, silicon(IV) oxide and diamond. All the outer energy level electrons of the carbon atoms are used to form covalent bonds, so there are no electrons available to enable diamond to conduct electricity.

It is possible to manufacture both allotropes of carbon. Diamond is made by heating graphite to about 300 °C at very high pressures. Diamond made by this method is known as industrial diamond. Graphite can be made by heating a mixture of coke and sand at a very high temperature in an electric arc furnace for about 24 hours.

The various uses of graphite and diamond result from their differing properties (Figure 4.24).

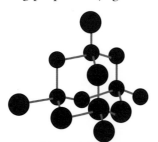

a small part of the structure

a view of a much larger part of the structure

a The structure of diamond.

b The Regent Diamond has been worn by Queen Elizabeth II.

Figure 4.24 Uses of graphite (as a pencil and in a squash racket) and diamond (as a toothed saw to cut marble and on a dentist's drill).

Figure 4.23

Buckminsterfullerene – an unusual form of carbon

In 1985 a new allotrope of carbon was obtained by Richard Smalley and Robert Curl of Rice University, Texas. It was formed by the action of a laser beam on a sample of graphite.

The structure of buckminsterfullerene can be seen in Figure 4.25.

Figure 4.25 Buckminsterfullerene – a 'bucky ball' (C_{60}).

This spherical structure is composed of 60 carbon atoms covalently bonded together. Further spherical forms of carbon, 'bucky balls', containing 70, 72 and 84 carbon atoms have been identified and the discovery has led to a whole new branch of inorganic carbon chemistry. It is thought that this type of molecule exists in chimney soot. Chemists have suggested that due to the large surface area of the bucky balls they may have uses as catalysts (Chapter 11, p. 166). Also they may have uses as superconductors.

Buckminsterfullerene is named after an American architect, Buckminster Fuller, who built complex geometrical structures (Figure 4.26).

Questions

1 Explain the difference between ionic and covalent bonding. Discuss in what ways the electronic structure of the noble gases is important in both of these theories of bonding.

2 Compare the structures of silicon(IV) oxide (p. 60) and diamond, and describe their physical properties in terms of these structures.

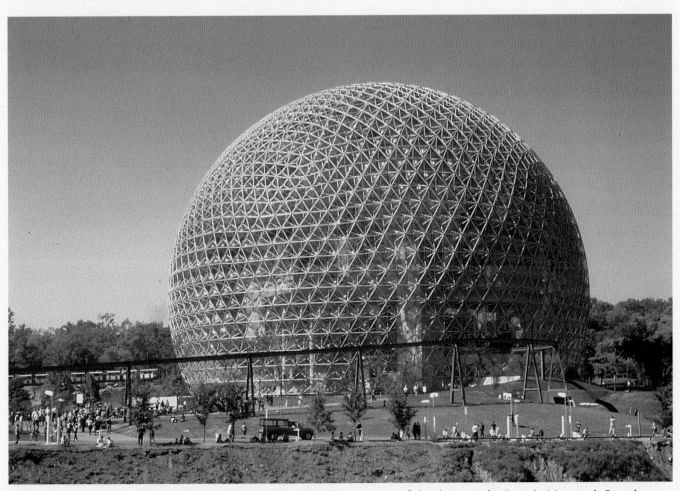

Figure 4.26 C_{60} has a structure similar to a football and to the structure of the dome at the Expo in Montreal, Canada.

■ *Glasses and ceramics*

Glasses

Glasses are irregular giant molecular structures held together by strong covalent bonds. Glass can be made by heating silicon(IV) oxide with other substances until a thick viscous liquid is formed. As this liquid cools, the atoms present cannot move freely enough to return to their arrangement within the pure silicon(IV) oxide structure. Instead they are forced to form a disordered arrangement as shown in Figure 4.27. Glass is called a **supercooled liquid**.

● O ● Si

Figure 4.27 Two-dimensional structure of silicon(IV) oxide.

The glass used in bottles and windows is **soda glass**. This type of glass is made by heating a mixture of sand (silicon(IV) oxide), soda (sodium carbonate) and lime (calcium oxide). Pyrex is a borosilicate glass (Figure 4.28). It is made by incorporating some boron oxide into the silicon(IV) oxide structure so that silicon atoms are replaced by boron atoms. This type of glass is tougher than soda glass and more resistant to temperature changes. It is, therefore, used in the manufacture of cooking utensils and laboratory glassware.

Figure 4.28 This glassware is made from Pyrex, a heat resistant glass.

Ceramics

The word **ceramic** comes from the Greek word **keramos** meaning pottery or 'burnt stuff'. Clay dug from the ground contains a mixture of several materials. The main one is a mineral called kaolinite, $Al_2Si_2O_5(OH)_4$, in which the atoms are arranged in layers in a giant structure. While wet, the clay can be moulded because the kaolinite crystals move over one another. However, when it is dry the clay becomes rigid because the crystals stick together.

During firing in a furnace, the clay is heated to a temperature of 1000 °C. A complicated series of chemical changes take place, new minerals are formed and some of the substances in the clay react to form a type of glass. The material produced at the end of the firing, the ceramic, consists of many minute mineral crystals bonded together with glass.

Modern ceramic materials now include zirconium oxide (ZrO_2), titanium carbide (TiC), and silicon nitride (SiN). There are now many more uses of these new ceramic materials. For example, vehicle components such as ceramic bearings do not need lubrication – even at high speeds. In space technology, ceramic tiles protect the Space Shuttle from intense heat during its re-entry into the Earth's atmosphere. In the power supply industry, they are used as insulators due to the fact that they do not conduct electricity (Figure 4.29).

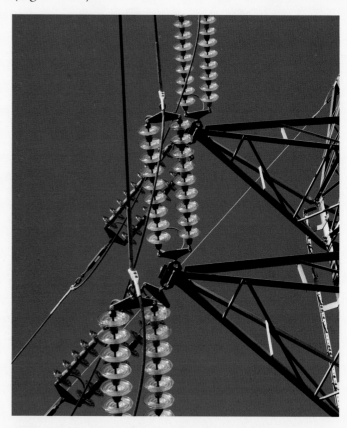

Figure 4.29 Ceramic discs prevent short-circuits in the national grid.

Questions

1 Draw up a table to summarise the properties of the different types of substances you have met in this chapter. Your table should include examples from ionic substances, covalent substances (simple and giant), ceramics and glasses.

2 Use your research skills, including suitable websites, to discover details of recently developed bioceramics as well as ceramics used as superconductors.

Metallic bonding

Another way in which atoms obtain a more stable electron structure is found in metals. The electrons in the outer energy level of the atom of a metal move freely throughout the structure (they are **delocalised** forming a mobile 'sea' of electrons (Figure 4.30)). When the metal atoms lose these electrons, they form a lattice of positive ions. Therefore, metals consist of positive ions embedded in moving clouds of electrons. The negatively charged electrons attract all the positive metal ions and bond them together with strong electrostatic forces of attraction as a single unit. This is the **metallic bond**.

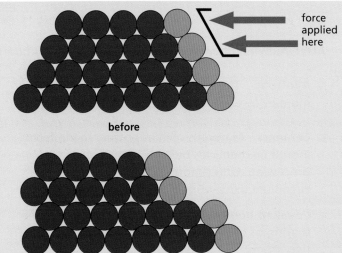

Figure 4.31 The positions of the positive ions in a metal before and after a force has been applied.

Figure 4.30 Metals consist of positive ions surrounded by a 'sea' of electrons.

Properties of metals

Metals have the following properties.

Figure 4.32 Arrangement of ions in the crystal lattice of a metal.

- They usually have high melting and boiling points due to the strong attraction between the positive metal ions and the mobile 'sea' of electrons.
- They conduct electricity due to the mobile electrons within the metal structure. When a metal is connected in a circuit, the electrons move towards the positive terminal while at the same time electrons are fed into the other end of the metal from the negative terminal.
- They are malleable and ductile. Unlike the fixed bonds in diamond, metallic bonds are not rigid but are still strong. If a force is applied to a metal, rows of ions can slide over one another. They reposition themselves and the strong bonds re-form as shown in Figure 4.31. Malleable means that metals can be hammered into different shapes. Ductile means that the metals can be pulled out into thin wires.
- They have high densities because the atoms are very closely packed in a regular manner as can be seen in Figure 4.32. Different metals show different types of packing and in doing so they produce the arrangements of ions shown in Figure 4.33.

close-packed hexagonal structure (e.g. magnesium, density = 1.74 g cm^{-3})

body-centred cubic structure (e.g. iron, density = 7.87 g cm^{-3})

face-centred cubic structure (e.g. copper, density = 8.92 g cm^{-3})

Figure 4.33 Relating different structures to the density of metal.

Questions

1 Explain the terms:
  a malleable
  b ductile.

2 Explain why metals are able to conduct heat and electricity.

3 Explain why the melting point of magnesium (649 °C) is much higher than the melting point of sodium (97.9 °C).

4 Suggest reasons why plastics and ceramics are used to manufacture handles for metallic cooking utensils.

■ *Checklist*

After studying Chapter 4 you should know and understand the following terms.

■ **Allotropy**   The existence of an element in two or more different forms in the same physical state.

■ **Ceramics**   Materials such as pottery made from inorganic chemicals by high-temperature processing. Other modern ceramics include zirconium oxide and silicon nitride.

■ **Covalent bond**   A chemical bond formed by the sharing of one or more pairs of electrons between two atoms.

■ **Giant ionic structure**   A lattice held together by the electrostatic forces of attraction between ions.

■ **Giant molecular substance**   A substance containing thousands of atoms per molecule.

■ **Glass**   A supercooled liquid which forms a hard, brittle substance that is usually transparent and resistant to chemical attack.

■ **Intermolecular bonds**   Attractive forces which act between molecules, for example van der Waals' forces.

■ **Intramolecular bonds**   Forces which act within a molecule, for example covalent bonds.

■ **Ionic (electrovalent) bond**   A strong electrostatic force of attraction between oppositely charged ions.

■ **Lattice**   A regular three-dimensional arrangement of atoms/ions in a crystalline solid.

■ **Metallic bond**   An electrostatic force of attraction between the mobile 'sea' of electrons and the regular array of positive metal ions within the solid metal.

■ **Simple molecular substance**   These substances possess between one and a few hundred atoms per molecule.

■ **Supercooled liquid**   One which has cooled below its freezing point without solidification.

■ **Valency**   The combining power of an atom or group of atoms. The valency of an ion is equal to its charge. The charge is often referred to as the atom's **oxidation state**.

■ **X-ray diffraction**   A technique often used to study crystal structures.

Bonding and structure
Additional questions

1 Draw diagrams to show the bonding in each of the following compounds:
   a calcium fluoride (CaF_2)
   b oxygen (O_2)
   c magnesium chloride ($MgCl_2$)
   d tetrachloromethane (CCl_4).

2 Use the information given in Table 4.1 on p. 55 to work out the formula for:
   a silver oxide
   b zinc chloride
   c potassium sulfate
   d calcium nitrate
   e iron(II) nitrate
   f copper(II) carbonate
   g iron(III) hydroxide
   h aluminium fluoride.

3 Atoms of elements **X**, **Y** and **Z** have 16, 17 and 19 electrons, respectively. Atoms of argon have 18 electrons.
   a Determine the formulae of the compounds formed by the combination of the atoms of the elements:
      (i) **X** and **Z**
      (ii) **Y** and **Z**
      (iii) **X** with itself.
   b In each of the cases shown in **a**(i)–(iii) above, name the type of chemical bond formed.
   c Give two properties you would expect to be shown by the compounds formed in **a**(ii) and **a**(iii).

4 The diagram shows the arrangement of the outer electrons only in a molecule of ethanoic acid.

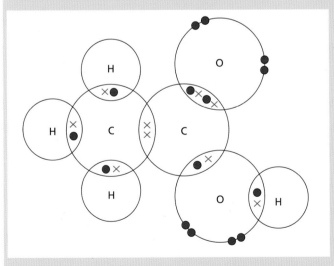

   a Name the different elements found in this compound.
   b What is the total number of atoms present in this molecule?
   c Between which two atoms is there a double covalent bond?
   d How many single covalent bonds does each carbon atom have?
   e Write a paragraph explaining the sorts of properties you would expect this substance to have.

5 Make a summary table of the properties of substances with covalent structures. Your table should include examples of both simple molecular and giant molecular substances.

6 The elements sodium and chlorine react together to form the compound sodium chloride, which has a giant ionic lattice structure.
   a What type of structure do the elements (i) sodium and (ii) chlorine have?
   b Draw a diagram to represent how the ions are arranged in the crystal lattice of sodium chloride.
   c Explain how the ions are held together in this crystal lattice.
   d Draw diagrams to show how the electrons are arranged in a sodium ion and a chloride ion. (The proton numbers of sodium and chlorine are 11 and 17, respectively.)
   e Make a table showing the properties of the three substances sodium, chlorine and sodium chloride. Include in your table:
      (i) the physical state at room temperature
      (ii) solubility in (or reaction with) water
      (iii) colour
      (iv) electrical conductivity.

7 Explain the following.
   a Ammonia is a gas at room temperature.
   b The melting points of sodium chloride and iodine are very different.
   c Metals generally are good conductors of electricity.
   d Buckminsterfullerene is an allotrope of carbon.
   e Metals usually have high melting and boiling points.

8 Discuss the following with reference to diamond and graphite.
   a Diamond is one of the hardest substances known.
   b Graphite is a good lubricating agent.
   c Graphite conducts electricity but diamond does not.

5 *Chemical calculations*

Relative atomic mass

Reacting masses

Calculating moles
Moles and elements
Moles and compounds
Moles and gases
Moles and solutions

Calculating formulae
Finding the formula

Moles and chemical equations

Checklist

Additional questions

■ *Relative atomic mass*

There are at present 117 different elements known. The atoms of these elements differ in mass because of the different numbers of protons, neutrons and electrons they contain. The actual mass of one atom is very small. For example, the mass of a single atom of sulfur is around:

$$0.000\,000\,000\,000\,000\,000\,000\,000\,053\,16\,g$$

This small quantity is not easy to work with so, as you saw in Chapter 3, a scale called the **relative atomic mass scale** is used. In this scale an atom of carbon is given a relative atomic mass A_r, of 12.00. All other atoms of the other elements are given a relative atomic mass compared to that of carbon.

An H atom is $\frac{1}{12}$ the mass of a C atom   An Mg atom is twice the mass of a C atom

H	C	Mg	S	Ca
1	12 fixed	24	32	40

Figure 5.1 The relative atomic masses of the elements H, C, Mg, S and Ca.

■ *Reacting masses*

Chemists often need to be able to show the relative masses of the atoms involved in a chemical process, for example, what mass of carbon dioxide would be produced if 6 g of carbon was completely combusted?

$$C + O_2 \rightarrow CO_2$$

Instead of using the actual masses of atoms we use the relative atomic mass to help us answer this type of question.

In the example above we can work out the **relative formula mass** of molecules such as O_2 and CO_2 using the relative atomic masses of the atoms they are made from. The RFM is the sum of the relative atomic masses of all those elements shown in the formula of the substance.

O_2 has a relative formula mass of $2 \times 16 = 32$
CO_2 has a relative formula mass of $12 + (2 \times 16) = 44$

So we can now use the equation to answer the question asked earlier.

$$
\begin{array}{ccc}
C & + & O_2 & \rightarrow & CO_2 \\
12 & & 32 & & 44
\end{array}
$$

Converting these relative masses to actual masses by adding mass units, g, would give:

$$
\begin{array}{ccc}
C & + & O_2 & \rightarrow & CO_2 \\
12 & & 32 & & 44 \\
12\,g & & 32\,g & & 44\,g
\end{array}
$$

The above calculation shows that if 12 g of carbon were burned completely then 44 g of carbon dioxide gas would be formed. So 6 g of carbon burning would result in the formation of 22 g of carbon dioxide gas.

Let us look at another example. What mass of hydrogen gas would be produced if 46 g of sodium was reacted with water? First of all write down the balanced chemical equation:

$$2Na + 2H_2O \rightarrow 2NaOH + H_2$$

Next find the relative atomic mass of sodium from the periodic table and work out the relative formula masses of water, sodium hydroxide and hydrogen gas.

Relative atomic mass of sodium is 23.
Relative formula mass of water, H_2O, is $(2 \times 1) + 16 = 18$
Relative formula mass of sodium hydroxide is $23 + 16 + 1 = 40$
Relative formula mass of hydrogen gas, H_2, is $2 \times 1 = 2$

Now you can write these masses under the balanced chemical equation taking into account the numbers used to balance the equation.

$$
\begin{array}{ccccc}
2Na & + & 2H_2O & \rightarrow & 2NaOH & + & H_2 \\
2 \times 23 = 46 & & 2 \times 18 = 36 & & 2 \times 40 = 80 & & 2
\end{array}
$$

These relative masses can now be converted into actual or reacting masses by putting in mass units, for example, grams.

$$2Na \quad + \quad 2H_2O \quad \rightarrow \quad 2NaOH \quad + \quad H_2$$

| 2 × 23 = 46 | 2 × 18 = 36 | 2 × 40 = 80 | 2 |
| 46 g | 36 g | 80 g | 2 g |

So the answer to the question of what mass of hydrogen would be produced if 46 g of sodium was reacted with water is 2 g.

Questions

1 What mass of carbon dioxide gas would be produced if 10 g of calcium carbonate reacted with an excess of hydrochloric acid?

2 What mass of sulfur dioxide would be produced if 64 tonnes of sulfur was completely reacted with oxygen gas?

Chemists often need to know how much of a substance has been formed or used up during a chemical reaction (Figure 5.2). This is particularly important in the chemical industry, where the substances being reacted (the **reactants**) and the substances being produced (the **products**) are worth thousands of pounds. Waste costs money!

To solve this problem they need a way of counting atoms, ions or molecules. Atoms, ions and molecules are very tiny particles and it is impossible to measure out a dozen or even a hundred of them. Instead, chemists weigh out a very large number of particles.

This number is 6×10^{23} atoms, ions or molecules and is called **Avogadro's constant** after the famous Italian scientist Amedeo Avogadro (1776–1856). An amount of substance containing 6×10^{23} particles is called a **mole** (often abbreviated to mol).

So, a mole of the element magnesium is 6×10^{23} atoms of magnesium and a mole of the element carbon is 6×10^{23} atoms of carbon (Figure 5.3).

a A mole of magnesium.

Figure 5.2 The chemists at paint manufacturers need to know how much pigment is going to be produced.

b A mole of carbon.

Figure 5.3

■ *Calculating moles*

You have already seen in Chapter 3 how we can compare the masses of all the other atoms with the mass of carbon atoms. This is the basis of the **relative atomic mass scale**. Chemists have found by experiment that if you take the relative atomic mass of an element in grams, it always contains 6×10^{23} or one mole of its atoms.

Moles and elements

For example, the relative atomic mass (A_r) of iron is 56, so one mole of iron is 56 g. Therefore, 56 g of iron contains 6×10^{23} atoms.

The A_r for aluminium is 27. In 27 g of aluminium it is found that there are 6×10^{23} atoms. Therefore, 27 g of aluminium is one mole of aluminium atoms.

The mass of a substance present in any number of moles can be calculated using the relationship:

$$\begin{array}{ccc} \text{mass} & = & \text{number} \times \text{mass of 1 mole} \\ \text{(in grams)} & & \text{of moles} \quad \text{of the element} \end{array}$$

Example 1

Calculate the mass of (a) 2 moles and (b) 0.25 mole of iron (A_r: Fe = 56).

a mass of 2 moles of iron
= number of moles 3 relative atomic mass (A_r)
= 2 × 56
= 112 g

b mass of 0.25 mole of iron
= number of moles 3 relative atomic mass (A_r)
= 0.25 × 56
= 14 g

If we know the mass of the element then it is possible to calculate the number of moles of that element using:

$$\frac{\text{number of}}{\text{moles}} = \frac{\text{mass of the element}}{\text{mass of 1 mole of that element}}$$

Example 2

Calculate the number of moles of aluminium present in (a) 108 g and (b) 13.5 g of the element (A_r: Al = 27).

a number of moles of aluminium
$$= \frac{\text{mass of aluminium}}{\text{mass of 1 mole of aluminium}}$$
$$= \frac{108}{27} = 4 \text{ moles}$$

b number of moles of aluminium
$$= \frac{\text{mass of aluminium}}{\text{mass of 1 mole of aluminium}}$$
$$= \frac{13.5}{27} = 0.5 \text{ mole}$$

Moles and compounds

The idea of the mole has been used so far only with elements and atoms. However, it can also be used with compounds (Figure 5.4).

What is the mass of 1 mole of water (H_2O) molecules? (A_r: H = 1; O = 16)

From the formula of water, H_2O, you will see that 1 mole of water molecules contains 2 moles of hydrogen (H) atoms and 1 mole of oxygen (O) atoms. The mass of 1 mole of water molecules is therefore:

$$(2 \times 1) + (1 \times 16) = 18 \text{ g}$$

The mass of 1 mole of a compound is called its molar mass. If you write the molar mass of a compound without any units then it is the relative formula mass, often called the **relative molecular mass** (M_r). So the relative formula mass of water is 18.

Now follow these examples to help you learn and understand more about moles and compounds.

Figure 5.4 One mole of water (H_2O) (left) and one mole of ethanol (C_2H_5OH) (right) in separate measuring cylinders.

Example 1

What is (a) the mass of 1 mole and (b) the relative formula mass (RFM) of ethanol, C_2H_5OH? (A_r: H = 1; C = 12; O = 16)

a One mole of C_2H_5OH contains 2 moles of carbon atoms, 6 moles of hydrogen atoms and 1 mole of oxygen atoms. Therefore:
mass of 1 mole of ethanol
$= (2 \times 12) + (6 \times 1) + (1 \times 16)$
$= 46\,g$

b The RFM of ethanol is 46.

Example 2

What is (a) the mass of 1 mole and (b) the RFM of nitrogen gas, N_2? (A_r: N = 14)

a Nitrogen is a diatomic gas. Each nitrogen molecule contains two atoms of nitrogen. Therefore:
mass of 1 mole of N_2
$= 2 \times 14$
$= 28\,g$

b The RFM of N_2 is 28.

The mass of a compound found in any number of moles can be calculated using the relationship:

mass of compound = number of moles of the compound $\times$ mass of 1 mole of the compound

Example 3

Calculate the mass of (a) 3 moles and (b) 0.2 moles of carbon dioxide gas, CO_2. (A_r: C = 12; O =16)

a One mole of CO_2 contains 1 mole of carbon atoms and 2 moles of oxygen atoms. Therefore:
mass of 1 mole of CO_2
$= (1 \times 12) + (2 \times 16)$
$= 44\,g$

mass of 3 moles of CO_2
$=$ number of moles $\times$ mass of 1 mole of CO_2
$= 3 \times 44$
$= 132\,g$

b mass of 0.2 mole of CO_2
$=$ number of moles $\times$ mass of 1 mole of CO_2
$= 0.2 \times 44$
$= 8.8\,g$

If we know the mass of the compound then we can calculate the number of moles of the compound using the relationship:

$$\text{number of moles of compound} = \frac{\text{mass of compound}}{\text{mass of 1 mole of the compound}}$$

Example 4

Calculate the number of moles of magnesium oxide, MgO, in (a) 80 g and (b) 10 g of the compound. (A_r: O = 16; Mg = 24)

a One mole of MgO contains 1 mole of magnesium atoms and 1 mole of oxygen atoms. Therefore:
mass of 1 mole of MgO
$= (1 \times 24) + (1 \times 16)$
$= 40\,g$

number of moles of MgO in 80 g
$$= \frac{\text{mass of MgO}}{\text{mass of 1 mole of MgO}}$$
$$= \frac{80}{40}$$
$= 2\,moles$

b number of moles of MgO in 10 g
$$= \frac{\text{mass of MgO}}{\text{mass of 1 mole of MgO}}$$
$$= \frac{10}{40}$$
$= 0.25\,mole$

Moles and gases

Many substances exist as gases. If we want to find the number of moles of a gas we can do this by measuring the volume rather than the mass.
Chemists have shown by experiment that:

One mole of any gas occupies a volume of approximately $24\,dm^3$ (24 l) at room temperature and pressure (rtp). This quantity is also known as the molar gas volume, V_m.

Therefore, it is relatively easy to convert volumes of gases into moles and moles of gases into volumes using the following relationship:

$$\text{number of moles of a gas} = \frac{\text{volume of the gas (in } dm^3 \text{ at rtp)}}{24\,dm^3}$$

or

$$\text{volume of a gas} = \text{number of moles of gas} \times 24\,dm^3 \text{ (in } dm^3 \text{ at rtp)}$$

Example 1

Calculate the number of moles of ammonia gas, NH_3, in a volume of $72 \, dm^3$ of the gas measured at rtp.

$$\text{number of moles of ammonia} = \frac{\text{volume of ammonia in } dm^3}{24 \, dm^3}$$

$$= \frac{72}{24}$$

$$= 3$$

Example 2

Calculate the volume of carbon dioxide gas, CO_2, occupied by (a) 5 moles and (b) 0.5 mole of the gas measured at rtp.

a volume of CO_2
= number of moles of $CO_2 \times 24 \, dm^3$
$= 5 \times 24$
$= 120 \, dm^3$

b volume of CO_2
= number of moles of $CO_2 \times 24 \, dm^3$
$= 0.5 \times 24$
$= 12 \, dm^3$

The volume occupied by one mole of any gas must contain 6×10^{23} molecules. Therefore, it follows that equal volumes of all gases measured at the same temperature and pressure must contain the same number of molecules. This idea was also first put forward by Amedeo Avogadro and is called **Avogadro's Law**.

Moles and solutions

Chemists often need to know the concentration of a solution. Sometimes it is measured in grams per cubic decimetre $(g \, dm^{-3})$ but more often concentration is measured in **moles per cubic decimetre ($mol \, dm^{-3}$)**. When 1 mole of a substance is dissolved in water and the solution is made up to $1 \, dm^3$ (1000 cm^3), a **1 molar ($1 \, mol \, dm^{-3}$)** solution is produced. Chemists do not always need to make up such large volumes of solution. A simple method of calculating the concentration is by using the relationship:

$$\text{concentration (in } mol \, dm^{-3}) = \frac{\text{number of moles}}{\text{volume (in } dm^3)}$$

Example 1

Calculate the concentration (in $mol \, dm^{-3}$) of a solution of sodium hydroxide, NaOH, which was made by dissolving $10 \, g$ of solid sodium hydroxide in water and making up to $250 \, cm^3$. (A_r: H = 1; O = 16; Na = 23)

1 mole of NaOH contains 1 mole of sodium, 1 mole of oxygen and 1 mole of hydrogen. Therefore:

mass of 1 mole of NaOH $= (1 \times 23) + (1 \times 16) + (1 \times 1)$
$= 40 \, g$

number of moles of NaOH in $10 \, g$

$$= \frac{\text{mass of NaOH}}{\text{mass of 1 mole of NaOH}}$$

$$= \frac{10}{40}$$

$$= 0.25$$

$$\left(250 \, cm^3 = \frac{250}{1000} \, dm^3 = 0.25 \, dm^3 \right)$$

concentration of the NaOH solution

$$= \frac{\text{number of moles of NaOH}}{\text{volume of solution (} dm^3)}$$

$$= \frac{0.25}{0.25}$$

$$= 1 \, mol \, dm^{-3}$$

Sometimes chemists need to know the mass of a substance that has to be dissolved to prepare a known volume of solution at a given concentration. A simple method of calculating the number of moles and so the mass of substance needed is by using the relationship:

$$\begin{array}{cc} \text{number of} & = \text{concentration} \times \text{volume of solution} \\ \text{moles} & (\text{in } mol \, dm^{-3}) \qquad (\text{in } dm^3) \end{array}$$

Example 2

Calculate the mass of potassium hydroxide, KOH, that needs to be used to prepare $500 \, cm^3$ of a $2 \, mol \, dm^{-3}$ solution in water. (A_r: H = 1; O = 16; K = 39)

number of moles of KOH

$$= \begin{array}{c} \text{concentration of solution} \\ (mol \, dm^{-3}) \end{array} \times \begin{array}{c} \text{volume of solution} \\ (dm^3) \end{array}$$

$$= 2 \times \frac{500}{1000}$$

$$= 1$$

1 mole of KOH contains 1 mole of potassium, 1 mole of oxygen and 1 mole of hydrogen. Therefore:

mass of 1 mole of KOH $= (1 \times 39) + (1 \times 16) + (1 \times 1)$
$= 56 \, g$

Therefore:

mass of KOH in 1 mole
= number of moles $\times$ mass of 1 mole
$= 1 \times 56$
$= 56 \, g$

Questions

Use the values of A_r which follow to answer the questions below:

H = 1; C = 12; N = 14; O = 16; Ne = 20; Na = 23; Mg = 24; S = 32; K = 39; Fe = 56; Cu = 63.5; Zn = 65.

One mole of any gas at rtp occupies 24 dm³.

1 Calculate the number of moles in:
  a 2 g of neon atoms
  b 4 g of magnesium atoms
  c 24 g of carbon atoms.

2 Calculate the mass of:
  a 0.1 mole of oxygen molecules
  b 5 moles of sulfur atoms
  c 0.25 mole of sodium atoms.

3 Calculate the number of moles in:
  a 9.8 g of sulfuric acid (H_2SO_4)
  b 40 g of sodium hydroxide (NaOH)
  c 720 g of iron(II) oxide (FeO).

4 Calculate the mass of:
  a 2 moles of zinc oxide (ZnO)
  b 0.25 mole of hydrogen sulfide (H_2S)
  c 0.35 mole of copper(II) sulfate ($CuSO_4$).

5 Calculate the number of moles at rtp in:
  a 2 dm³ of carbon dioxide (CO_2)
  b 240 dm³ of sulfur dioxide (SO_2)
  c 20 cm³ of carbon monoxide (CO).

6 Calculate the volume of:
  a 0.3 mole of hydrogen chloride (HCl)
  b 4.4 g of carbon dioxide
  c 34 g of ammonia (NH_3).

7 Calculate the concentration of solutions containing:
  a 0.2 mole of sodium hydroxide dissolved in water and made up to 100 cm³
  b 9.8 g of sulfuric acid dissolved in water and made up to 500 cm³.

8 Calculate the mass of:
  a copper(II) sulfate ($CuSO_4$) which needs to be used to prepare 500 cm³ of a 0.1 mol dm⁻³ solution
  b potassium nitrate (KNO_3) which needs to be used to prepare 200 cm³ of a 2 mol dm⁻³ solution.

Calculating formulae

If we have 1 mole of a compound, then the formula shows the number of moles of each element in that compound. For example, the formula for lead(II) bromide is $PbBr_2$. This means that 1 mole of lead(II) bromide contains 1 mole of lead ions and 2 moles of bromide ions. If we do not know the formula of a compound, we can find the masses of the elements present experimentally and these masses can be used to work out the formula of that compound.

Finding the formula

Magnesium oxide
When magnesium ribbon is heated strongly, it burns very brightly to form the white powder called magnesium oxide.

magnesium + oxygen → magnesium oxide

Figure 5.5 Apparatus used to determine magnesium oxide's formula.

Table 5.1 Data from experiment shown in Figure 5.5.

Mass of crucible	14.63 g
Mass of crucible and magnesium	14.87 g
Mass of crucible and magnesium oxide	15.03 g
Mass of magnesium used	0.24 g
Mass of oxygen which has reacted with the magnesium	0.16 g

The data shown in Table 5.1 were obtained from an experiment using the apparatus shown in Figure 5.5 to find the formula for this white powder, magnesium oxide. From these data we can calculate the number of moles of each of the reacting elements. (A_r: O = 16; Mg = 24)

	Mg	O
Masses reacting (g)	0.24	0.16
Number of moles	$\dfrac{0.24}{24}$	$\dfrac{0.16}{16}$
	= 0.01	= 0.01
Ratio of moles	1	1
Formula	MgO	

This formula is the **empirical formula** of the compound. It shows the simplest ratio of the atoms present.

Unknown compound 1

In another experiment an unknown organic compound was found to contain 0.12 g of carbon and 0.02 g of hydrogen. Calculate the empirical formula of the compound. (A_r: H = 1; C = 12)

	C	H
Masses (g)	0.12	0.02
Number of moles	$\dfrac{0.12}{12}$	$\dfrac{0.02}{1}$
	= 0.01	= 0.02
Ratio of moles	1	2
Empirical formula	CH_2	

From our knowledge of bonding (Chapter 4, p. 56) we know that a molecule of this formula cannot exist. However, molecules with the following formulae do exist: C_2H_4, C_3H_6, C_4H_8 and C_5H_{10}. All of these formulae show the same ratio of carbon atoms to hydrogen atoms, CH_2, as our unknown. To find out which of these formulae is the actual formula for the unknown organic compound, we need to know the mass of one mole of the compound.

Using a mass spectrometer, the relative molecular mass (M_r) of this organic compound was found to be 56. We need to find out the number of empirical formulae units present:

M_r of the empirical formula unit
= $(1 \times 12) + (2 \times 1)$
= 14

Number of empirical formula units present

= $\dfrac{M_r \text{ of compound}}{M_r \text{ of empirical formula unit}}$

= $\dfrac{56}{14} = 4$

Therefore, the actual formula of the unknown organic compound is $4 \times CH_2 = C_4H_8$.

This substance is called butene. C_4H_8 is the **molecular formula** for this substance and shows the **actual** numbers of atoms of each element present in one molecule of the substance.

Sometimes the composition of a compound is given as a percentage by mass of the elements present. In cases such as this the procedure shown in the next example is followed.

Unknown compound 2

Calculate the empirical formula of an organic compound containing 92.3% carbon and 7.7% hydrogen by mass. The M_r of the organic compound is 78. What is its molecular formula? (A_r: H = 1; C = 12)

	C	H
% by mass	92.3	7.7
Masses in 100 g	92.3 g	7.7 g
Number of moles	$\dfrac{92.3}{12}$	$\dfrac{7.7}{1}$
	= 7.7	= 7.7
Ratio of moles	1	1
Empirical formula	CH	

M_r of the empirical formula unit CH
= 12 + 1
= 13

Number of empirical formula units present

= $\dfrac{M_r \text{ of compound}}{M_r \text{ of empirical formula unit}}$

= $\dfrac{78}{13}$

= 6

The molecular formula of the organic compound is $6 \times CH = C_6H_6$. This is a substance called benzene.

Questions

Use the following values of A_r to answer the questions below: H = 1; C = 12; O = 16; Ca = 40.

1 Determine the empirical formula of an oxide of calcium formed when 0.4 g of calcium reacts with 0.16 g of oxygen.

2 Determine the empirical formula of an organic hydrocarbon compound which contains 80% by mass of carbon and 20% by mass of hydrogen. If the M_r of the compound is 30, what is its molecular formula?

Moles and chemical equations

When we write a balanced chemical equation we are indicating the numbers of moles of reactants and products involved in the chemical reaction. Consider the reaction between magnesium and oxygen.

$$\text{magnesium} + \text{oxygen} \rightarrow \text{magnesium oxide}$$
$$2Mg(s) + O_2(g) \rightarrow 2MgO(s)$$

This shows that 2 moles of magnesium react with 1 mole of oxygen to give 2 moles of magnesium oxide.

Using the ideas of moles and masses we can use this information to calculate the quantities of the different chemicals involved.

$$2Mg(s) + O_2(g) \rightarrow 2MgO(s)$$

2 moles	1 mole	2 moles
2×24	$1 \times (16 \times 2)$	$2 \times (24 + 16)$
$= 48\,g$	$= 32\,g$	$= 80\,g$

You will notice that the total mass of reactants is equal to the total mass of product. This is true for any chemical reaction and it is known as the **Law of conservation of mass**. This law was understood by the Greeks but was first clearly formulated by Antoine Lavoisier in 1774. Chemists can use this idea to calculate masses of products formed and reactants used in chemical processes before they are carried out.

Example using a solid

Lime (calcium oxide, CaO) is used in the manufacture of mortar. It is manufactured in large quantities by a company called Tilcon at their Swindon quarry in North Yorkshire in the UK (see Figure 5.6) by heating limestone (calcium carbonate, $CaCO_3$).

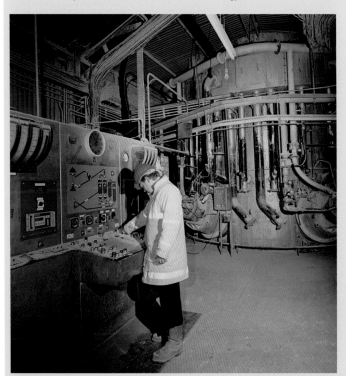

Figure 5.6 Calcium oxide production at Swindon quarry, UK.

The equation for the process is:

$$CaCO_3(s) \rightarrow CaO(s) + CO_2(g)$$

1 mole	1 mole	1 mole
$(40 + 12 + (3 \times 16))$	$40 + 16$	$12 + (2 \times 16)$
$= 100\,g$	$= 56\,g$	$= 44\,g$

Calculate the amount of lime produced when 10 tonnes of limestone are heated (Figure 5.7). (A_r: C = 12; O = 16; Ca = 40)

$$1 \text{ tonne (t)} = 1000\,kg$$
$$1\,kg = 1000\,g$$

Figure 5.7 How much lime is produced?

From this relationship between grams and tonnes we can replace the masses in grams by masses in tonnes.

$$CaCO_3(s) \rightarrow CaO(s) + CO_2(g)$$

	100 t	56 t	44 t
Hence	10 t	5.6 t	4.4 t

The equation now shows that 100 t of limestone will produce 56 t of lime. Therefore, 10 t of limestone will produce 5.6 t of lime.

Example using a gas

Many chemical processes involve gases. The volume of a gas is measured more easily than its mass. This example shows how chemists work out the volumes of gaseous reactants and products needed using Avogadro's Law and the idea of moles.

Some rockets use hydrogen gas as a fuel. When hydrogen burns in oxygen it forms water vapour. Calculate the volumes of (a) $O_2(g)$ used and (b) water, $H_2O(g)$, produced if $960\,dm^3$ of hydrogen gas, $H_2(g)$, were burned in oxygen. (A_r: H = 1; O = 16) Assume 1 mole of any gas occupies a volume of $24\,dm^3$.

$2H_2(g)$	+ $O_2(g)$	$\rightarrow 2H_2O(g)$
2 moles	1 mole	2 moles
2×24	1×24	2×24
$= 48\,dm^3$	$= 24\,dm^3$	$= 48\,dm^3$

Therefore:

(×2)	$96\,dm^3$	$48\,dm^3$	$96\,dm^3$
(×10)	$960\,dm^3$	$480\,dm^3$	$960\,dm^3$

When $960\,dm^3$ of hydrogen are burned in oxygen:

a $480\,dm^3$ of oxygen are required and
b $960\,dm^3$ of $H_2O(g)$ are produced.

Example 1 using a solution

Chemists usually carry out reactions using solutions. If they know the concentration of the solution(s) they are using they can find out the quantities reacting.

Calculate the volume of $1\,mol\,dm^{-3}$ solution of H_2SO_4 required to react completely with $6\,g$ of magnesium. (A_r: Mg = 24)

number of moles of magnesium

$$= \frac{\text{mass of magnesium}}{\text{mass of 1 mole of magnesium}}$$

$$= \frac{6}{24}$$

$$= 0.25$$

$Mg(s)$	+ $H_2SO_4(aq)$	$\rightarrow MgSO_4(aq)$ +	$H_2(g)$
1 mole	1 mole	1 mole	1 mole
0.25 mol	0.25 mol	0.25 mol	0.25 mol

We can see that $0.25\,mol$ of $H_2SO_4(aq)$ is required. Using:

volume of $H_2SO_4(aq)$ (dm^3)

$$= \frac{\text{moles of } H_2SO_4}{\text{concentration of } H_2SO_4 \text{ (mol\,dm}^{-3})}$$

$$= \frac{0.25}{1}$$

$$= 0.25\,dm^3 \text{ or } 250\,cm^3$$

Example 2 using a solution

What is the concentration of sodium hydroxide solution used in the following neutralisation reaction? $40\,cm^3$ of $0.2\,mol\,dm^{-3}$ solution of hydrochloric acid just neutralised $20\,cm^3$ of sodium hydroxide solution.

number of moles of HCl used
= concentration $(mol\,dm^{-3}) \times$ volume (dm^3)
$= 0.2 \times 0.04$
$= 0.08$

$HCl(aq)$	+ $NaOH(aq)$	$\rightarrow NaCl(aq)$ +	$H_2O(l)$
1 mole	1 mole	1 mole	1 mole
0.08 mol	0.08 mol	0.08 mol	0.08 mol

You will see that $0.08\,mole$ of NaOH would have been present. The concentration of the NaOH(aq) will be given by:

concentration of NaOH $(mol\,dm^{-3})$

$$= \frac{\text{number of moles of NaOH}}{\text{volume of NaOH } (dm^3)}$$

(volume of NaOH in $dm^3 = \frac{20}{1000} = 0.02$)

$$= \frac{0.08}{0.02}$$

$$= 4\,mol\,dm^{-3}$$

Percentage yield

Chemical reactions rarely produce the predicted amount of product from the masses of reactants in the reaction, they are not 100% efficient.

An example of this is the reaction of carbon with oxygen to produce carbon dioxide gas.

$$C(s) + O_2(g) \rightarrow CO_2(g)$$

From the equation for the reaction you can see that one mole of carbon reacts with oxygen to give one mole of carbon dioxide gas.

If you burn 12 grams, 1 mole, of carbon to make CO_2, then amount of carbon dioxide expected is 44 grams, 1 mole of CO_2, The **theoretical yield** of carbon dioxide from this reaction is 44 grams. This only occurs, however, if the reaction is 100% efficient.

The mas sof carbon dioxide you will get will be less than 44 grams because another reaction can also occur between carbon and oxygen. Some carbon reacts to make carbon monoxide, CO.

$$2\,C(s)\,1\,O_2(g) \rightarrow 2\,CO(g)$$

The percentage yield of the reaction is based on the amount of carbon dioxide that is actually produced against what should have been produced, if the reaction was 100% efficient.

If 12 grams of carbon was burned in excess oxygen and only 28 grams of carbon dioxide was produced the percentage yield can be worked out as shown below.

$$\text{Percentage yield of carbon dioxide} = \frac{28}{44} \times 100$$

$$= 63.6\%$$

In general:

$$\text{percentage yield} = \frac{\text{actual yield}}{\text{theoretical yield}} \times 100$$

Questions

Use the following A_r values to answer the questions below: O = 16; Mg = 24; S = 32; K = 39; Cu = 63.5.

1 Calculate the mass of sulfur dioxide produced by burning 16 g of sulfur in an excess of oxygen in the Contact process (see p. 227).

2 Calculate the mass of sulfur which, when burned in excess oxygen, produces 640 g of sulfur dioxide in the Contact process.

3 Calculate the mass of copper required to produce 159 g of copper(II) oxide when heated in excess oxygen.

4 In the rocket mentioned previously in which hydrogen is used as a fuel, calculate the volume of hydrogen used to produce 24 dm³ of water ($H_2O(g)$).

5 Calculate the volume of 2 mol dm⁻³ solution of sulfuric acid required to react with 24 g of magnesium.

6 What is the concentration of potassium hydroxide solution used in the following neutralisation reaction? 20 cm³ of 0.2 mol dm⁻³ solution of hydrochloric acid just neutralised 15 cm³ of potassium hydroxide solution.

■ *Checklist*

After studying Chapter 5 you should know and understand the following terms.

■ **Avogadro's Law** Equal volumes of all gases measured under the same conditions of temperature and pressure contain equal numbers of molecules.

■ **Calculating moles of compounds**

mass of compound = number of × mass of 1 mole
(in grams)   moles   of compound

$$\text{number of moles} = \frac{\text{mass of compound}}{\text{mass of 1 mole of compound}}$$

■ **Calculating moles of elements**

mass of element = number of × mass of 1 mole
(in grams)   moles   of the element

$$\text{number of moles} = \frac{\text{mass of the element}}{\text{mass of 1 mole of that element}}$$

■ **Calculating moles of gases**

1 mole of any gas occupies 24 dm³ (litres) at room temperature and pressure (rtp).

$$\text{number of moles of gas} = \frac{\text{volume of the gas (in dm}^3\text{ at rtp)}}{24\,\text{dm}^3}$$

■ **Calculating moles of solutions**

$$\text{concentration of a solution (in mol dm}^{-3}) = \frac{\text{number of moles of solute}}{\text{volume (in dm}^3)}$$

number of = concentration × volume of solution
moles   (in mol dm⁻³)   (in dm³)

■ **Empirical formula** A formula showing the simplest ratio of atoms present.

■ **Mole** The amount of substance which contains 6×10^{23} atoms, ions or molecules. This number is called Avogadro's constant.

Atoms – 1 mole of atoms has a mass equal to the relative atomic mass (A_r) in grams.

Molecules – 1 mole of molecules has a mass equal to its relative molecular mass (M_r) in grams.

■ **Molecular formula** A formula showing the actual number of atoms of each element present in one molecule.

■ **Relative formula mass (RFM)** The sum of the relative atomic masses of all those elements shown in the formula of the substance. This is often referred to as the **relative molecular mass** (M_r).

■ **Percentage yield**

$$\text{percentage yield} = \frac{\text{actual yield}}{\text{theoretical yield}} \times 100$$

Chemical calculations
Additional questions

Use the data in the table below to answer the questions which follow:

Element	A_r
H	1
C	12
N	14
O	16
Na	23
Mg	24
Si	28
S	32
Cl	35.5
Fe	56

1 Calculate the mass of:
 a 1 mole of:
   (i) chlorine molecules
   (ii) iron(III) oxide.
 b 0.5 mole of:
   (i) magnesium nitrate
   (ii) ammonia.

2 Calculate the volume occupied, at rtp, by the following gases. (One mole of any gas occupies a volume of 24 dm³ at rtp.)
 a 12.5 moles of sulfur dioxide gas.
 b 0.15 mole of nitrogen gas.

3 Calculate the number of moles of gas present in the following:
 a 36 cm³ of sulfur dioxide
 b 144 dm³ of hydrogen sulfide.

4 Use the following experimental information to determine the empirical formula of an oxide of silicon.
 Mass of crucible   18.20 g
 Mass of crucible + silicon   18.48 g
 Mass of crucible + oxide of silicon   18.80 g

5 a Calculate the empirical formula of an organic liquid containing 26.67% of carbon, 2.22% of hydrogen with the rest being oxygen.
 b The M_r of the liquid is 90. What is its molecular formula?

6 Iron is extracted from its ore, haematite, in the blast furnace. The main extraction reaction is:

$$Fe_2O_3(s) + 3CO(g) \rightarrow 2Fe(s) + 3CO_2(g)$$

 a Name the reducing agent in this process.
 b Name the oxide of iron shown in the equation.
 c Explain why this is a **redox** reaction.
 d Calculate the mass of iron which will be produced from 640 tonnes of haematite.

7 Consider the following information about the newly discovered element, vulcium, whose symbol is Vu. 'Vulcium is a solid at room temperature. It is easily cut by a penknife to reveal a shiny surface which tarnishes quite rapidly. It reacts violently with water, liberating a flammable gas and forms a solution with a pH of 13. When vulcium reacts with chlorine, it forms a white crystalline solid containing 29.5% chlorine.'
 (A_r: Vu = 85)
 a Calculate the empirical formula of vulcium chloride.
 b To which group of the periodic table should vulcium be assigned?
 c Write a word and balanced chemical equation for the reaction between vulcium and chlorine.
 d What other information in the description supports the assignment of group you have given to vulcium?
 e What type of bonding is present in vulcium chloride?
 f Write a word and balanced chemical equation for the reaction between vulcium and water.
 g Write the formulae for:
  (i) vulcium sulfate
  (ii) vulcium carbonate
  (iii) vulcium hydroxide.
  Look at the periodic table (p. 38) to find out the real name of vulcium.

8 0.048 g of magnesium was reacted with excess dilute hydrochloric acid at room temperature and pressure. The hydrogen gas given off was collected.
 a Write a word and balanced symbol equation for the reaction taking place.
 b Draw a diagram of an apparatus which could be used to carry out this experiment and collect the hydrogen gas.
 c How many moles of magnesium were used?
 d Using the equation you have written in your answer to **a**, calculate the number of moles of hydrogen and hence the volume of this gas produced.
 e Calculate the volume of a solution containing 0.1 mol dm⁻³ hydrochloric acid which would be needed to react exactly with 0.048 g of magnesium.

6 Electrolysis and its uses

Electrolysis of lead(II) bromide

Electrolysis of aluminium oxide
Anodising

Extraction of sodium

Electrolysis of aqueous solutions
Electrolysis of water
Chlor-alkali industry

Electrolysis of copper(II) sulfate solution

Purification of copper

Electrolysis guidelines

Electroplating
Plating plastics

Calculations in electrolysis

Checklist

Additional questions

What do all the items in the photographs shown in Figure 6.1 have in common? They all involve electricity through a process known as **electrolysis**. Electrolysis is the process of splitting up (decomposing) substances by passing an electric current through them. The substance which is decomposed is called the **electrolyte** (Figure 6.2, opposite). An electrolyte is a substance that conducts electricity when in the molten state or in solution. The electricity is carried through the electrolyte by **ions**. In the molten state and in solution the ions are free to move to the appropriate electrodes due to weakened forces of attraction between them.

- Substances that do not conduct electricity when in the molten state or in solution are called **non-electrolytes**.
- Substances that conduct electricity **to a small extent** in the molten state or in solution are called **weak electrolytes**.

The electric currrent enters and leaves the electrolyte through **electrodes**, which are usually made of unreactive metals such as platinum or of the non–metal carbon. These are said to be **inert** electrodes because they do not react with the products of electrolysis. The names given to the two electrodes are **cathode**, the negative electrode which attracts **cations** (positively charged ions), and **anode**, the positive electrode which attracts **anions** (negatively charged ions).

Note that the conduction which takes place in the electrodes is due to the movement of delocalised electrons (pp. 61 and 65) whereas in the electrolyte, as stated earlier, the charge carriers are ions.

Electrolysis is very important in industry. To help you to understand what is happening in the process shown in the photographs, we will first consider the electrolysis of lead(II) bromide.

a This watch has a thin coating of gold over steel; the thin coating is produced by electrolysis.

b This picture frame has been silver plated using an electroplating process involving electrolysis.

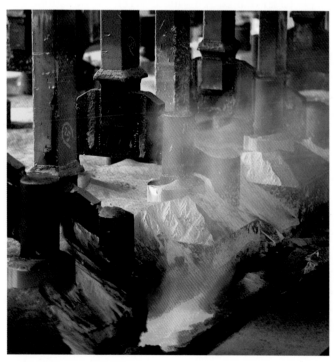

c Aluminium is produced by electrolysis.

Figure 6.1

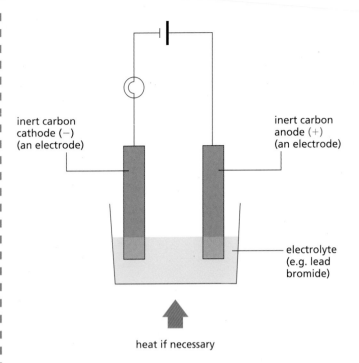

Figure 6.2 The important terms used in electrolysis.

■ *Electrolysis of lead(II) bromide*

Consider Figure 6.3, which shows solid lead(II) bromide (PbBr$_2$) in a crucible with two carbon electrodes in contact with it. When the electrodes are first connected, the bulb does not light, because the solid compound does not allow electricity to pass through it. However, when the compound is heated until it is molten, the bulb does light. The lead(II) bromide is now behaving as an electrolyte. When this happens an orange-red vapour is seen at the anode and lead metal is produced at the cathode.

The break-up (decomposition) of lead(II) bromide into its constituent elements by the passage of an electric current is called **electrolysis**.

$$\text{molten lead(II) bromide} \rightarrow \text{bromine} + \text{lead}$$
$$\text{PbBr}_2\text{(l)} \rightarrow \text{Br}_2\text{(g)} + \text{Pb(l)}$$

For lead metal to be formed, or deposited, at the cathode, the lead ions must be attracted towards the electrode (Figure 6.4). To produce lead metal atoms these lead ions must each collect two electrons at the cathode:

$$\text{lead ion} + \text{electrons} \rightarrow \text{lead atom}$$
$$\text{Pb}^{2+}\text{(l)} + 2e^- \rightarrow \text{Pb(l)}$$

This process of gaining electrons is called **reduction** (p. 15).

To form bromine molecules each bromide ion must first of all lose its extra negative charge at the anode and so form a neutral bromine atom:

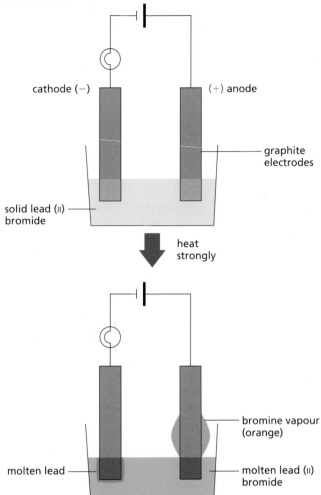

Figure 6.3 The electrolysis of molten lead(II) bromide.

Figure 6.4 The lead ions (Pb^{2+}) are attracted to the cathode and the bromide ions (Br$^-$) are attracted to the anode.

$$\text{bromide ion} \rightarrow \text{bromine atom} + \text{electron}$$
$$\text{Br}^-\text{(l)} \rightarrow \text{Br} + e^-$$

Two bromine atoms then combine to form a bromine molecule:

$$\text{bromine atoms} \rightarrow \text{bromine molecule}$$
$$2\text{Br} \rightarrow \text{Br}_2\text{(g)}$$

Question

1 Predict the products of the electrolysis of molten potassium chloride.

Electrolysis of aluminium oxide

Aluminium is the most abundant metallic element in the Earth's crust. It was discovered in 1825 by Hans Christian Oersted in Copenhagen, Denmark, but was first isolated by Friedrich Wöhler in 1827. It makes up 8% of the crust and is found in the minerals bauxite (Figure 6.5), cryolite and mica, as well as in clay.

In the nineteenth century Napoleon III owned a very precious dinner service. It was said to be made of a metal more precious than gold. That metal was aluminium. The reason it was precious was that it was very rarely found as the pure metal. Aluminium is a reactive metal and as such was very difficult to extract from its ore. Reactive metals hold on tightly to the element(s) they have combined with and many are extracted from their ores by electrolysis.

Today we use aluminium in very large quantities. The annual production in the world is 19.5 million tonnes. The commercial extraction of aluminium has been made possible by two scientists, working independently of each other, who discovered a method using electrolysis. The two scientists were Charles Martin Hall (USA), who discovered the process in 1886, and the French chemist Paul Héroult, who discovered the process independently in the same year. The process they developed, often called the Hall–Héroult process, involves the electrolysis of aluminium oxide (alumina). The process involves the following stages.

- Bauxite, an impure form of aluminium oxide, is first treated with sodium hydroxide to obtain pure aluminium oxide, removing impurities such as iron(III) oxide and sand. This improves the conductivity of the molten aluminium oxide.
- The purified aluminium oxide is then dissolved in molten cryolite (Na_3AlF_6). Cryolite, a mineral found naturally in Greenland, is used to reduce the working temperature of the Hall–Héroult cell from 2017 °C (the melting point of pure aluminium oxide) to between 800 and 1000 °C. Therefore, the cryolite provides a considerable saving in the energy requirements of the process. In recent years it has become necessary to manufacture the cryolite.
- The molten mixture is then electrolysed in a cell similar to that shown in Figure 6.6.

Figure 6.5 Bauxite mining.

crust of solid electrolyte

graphite anodes

aluminium out

4m wide

1m deep

syphon

8m long

molten aluminium collecting on floor of cell

tank lined with graphite cathode

molten electrolyte

Figure 6.6 The Hall–Héroult cell is used in industry to extract aluminium by electrolysis.

The anodes of this process are blocks of graphite which are lowered into the molten mixture from above. The cathode is the graphite lining of the steel vessel containing the cell.

Aluminium oxide is an ionic compound. When it is melted the ions become mobile, as the strong electrostatic forces of attraction between them are broken by the input of heat energy. During electrolysis the negatively charged oxide ions are attracted to the anode (the positive electrode), where they lose electrons to form oxygen gas.

oxide ions → oxygen molecules + electrons
$$2O^{2-}(l) \rightarrow O_2(g) + 4e^-$$

The positive aluminium ions are attracted to the cathode (the negative electrode). They gain electrons to form molten aluminium metal.

aluminium ions + electrons → aluminium metal
$$Al^{3+}(l) + 3e^- \rightarrow Al(l)$$

The gain of electrons which takes place at the cathode is called **reduction**. The loss of electrons which takes place at the anode is called **oxidation**. This is a further way of considering oxidation and reduction (Chapter 2, p. 15). A handy way of remembering it is **oil rig** (**o**xidation **i**s **l**oss, **r**eduction **i**s **g**ain of electrons).

The overall reaction which takes place in the cell is:

aluminium oxide $\xrightarrow{\text{electrolysis}}$ aluminium + oxygen
$$2Al_2O_3(l) \xrightarrow{\hspace{2cm}} 4Al(l) + 3O_2(g)$$

The molten aluminium collects at the bottom of the cell and it is siphoned out at regular intervals. No problems arise with other metals being deposited, since the cryolite is largely 'unaffected' by the flow of electricity. Problems do arise, however, with the graphite anodes. At the working temperature of the cell, the oxygen liberated reacts with the graphite anodes, producing carbon dioxide.

carbon + oxygen → carbon dioxide
$$C(s) + O_2(g) \rightarrow CO_2(g)$$

The anodes burn away and have to be replaced on a regular basis.

The electrolysis of aluminium oxide is a continuous process in which vast amounts of electricity are used. Approximately 15kWh of electricity are used to produce 1kg of aluminium. In order to make the process an economic one, a cheap form of electricity is required. Hydroelectric power (HEP) is usually used for this process. The plant shown in Figure 6.7 uses an HEP scheme to provide some of the electrical energy required for this process. Further details about HEP are given in Chapter 13, p. 188.

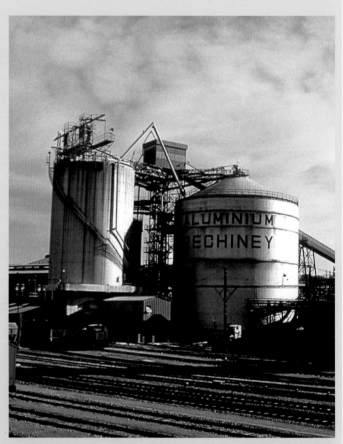

Figure 6.7 An aluminium smelting plant.

Using cheap electrical energy has allowed aluminium to be produced in such large quantities that it is the second most widely used metal after iron. It is used in the manufacture of electrical cables because of its low density, chemical inertness and good electrical conductivity. Owing to the first two of these properties, it is also used for making cars, bikes and cooking foil as well as in alloys (Chapter 9, p. 141) such as duralumin, which is used in the manufacture of aeroplane bodies (Figure 6.8).

Figure 6.8 Aluminium is used in the manufacture of aeroplane bodies.

Figure 6.9 Bauxite pollution of the Amazon in Brazil.

Environmental problems associated with the location of aluminium plants are concerned with:

- the effects of the extracted impurities, which form a red mud (Figure 6.9)
- the fine cryolite dust, which is emitted through very tall chimneys so as not to affect the surrounding area
- the claimed link between environmental aluminium and a degenerative brain disorder called Alzheimer's disease – it is thought that aluminium is a major influence on the early onset of this disease. However, the evidence is still inconclusive.

Anodising

This is a process in which the surface coating of oxide on aluminium (Al_2O_3) is made thicker. In this process the aluminium object is made the anode in a cell in which the electrolyte is dilute sulphuric acid. During the electrolysis process, oxygen is produced at the anode and combines with the aluminium. The oxide layer on the surface of the aluminium therefore increases. Dyes can be mixed with the electrolyte and so the new thicker coating of oxide is colourful and also decorative (Figure 6.10).

Figure 6.10 The oxide layer on the surface of these aluminium cups has been thickened, and dyes added to obtain the vibrant colours.

Questions

1 Produce a flow chart to summarise the processes involved in the extraction of aluminium metal from bauxite.

2 Explain why the mixture of gases formed at the anode contains oxygen, carbon dioxide and traces of fluorine.

Extraction of sodium

Sodium, a very reactive metal, was first discovered by Sir Humphry Davy in 1807. It is now extracted by electrolysing molten sodium chloride in the Down's cell. Sodium chloride may be obtained from the evaporation of sea water or mined as rock salt (Figure 6.11). The rock salt is mined from large underground caverns. Rock salt for the chemical industry is obtained by dissolving the salt and pumping the brine to the surface.

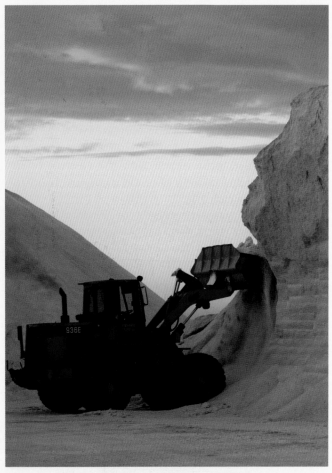

Figure 6.11 Rock salt mining in Spain.

The first economic extraction of sodium was developed by Down in 1921 in North America using hydroelectric power generated by the Niagara Falls. This ensured the availability of cheap electricity.

The process involves the following.

• Sodium chloride, obtained from rock salt or sea water, is mixed with calcium chloride and melted. The calcium chloride is added to the sodium chloride electrolyte to reduce the working temperature of the cell from 801 °C (the melting point of sodium chloride) to 600 °C (the melting point of the mixture). This saves electrical energy and, therefore, makes the process more economical.
• The mixture is then electrolysed in a cell similar to the one in Figure 6.12.

Figure 6.12 The Down's cell used for the electrolysis of sodium chloride.

The chloride ions are attracted to the anode, where they lose electrons and form chlorine gas.

chloride ions → chlorine molecules + electrons
$$2Cl^-(l) \rightarrow Cl_2(g) + 2e^-$$

The positive sodium ions are attracted to the cathode. They gain electrons to form molten sodium metal.

sodium ions + electrons → sodium metal
$$Na^+(l) + e^- \rightarrow Na(l)$$

The overall reaction which takes place in the cell is:

sodium chloride $\xrightarrow{\text{electrolysis}}$ sodium + chlorine
$$2NaCl(l) \longrightarrow 2Na(l) + Cl_2(g)$$

The cathode is a circle of steel around the graphite anode. At 600 °C sodium and chlorine would react violently together to re-form sodium chloride. To prevent this from happening, the Down's cell contains a steel gauze around the graphite anode to keep it and the cathode apart. The molten sodium floats on the electrolyte and is run off for storage.

A problem arises, however, in that calcium ions are also attracted to the cathode, where they form calcium metal. Therefore, the sodium which is run off contains a significant proportion of calcium. Fortunately, the calcium crystallises out when the mixture cools and relatively pure sodium metal remains (Figure 6.13, overleaf).

Figure 6.13 Molten sodium – the result of electrolysis of sodium chloride.

The worldwide production of sodium each year is 60 000–80 000 tonnes.

Sodium is used as a liquid coolant in nuclear power stations as well as in street lighting and in the production of metals such as titanium. Chlorine gas (the co-product) is sold to make the process more economical (see overleaf and p. 43 for the properties of chlorine).

■ *Electrolysis of aqueous solutions*

Other industrial processes involve the electrolysis of aqueous solutions. To help you to understand what is happening in these processes, we will first consider the electrolysis of water.

Electrolysis of water

Pure water is a very poor conductor of electricity because there are so few ions in it. However, it can be made to decompose if an electric current is passed through it in a Hofmann voltameter, as in Figure 6.14.

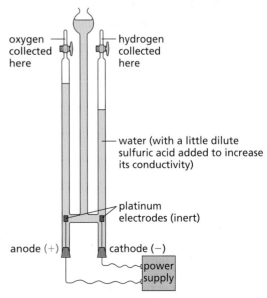

oxygen collected here — — hydrogen collected here

— water (with a little dilute sulfuric acid added to increase its conductivity)

— platinum electrodes (inert)

anode (+) — — cathode (−)

power supply

Questions

1 Use your research skills, including the Internet, to find uses, other than those given in the text, for sodium.

2 A student carries out the electrolysis of molten lead(ii) bromide in a fume cupboard.
   a Draw a diagram to show a suitable apparatus the student could use to carry out this experiment.
   b Write anode and cathode reactions to represent the processes taking place during the electrolysis.
   c Why does this experiment need to be carried out in a fume cupboard?
   d Find uses for the anode product of this cell.

Figure 6.14 A Hofmann voltameter used to electrolyse water. ■ **85**

To enable water to conduct electricity better, some dilute sulfuric acid (or sodium hydroxide solution) is added. When the power is turned on and an electric current flows through this solution, gases can be seen to be produced at the two electrodes and they are collected in the side arms of the apparatus. After about 20 minutes, roughly twice as much gas is produced at the cathode as at the anode.

The gas collected at the cathode burns with a squeaky pop, showing it to be hydrogen gas. For hydrogen to be collected in this way, the positively charged hydrogen ions must have moved to the cathode.

$$\text{hydrogen ions} + \text{electrons} \rightarrow \text{hydrogen molecules}$$
$$4H^+{}_{(aq)} + 4e^- \rightarrow 2H_2{}_{(g)}$$

If during this process the water molecules lose $H^+{}_{(aq)}$, then the remaining portion must be hydroxide ions, $OH^-{}_{(aq)}$. These ions are attracted to the anode. The gas collected at the anode relights a glowing splint, showing it to be oxygen.

This gas is produced in the following way.

$$\begin{array}{ccccc} \text{hydroxide} & \rightarrow & \text{water} & + & \text{oxygen} & + & \text{electrons} \\ \text{ions} & & \text{molecules} & & \text{molecules} \end{array}$$
$$4OH^-{}_{(aq)} \rightarrow 2H_2O_{(l)} + O_2{}_{(g)} + 4e^-$$

This experiment was first carried out by Sir Humphry Davy. It confirmed that the formula for water was H_2O.

Chlor-alkali industry

The electrolysis of saturated sodium chloride solution (brine) is the basis of a major industry. In countries where rock salt (sodium chloride) is found underground it is mined. In other countries it can be obtained by evaporation of sea water in large shallow lakes. Three very important substances are produced in this electrolysis process – chlorine, sodium hydroxide and hydrogen. The electrolytic process is a very expensive one, requiring vast amounts of electricity. The process is economical only because all three products have a large number of uses (Figure 6.15).

Figure 6.15 The chlor-alkali industry.

There are two well-established methods for electrolysing brine, the **diaphragm cell** and the **mercury cell**. However, recent developments in electrolysis technology, by chemical engineers, have produced the **membrane cell** (Figure 6.16). This method is now preferred to the other two because it produces a purer product, it causes less pollution and it is cheaper to run.

The brine is first purified to remove calcium, strontium and magnesium compounds by a process of ion exchange (see Chapter 8, p. 121).

a A section through the membrane cell.

b A diagrammatic representation of the reactions going on inside the cell.

Figure 6.16

The membrane cell is used continuously, with fresh brine flowing into the cell as the process breaks up the brine. The cell has been designed to ensure that the products do not mix. The ions in this concentrated sodium chloride solution are:

from the water: H^+(aq) OH^-(aq)
from the sodium chloride: Na^+(aq) Cl^-(aq)

When the current flows, the chloride ions, Cl^-(aq), are attracted to the anode. Chlorine gas is produced by the electrode process.

$$\text{chloride ions} \xrightarrow{\text{oxidation}} \text{chlorine molecules + electrons}$$
$$2Cl^-(aq) \longrightarrow Cl_2(g) + 2e^-$$

This leaves a high concentration of sodium ions, Na^+(aq), around the anode.

The hydrogen ions, H^+(aq), are attracted to the cathode and hydrogen gas is produced.

$$\text{hydrogen ions + electrons} \xrightarrow{\text{reduction}} \text{hydrogen molecules}$$
$$2H^+(aq) + 2e^- \longrightarrow H_2(g)$$

This leaves a high concentration of hydroxide ions, OH^-(aq), around the cathode. The sodium ions, Na^+(aq), are drawn through the membrane, where they combine with the OH^-(aq) to form sodium hydroxide, NaOH, solution. The annual production worldwide is 41 million tonnes.

Tests for aqueous cations

Aqueous sodium hydroxide can be used to identify salts of Al^{3+}, Ca^{2+}, Cu^{2+}, Fe^{2+}, Fe^{3+}, Zn^{2+}. The colour of the precipitate and its behaviour in excess sodium hydroxide solution will help identify the metal present. In the case of ammonium salts, ammonia gas is produced on warming.

Table 6.1

Cation	Effect of aqueous sodium hydroxide
Aluminium (Al^{3+})	White precipitate, soluble in excess giving a colourless solution
Ammonium (NH_4^+)	Ammonia produced on warming
Calcium (Ca^{2+})	White precipitate, insoluble in excess
Copper (Cu^{2+})	Light blue precipitate, insoluble in excess
Iron(II) (Fe^{2+})	Green precipitate, insoluble in excess
Iron(III) (Fe^{3+})	Red-brown precipitate, insoluble in excess
Zinc (Zn^{2+})	White precipitate, soluble in excess, giving a colourless solution

If the nitrate ion is present in solution, then upon addition of aqueous sodium hydroxide and a little aluminium foil, ammonia gas is produced on warming.

Questions

1 Suggest a reason for only 'roughly' twice as much hydrogen gas being produced at the cathode as oxygen gas at the anode in the electrolysis of water.

2 Account for the following observations which were made when concentrated sodium chloride solution, to which a little universal indicator had been added, was electrolysed in the laboratory in a Hofmann voltameter.

   a The universal indicator initially turns red in the region of the anode, but as the electrolysis proceeds it loses its colour.

   b The universal indicator turns blue in the region of the cathode.

3 Why is it important to remove compounds of calcium, strontium and magnesium before brine is electrolysed?

4 The uses of sodium hydroxide can be separated on a percentage basis as follows:

Neutralisation	5%
Paper manufacture	5%
Oil refining	5%
Soap/detergents	5%
Manufacture of rayon and acetate fibres	16%
Manufacture of chemicals	30%
Miscellaneous uses	34%

Use a graph-plotting program to create a 'pie' chart of these data.

Electrolysis of copper(II) sulfate solution

Copper(II) sulfate solution ($CuSO_4(aq)$) may be electrolysed using inert graphite electrodes in a cell similar to that shown in Figure 6.17. When the solution is electrolysed, oxygen gas and copper metal are formed at the anode and cathode respectively. Four ions are present in solution:

from the water: $H^+(aq)$ $OH^-(aq)$
from the copper(II) sulfate: $Cu^{2+}(aq)$ $SO_4^{2-}(aq)$

Figure 6.17 The electrolysis of copper(II) sulfate solution using inert electrodes.

$H^+(aq)$ and $Cu^{2+}(aq)$ ions are both attracted to the cathode, the Cu^{2+} ions accepting electrons more readily than the H^+ ions (preferential discharge). Copper metal is therefore deposited at the cathode (Figure 6.17).

$$copper\ ions + electrons \rightarrow copper\ atoms$$
$$Cu^{2+}(aq)\ +\ 2e^-\ \rightarrow\ Cu(s)$$

$OH^-(aq)$ and $SO_4^{2-}(aq)$ ions are both attracted to the anode. The OH^- ions release electrons more easily than the SO_4^{2-} ions, so oxygen gas and water are produced at the anode (Figure 6.18).

$$hydroxide\ ions \rightarrow oxygen + water + electrons$$
$$4OH^-(aq)\ \rightarrow\ O_2(g)\ + 2H_2O(l) +\ 4e^-$$

Figure 6.18 Oxygen is given off at the anode and copper is deposited at the cathode.

Purification of copper

Because copper is a very good conductor of electricity, it is used for electrical wiring and cables (Figure 6.19).

Figure 6.19a The copper used in electrical wiring has to be very pure.

Figure 6.19b Due to the high density of copper and its cost, steel-cored aluminium cables are used for electrical transmission through national grids.

Figure 6.21 The movement of ions in the purification of copper by electrolysis.

However, even small amounts of impurities cut down this conductivity quite noticeably whether in fine wires or larger cables. The metal must be 99.99% pure to be used in this way. To ensure this level of purity, the newly extracted copper has to be purified by electrolysis (Figure 6.20).

The impure copper is used as the anode and is typically 1m square, 35–50 mm thick and 330 kg in weight. The cathode is a 1 mm thick sheet and weighs about 5 kg; it is made from very pure copper. The electrolyte is a solution of copper(II) sulfate $(0.3\,mol\,dm^{-3})$ acidified with a $2\,mol\,dm^{-3}$ solution of sulfuric acid to help the solution conduct electricity.

When the current flows, the copper moves from the impure anode to the pure cathode. Any impurities fall to the bottom of the cell and collect below the anode in the form of a slime. This slime is rich in precious metals and the recovery of these metals is an important aspect of the economics of the process. The electrolysis proceeds for about three weeks until the anodes are reduced to about 10% of their original size and the cathodes weigh between 100 and 120 kg. A potential of 0.25V and a current density of $200\,Am^{-2}$ are usually used.

Figure 6.20 Copper purification process.

The ions present in the solution are:

from the water:	H^+(aq)	OH^-(aq)
from the copper(II) sulfate:	Cu^{2+}(aq)	SO_4^{2-}(aq)

During the process the impure anode loses mass because the copper atoms lose electrons and become copper ions, Cu^{2+}(aq) (Figure 6.21, p. 89).

$$\text{copper atoms} \rightarrow \text{copper ions} + \text{electrons}$$
$$Cu(s) \rightarrow Cu^{2+}(aq) + 2e^-$$

The electrons released at the anode travel around the external circuit to the cathode. There the electrons are passed on to the copper ions, Cu^{2+}(aq), from the copper(II) sulfate solution and the copper is deposited or copper plated on to the cathode.

$$\text{copper ions} + \text{electrons} \rightarrow \text{copper atoms}$$
$$Cu^{2+}(aq) + 2e^- \rightarrow Cu(s)$$

The annual production of copper worldwide is 12 million tonnes. However, a large amount of the copper we need is obtained by recycling. This way of obtaining copper is increasing in importance as it becomes more difficult and expensive to locate and extract the copper ore.

■ Electrolysis guidelines

The following points may help you work out the products of electrolysis in unfamiliar situations. They will also help you remember what happens at each electrode.

- Non-metals are produced at the anode whereas metals and hydrogen gas are produced at the cathode.
- At the inert anode, chlorine, bromine and iodine (the halogens) are produced in preference to oxygen.
- At the inert cathode, hydrogen is produced in preference to metals unless unreactive metals such as copper and nickel are present.

Questions

1 Why do you think it is advantageous to use inert electrodes in the electrolysis processes?

2 Predict the products of electrolysis of a solution of copper(II) sulfate if carbon electrodes are used instead of those made from copper as referred to in the purification of copper section.

3 Predict the products of the electrolysis of concentrated hydrochloric acid using platinum electrodes.

4 Using your knowledge of electrolysis, predict the likely products of the electrolysis of copper(II) chloride solution, using platinum electrodes. Write electrode equations for the formation of these products.

5 Suggest reasons why:
   a copper and aluminium are used in electrical cables
   b plastics and ceramics are used to manufacture handles for metallic cooking utensils.

■ Electroplating

Electroplating is the process involving electrolysis to plate, or coat, one metal with another or a plastic with a metal. Often the purpose of electroplating is to give a protective coating to the metal beneath. For example, bath taps are chromium plated to prevent corrosion, and at the same time are given a shiny, more attractive finish (Figure 6.22).

Figure 6.22 This tap has been chromium plated.

The electroplating process is carried out in a cell such as the one shown in Figure 6.23a. This is often known as the 'plating bath' and it contains a suitable electrolyte, usually a solution of a metal salt.

For silver plating the electrolyte is a solution of a silver salt. The article to be plated is made the cathode in the cell so that the metal ions move to it when the current is switched on. The cathode reaction in this process is:

$$\text{silver ions} + \text{electrons} \rightarrow \text{silver atoms}$$
$$Ag^+(aq) + e^- \rightarrow Ag(s)$$

anode (+)　　　cathode (−)

object to be plated,
e.g. spoon

anode made
from the metal
being used
for plating,
e.g. silver

electrolyte
containing the
metal being used
for plating,
e.g. silver

a Silver plating a spoon.

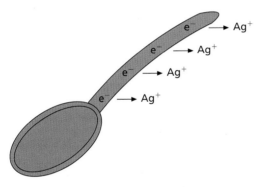

The Ag⁺ ions are attracted to the
cathode, where they gain electrons

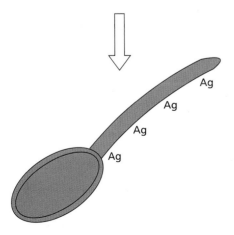

A coating of silver forms
on the spoon at the cathode

b Explaining silver plating.

Figure 6.23

Plating plastics

Nowadays it is not only metals that are electroplated. Plastics have been developed that are able to conduct electricity. For example, the plastic poly(pyrrole) can be electroplated in the same way as the metals we have discussed above (Figure 6.24).

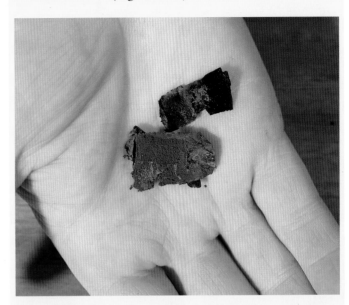

Figure 6.24 This plastic has been coated with copper by electrolysis.

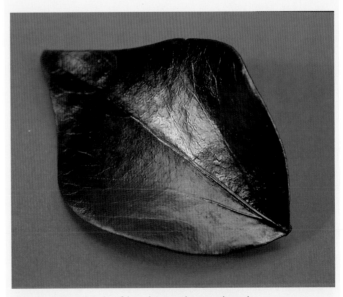

Figure 6.25 This leaf has been electroplated.

Questions

1 The leaf in Figure 6.25 has been copper plated. Suggest a suitable method for copper plating the leaf.

2 Explain why copper(II) chloride solution would not be used as an electrolyte in the electrolyte cell used for copper plating.

Calculations in electrolysis

The quantity of electricity flowing through an electrolysis cell is measured in **coulombs** (C). If one **ampere** (A) is passed for one second, the quantity of electricity is said to be 1 coulomb.

$$\text{quantity of electricity} = \text{current} \times \text{time}$$
$$\text{(coulombs)} \quad\quad \text{(amperes)} \quad \text{(seconds)}$$

Therefore, if 2 amperes flow for 5 seconds then the quantity of electricity passed is 10 coulombs.

In the purification of copper you saw that copper was deposited at the cathode. The electrode equation is:

$$Cu^{2+}(aq) + 2e^- \rightarrow Cu(s)$$
$$1\,\text{mole} \quad 2\,\text{moles} \quad 1\,\text{mole}$$

This equation tells us that 1 mole of copper(II) ions combines with 2 moles of electrons to produce 1 mole of copper metal atoms (63.5 g).

An old name for a mole of electrons was the **faraday**. The unit was named after an English scientist, Michael Faraday (Figure 6.26), who carried out many significant experiments while investigating the nature of magnetism and electricity.

Figure 6.26 In 1883 Michael Faraday was the first scientist to measure the masses of elements produced during electrolysis.

From accurate electrolysis experiments it has been found that:

$$1 \text{ mole of electrons} = 96\,500 \text{ coulombs}$$

Therefore, the quantity of electricity required to deposit 1 mole of copper atoms (63.5 g) is:

$$2 \times 96\,500 = 193\,000 \text{ coulombs}$$

Example 1

Calculate the number of moles of electrons required to deposit 10 g of silver on the surface of a fork during an electroplating process. (A_r: Ag = 108)

$$\text{Electrode equation: } Ag^+(aq) + e^- \rightarrow Ag(s)$$

1 mole of silver is deposited by 1 mole of electrons. Therefore, 108 g of silver is deposited by 1 mole of electrons.
Hence 1 g of silver is deposited by $\frac{1}{108}$ mole of electrons.
Therefore, 10 g of silver are deposited by

$$\frac{1}{108} \times 10 = 0.093 \text{ moles of electrons}$$

Example 2

Calculate the volume of oxygen gas, measured at room temperature and pressure (rtp), liberated at the anode in the electrolysis of acidified water by 2 moles of electrons.
(1 mole of oxygen at rtp occupies a volume of 24 dm³.)

$$\text{Electrode equation: } 4OH^-(aq) \rightarrow 2H_2O(l) + O_2(g) + 4e^-$$

1 mole of oxygen gas is liberated by 4 moles of electrons.
Therefore, 24 dm³ of oxygen is liberated by 4 moles of electrons.
Hence 12 dm³ of oxygen would be liberated by 2 moles of electrons.

Example 3

The industrial production of aluminium uses a current of 25 000 amperes. Calculate the time required to produce 10 kg of aluminium from the electrolysis of molten aluminium oxide.

$$\text{Electrode equation: } Al^{3+}(l) + 3e^- \rightarrow Al(l)$$

1 mole of aluminium is produced by 3 moles of electrons.
27 g of aluminium is produced by 3 moles of electrons.

Hence 1 g of aluminium would be produced by $\frac{3}{27}$ moles of electrons.

Therefore, 10 000 g of aluminium would be produced by

$$\frac{3}{27} \times 10\,000 \text{ moles of electrons}$$

$$= 1111.1 \text{ moles of electrons}$$

So, the quantity of electricity

$$= \text{number of moles of electrons} \times 96\,500$$
$$= 1111.1 \times 96\,500$$
$$= 1.07 \times 10^8\,C$$

Time (seconds)

$$= \frac{\text{coulombs}}{\text{amperes}}$$

$$= \frac{1.07 \times 10^8}{25\,000}$$

$$= 4280 \text{ seconds (71.3 minutes)}$$

Questions

1 Write equations which represent the discharge at the cathode of the following ions:
   a K^+
   b Pb^{2+}
   c Al^{3+}
   and at the anode of:
   d Br^-
   e O^{2-}
   f F^-

2 How many moles of electrons are required to discharge 1 mole of the following ions?
   a Mg^{2+}
   b K^+
   c Al^{3+}
   d O^{2-}

3 How many coulombs are required to discharge 1 mole of the following ions?
   (1 mole of electrons = 96 500 coulombs)
   a Cu^{2+}
   b Na^+
   c Pb^{2+}

4 Calculate the number of moles of electrons required to deposit 6.35 g of copper on a metal surface.
   (A_r: Cu = 63.5)

■ *Checklist*

After studying Chapter 6 you should know and understand the following terms.

■ **Anode**   The positive electrode. It is positively charged because electrons are drawn away from it.

■ **Cathode**   The negative electrode. It is negatively charged because an excess of electrons move towards it.

■ **Coulombs**   The unit used to measure the quantity of electricity:

$$1 \text{ coulomb} = 1 \text{ ampere} \times 1 \text{ second}$$

■ **Down's cell**   The electrolysis cell in which sodium is extracted from molten sodium chloride to which 60% calcium chloride has been added to reduce the working temperature to 600 °C. This cell has a graphite anode surrounded by a cylindrical steel cathode.

■ **Electrode**   A point where the electric current enters and leaves the electrolytic cell. An inert electrode is usually made of platinum or carbon and does not react with the electrolyte or the substances produced at the electrodes themselves.

■ **Electrolysis**   A process in which a chemical reaction is caused by the passage of an electric current.

■ **Electrolyte**   A substance which will carry electric current only when it is molten or dissolved.

■ **Electroplating**   The process of depositing metals from solution in the form of a layer on other surfaces such as metal or plastic.

■ **Hall–Héroult cell**   The electrolysis cell in which aluminium is extracted from purified bauxite dissolved in molten cryolite at 900 °C. This cell has both a graphite anode and a graphite cathode.

■ **Membrane cell**   An electrolytic cell used for the production of sodium hydroxide, hydrogen and chlorine from brine in which the anode and cathode are separated by a membrane.

■ **One mole of electrons**   This is the same as one Faraday and is equivalent to 96 500 coulombs.

■ **Oxidation**   Takes place at the anode and involves a negative ion losing electrons.

■ **Reduction**   Takes place at the cathode and involves a positive ion gaining electrons.

Electrolysis and its uses
Additional questions

1 Explain the meaning of each of the following terms. Use a suitable example, in each case, to help with your explanation.
   a Anode.
   b Cathode.
   c Electrolysis.
   d Electrolyte.
   e Anion.
   f Cation.
   g Oxidation.
   h Reduction.

2 Copper is purified by electrolysis, as in the example shown below.

   a Name the materials used for the electrodes **A** and **B**.
   b Name the electrolyte **C** and substance **D**.
   c Why is substance **D** of economic importance in respect of this process?
   d Write equations for the reactions which take place at the cathode and anode during this process.
   e Draw a labelled diagram to show the cell after electrolysis has taken place.
   f Why has electrolyte **C** to be acidified with the dilute sulphuric acid?
   g Why does copper have to be 99.99% pure for use in electrical cables?

3 Copy and complete the table below, which shows the results of the electrolysis of four substances using inert electrodes.

Electrolyte	Product at anode (positive electrode)	Product at cathode (negative electrode)
Molten aluminium oxide		Aluminium
Concentrated sodium chloride solution	Chlorine	
Molten lithium chloride		
Silver nitrate solution		Silver

   a State what you understand by 'inert electrodes'.
   b Explain why the lithium chloride solution becomes progressively more alkaline during electrolysis.
   c Explain why solid lithium chloride is a non-conductor of electricity, whereas molten lithium chloride and lithium chloride solution are good conductors of electricity.

   d During the electrolysis of molten aluminium chloride ($AlCl_3$) the carbon anodes are burned away. Explain why this should happen and write balanced chemical equations for the reactions that take place.

4 The industrial production of sodium uses a current of 15 000 amperes. Calculate the time required to produce 100 kg of sodium from the electrolysis of molten sodium chloride. (1 mole of electrons = 96 500 coulombs. A_r: Na = 23)

5 Calculate the volume of hydrogen gas, measured at room temperature and pressure, liberated at the cathode in the electrolysis of acidified water by 18 moles of electrons. (One mole of any gas at rtp occupies a volume of 24 dm^3.)

6 Calculate the number of moles of electrons required to produce 10 g of gold deposited on the surface of some jewellery during an electroplating process. (A_r: Au = 197; note gold in solution is present as the Au^{3+} ion)

7 A pupil carried out an experiment in a fume cupboard to find out how electricity affected different substances. Some of the substances were in aqueous solution, others were in the molten state. Carbon electrodes were used in each experiment and she wrote down her results in a table with these headings.

Substance	What was formed at the cathode (−)	What was formed at the anode (+)

   Make a table like the one shown and fill it in with what you think happened for each of the substances below.
   a Molten lead iodide.
   b Sugar solution.
   c Silver nitrate solution.
   d Copper(II) sulfate solution.
   e Molten sodium bromide.
   f Ethanol solution.

8 Sodium hydroxide is made by the electrolysis of brine.
   a Draw and label a simplified diagram of the cell used in this process. Make certain that you have labelled on the diagram:
   (i) the electrolyte
   (ii) the material of the electrodes
   (iii) the material of the membrane.
   b Write equations for the reactions which take place at the cathode and anode. State clearly whether a reaction is oxidation or reduction.
   c Give two large-scale uses of the products of this electrolytic process.
   d Comment on the following statement: 'This electrolytic process is a very expensive one'.
   e Both the membrane cell and the older mercury cell make sodium hydroxide of high purity. Explain why the membrane cell is now the preferred way of making sodium hydroxide.

7 Acids, bases and salts

Acids and alkalis
Theories of acids and bases
The relative strengths of acids and bases
Neutralising an acid
Formation of salts
Methods of preparing soluble salts
Methods of preparing insoluble salts
More about salts
Testing for different salts

Crystal hydrates
Calculation of water of crystallisation
Solubility of salts in water
Calculating solubility
Solubility curves
Titration
Checklist
Additional questions

Acids and alkalis

All the substances shown in Figure 7.1 contain an **acid** of one sort or another. Acids are certainly all around us. What properties do these substances have which make you think that they are acids or contain acids?

The word acid means 'sour' and all acids possess this property. They are also:

• soluble in water
• corrosive.

Alkalis are very different from acids. They are the chemical 'opposite' of acids.

• They will remove the sharp taste from an acid.
• They have a soapy feel.

Some common alkaline substances are shown in Figure 7.2.

It would be too dangerous to taste a liquid to find out if it was acidic. Chemists use substances called **indicators** which change colour when they are added to acids or alkalis. Many indicators are dyes which have been extracted from natural sources. For example, litmus is a purple dye which has been extracted from lichens. Litmus turns red when it is added to an acid and turns blue when added to an alkali. Some other indicators are shown in Table 7.1, along with the colours they turn in acids and alkalis.

Figure 7.1 What do all these foods have in common?

Table 7.1 Indicators and their colours in acid and alkaline solution.

Indicator	Colour in acid solution	Colour in alkaline solution
Blue litmus	Red	Blue
Methyl orange	Pink	Yellow
Methyl red	Red	Yellow
Phenolphthalein	Colourless	Pink
Red litmus	Red	Blue

Figure 7.2 Some common alkaline substances.

These indicators tell chemists whether a substance is acid or alkaline (Figure 7.3). To obtain an idea of how acidic or alkaline a substance is, we use another indicator known as a **universal indicator**. This indicator is a mixture of many other indicators. The colour shown by this indicator can be matched against a **pH scale**. The pH scale was developed by a Scandinavian chemist called Søren Sørenson. The pH scale runs from below 0 to 14. A substance with a pH of less than 7 is an acid. One with a pH of greater than 7 is alkaline. One with a pH of 7 is said to be neither acid nor alkaline, that is neutral. Water is the most common example of a neutral substance. Figure 7.4 shows the universal indicator colour range along with everyday substances with their particular pH values.

Another way in which the pH of a substance can be measured is by using a pH meter (Figure 7.5). The pH electrode is placed into the solution and a pH reading is given on the digital display.

Figure 7.5 A digital pH meter.

Figure 7.3 Indicators tell you if a substance is acid or alkaline.

a The pH scale.

b Universal indicator in solution, showing the colour range.

Figure 7.4

Theories of acids and bases

There have been many attempts to define the difference between acids and bases. The first real attempt took place in 1777, when the Frenchman Antoine Lavoisier (Figure 7.6) suggested that acids were substances that contained oxygen.

Figure 7.6 Antoine Lavoisier (1714–1794) at work in his laboratory.

It was not long after this that the 'hydro-halic' acids (HCl, HBr, and so on) were discovered and they had no oxygen present in them. This produced a modified theory in 1810 when the English chemist Sir Humphry Davy (1778–1829) suggested that all acids contain hydrogen as the important element; however, it was pointed out that there were many hydrogen-containing substances that were not acids. The German chemist Justus von Liebig (Figure 7.7) then made the next useful proposal about acids (1838) when he suggested that acids were substances that can react with metals to produce hydrogen gas.

Figure 7.7 Justus von Liebig (1803–1873), who defined acids more closely.

This theory was followed in 1884 by the first really comprehensive theory of acids and bases, produced by the Swedish chemist Svante Arrhenius (1859–1927). He suggested that since these acid solutions were electrolytes (see Chapter 6) their solutions contained many ions. According to Arrhenius' theory, acids produce hydrogen ions (H^+) when they dissolve in water, whereas bases produce hydroxide ions (OH^-).

It was thus recognised that water plays an important part in the acidity of a substance. This led to the suggestion that the hydrogen ion cannot exist alone in aqueous solution. This was prompted by the fact that gaseous hydrogen chloride, $HCl(g)$, is not acidic but when it dissolves in water an acidic solution is produced.

These ideas were rather limiting since they only applied to aqueous solutions. There were situations where acid–base reactions were taking place in solvents other than water, or even in no solvent at all. This problem was addressed in 1923 by the Danish chemist Johannes Brønsted (1879–1947) and the English chemist Thomas Lowry (1874–1936) when they independently proposed a more general definition of acids and bases, and the study of acids and bases took a great step forward. This theory became known as the **Brønsted–Lowry** theory of acids and bases.

The Brønsted–Lowry theory

This theory defined:

- an acid as an H^+ ion (or proton) donor
- a base as an H^+ ion (or proton) acceptor.

The theory explains why a pure acid behaves differently from its aqueous solution, since for an acid to behave as an H^+ ion donor it must have another substance present to accept the H^+ ion. So the water, in the aqueous acid solution, is behaving as a Brønsted–Lowry base and accepting an H^+ ion. Generally:

$$HA(aq) + H_2O(l) \rightarrow H_3O^+(aq) + A^-(aq)$$
$$\text{acid} \qquad \text{base}$$

If a substance can behave both as a Brønsted–Lowry acid and as a Brønsted–Lowry base then it is called **amphoteric**. Water has this ability. As well as reacting with acids (above) it can also react with Brønsted–Lowry bases such as ammonia in the following way to form the base OH^-:

$$NH_3(aq) + H_2O(l) \rightarrow NH_4^+(aq) + OH^-(aq)$$
$$\text{base} \qquad \text{acid}$$

The reaction between hydrogen chloride gas and ammonia can be described as an acid–base reaction under this theory. The hydrogen chloride molecule acts as a proton donor and the ammonia molecule acts as the proton acceptor (Figure 7.8).

$$HCl(g) + NH_3(g) \rightarrow NH_4^+Cl^-(s)$$
$$\text{acid} \qquad \text{base}$$

Figure 7.8 The hydrogen chloride molecule (from concentrated hydrochloric acid) acts as a hydrogen ion donor. The ammonia molecule (from concentrated ammonia) acts as a hydrogen ion acceptor.

The relative strengths of acids and bases

The relative strength of an acid is found by comparing one acid with another. The strength of any acid depends upon how many molecules dissociate (or ionise) when the acid is dissolved in water. The relative strength of a base is found by comparing one base with another and is again dependent upon the dissociation of the base in aqueous solution.

Strong and weak acids

A typical strong acid is hydrochloric acid. It is formed by dissolving hydrogen chloride gas in water. In hydrochloric acid the ions formed separate completely.

$$\text{hydrogen chloride} \xrightarrow{\text{water}} \text{hydrogen ions} + \text{chloride ions}$$

$$HCl(g) \xrightarrow{\text{water}} H^+(aq) + Cl^-(aq)$$

For hydrochloric acid *all* the hydrogen chloride molecules break up to form H^+ ions and Cl^- ions. Any acid that behaves in this way is termed a **strong acid**. Both sulfuric acid and nitric acid also behave in this way and are therefore also termed strong acids. All these acids have a high concentration of hydrogen ions in solution ($H^+(aq)$) and have a low pH. Their solutions are good conductors of electricity and they react quickly with metals, bases and metal carbonates.

When strong acids are neutralised by strong alkalis the following reaction takes place between hydrogen ions and hydroxide ions.

$$H^+(aq) + OH^-(aq) \rightarrow H_2O(l)$$

A **weak acid** such as ethanoic acid, which is found in vinegar, produces few hydrogen ions when it dissolves in water compared with a strong acid of the same concentration. It is only partially ionised. Its solution has a higher pH than a strong acid, but still less than 7.

$$\text{ethanoic acid} \underset{\text{water}}{\rightleftharpoons} \text{hydrogen ions} + \text{ethanoate ions}$$

$$CH_3COOH(l) \underset{\text{water}}{\rightleftharpoons} H^+(aq) + CH_3COO^-(aq)$$

The $\rightleftharpoons$ sign means that the reaction is **reversible**. This means that if the ethanoic acid molecule breaks down to give hydrogen ions and ethanoate ions then they will react together to re-form the ethanoic acid molecule. The fact that fewer ethanoic acid molecules dissociate compared with a strong acid, and that the reaction is reversible, means that few oxonium ions are present in the solution. Other examples of weak acids are citric acid, found in oranges and lemons, carbonic acid, found in soft drinks, sulfurous acid (acid rain) (Figure 7.9) and ascorbic acid (vitamin C).

Solutions of weak acids are poorer conductors of electricity and have slower reactions with metals, bases and metal carbonates.

Figure 7.9 Sulfurous acid is found in acid rain. It is a weak acid and is oxidised to sulfuric acid (a strong acid). Acid rain damages the environment quite badly.

All acids when in aqueous solution produce hydrogen ions, $H^+(aq)$. To say an acid is a strong acid does not mean it is concentrated. The *strength* of an acid tells you how easily it dissociates (ionises) to produce hydrogen ions. The *concentration* of an acid indicates the proportions of water and acid present in aqueous solution. It is important to emphasise that a strong acid is still a strong acid even when it is in dilute solution and a weak acid is still a weak acid even if it is concentrated.

Strong and weak bases

An alkali is a base which produces hydroxide ions, $OH^-(aq)$, when dissolved in water. Sodium hydroxide is a **strong alkali** because when it dissolves in water its lattice breaks up completely to produce ions.

$$\text{sodium hydroxide} \xrightarrow{\text{water}} \text{sodium ions} + \text{hydroxide ions}$$

$$NaOH(s) \xrightarrow{} Na^+(aq) + OH^-(aq)$$

These substances which are strong alkalis produce large quantities of hydroxide ions and have a high pH. Other common, strong soluble bases include potassium hydroxide.

A **weak alkali**, such as ammonia, produces fewer hydroxide ions when it dissolves in water than a strong soluble base of the same concentration. It is only partially ionised. It has a lower pH than a strong base, but still above 7.

$$\text{ammonia} + \text{water} \rightleftharpoons \text{ammonium ions} + \text{hydroxide ions}$$

$$NH_3(g) + H_2O(l) \rightleftharpoons NH_4^+(aq) + OH^-(aq)$$

The ammonia molecules react with the water molecules to form ammonium ions and hydroxide ions. However, fewer ammonia molecules do this so only a low concentration of hydroxide ions is produced.

Neutralising an acid

A common situation involving neutralisation of an acid is when you suffer from indigestion. This is caused by a build-up of acid in your stomach. Normally you treat it by taking an indigestion remedy containing a substance which will react with and neutralise the acid.

In the laboratory, if you wish to neutralise a common acid such as hydrochloric acid you can use an alkali such as sodium hydroxide. If the pH of the acid is measured as some sodium hydroxide solution is added to it, the pH increases. If equal volumes of the same concentration of hydrochloric acid and sodium hydroxide are added to one another, the resulting solution is found to have a pH of 7. The acid has been neutralised and a neutral solution has been formed.

$$\text{hydrochloric} + \text{sodium} \rightarrow \text{sodium} + \text{water}$$
$$\text{acid} \qquad \text{hydroxide} \quad \text{chloride}$$
$$HCl_{(aq)} + NaOH_{(aq)} \rightarrow NaCl_{(aq)} + H_2O_{(l)}$$

As we have shown, when both hydrochloric acid and sodium hydroxide dissolve in water the ions separate completely. We may therefore write:

$$H^+_{(aq)}Cl^-_{(aq)} + Na^+_{(aq)}OH^-_{(aq)} \rightarrow Na^+_{(aq)} + Cl^-_{(aq)} + H_2O_{(l)}$$

You will notice that certain ions are unchanged on either side of the equation. They are called **spectator ions** and are usually taken out of the equation. The equation now becomes:

$$H^+_{(aq)} + OH^-_{(aq)} \rightarrow H_2O_{(l)}$$

This type of equation is known as an **ionic equation**. The reaction between any acid and alkali in aqueous solution can be summarised by this ionic equation. It shows the ion which causes acidity ($H^+_{(aq)}$) reacting with the ion which causes alkalinity ($OH^-_{(aq)}$) to produce neutral water ($H_2O_{(l)}$).

Questions

1 Complete the following equations:
 a $CH_3COOH + NaOH$
 b $H_2SO_4 + KOH$
 c $NH_3 + HBr$.
 In each case name the acid and the base. Also in parts **a** and **b** write the ionic equation for the reactions.

2 Explain the terms 'concentration' and 'strength' as applied to acids.

3 Explain what part water plays in the acidity of a solution.

4 Alongside the names of various chemicals below are shown their respective pH values in aqueous solution.

potassium hydroxide	pH 13
hydrogen bromide	pH 2
calcium hydroxide	pH 11
sodium chloride	pH 7
hydrogen chloride	pH 2
magnesium hydroxide	pH 10
citric acid	pH 4

Which of the substances is/are:
 a a strong acid?
 b a weak acid?
 c a strong alkali?
 d a weak alkali?
 e a neutral substance?
 In each case write a chemical equation to show the molecules/ions present in solution.

5 a Write a chemical equation to represent the neutralisation of sulfuric acid by sodium hydroxide.
 b Reduce this to an ionic equation.
 c Account for any difference you see between the ionic equation you have written and the one shown above for the reaction of hydrochloric acid and sodium hydroxide.

Formation of salts

In the example above, sodium chloride was produced as part of the neutralisation reaction. Compounds formed in this way are known as **normal salts**. A normal salt is a compound that has been formed when all the hydrogen ions of an acid have been replaced by metal ions or by the ammonium ion (NH_4^+).

Normal salts can be classified as those which are soluble in water or those which are insoluble in water. The following salts are soluble in cold water:

- all nitrates
- all common sodium, potassium and ammonium salts
- all chlorides except lead, silver and mercury
- all sulfates except lead, barium and calcium.

Salts are very useful substances, as you can see from Table 7.2 and Figure 7.10.

Table 7.2 Useful salts.

Salt	Use
Ammonium chloride	In torch batteries
Ammonium nitrate	In fertilisers
Calcium carbonate	Extraction of iron, making cement, glass making
Calcium chloride	In the extraction of sodium, drying agent (anhydrous)
Calcium sulfate	For making plaster boards, plaster casts for injured limbs
Iron(II) sulfate	In 'iron' tablets
Magnesium sulfate	In medicines
Potassium nitrate	In fertiliser and gunpowder manufacture
Silver bromide	In photography
Sodium carbonate	Glass making, softening water, making modern washing powders
Sodium chloride	Making hydrochloric acid, for food flavouring, hospital saline, in the Solvay process for the manufacture of sodium carbonate
Sodium stearate	In some soaps
Tin(II) fluoride	Additive to toothpaste

a Silver bromide is used in photography.

b Iron(II) sulfate is the iron containing substance in these iron and multivitamin tablets.

Figure 7.10 Some uses of salts.

If the acid being neutralised is hydrochloric acid, salts called **chlorides** are formed. Other types of salts can be formed with other acids. A summary of the different types of salt along with the acid they have been formed from is shown in Table 7.3.

Table 7.3 Types of salt and the acids they are formed from.

Acid	Type of salt	Example
Carbonic acid	Carbonates	Sodium carbonate (Na_2CO_3)
Ethanoic acid	Ethanoates	Sodium ethanoate (CH_3COONa)
Hydrochloric acid	Chlorides	Potassium chloride (KCl)
Nitric acid	Nitrates	Potassium nitrate (KNO_3)
Sulfuric acid	Sulphates	Sodium sulfate (Na_2SO_4)

Methods of preparing soluble salts

There are four general methods of preparing soluble salts:

Acid + metal

This method can only be used with the less reactive metals. It would be very dangerous to use a reactive metal such as sodium in this type of reaction. The metals usually used in this method of salt preparation are the **MAZIT** metals, that is, **m**agnesium, **a**luminium, **z**inc, **i**ron and **t**in. A typical experimental method is given below.

Excess magnesium ribbon is added to dilute nitric acid. During this addition an effervescence is observed due to the production of hydrogen gas. In this reaction the hydrogen ions from the nitric acid gain electrons from the metal atoms as the reaction proceeds.

$$\text{hydrogen ions} + \text{electrons} \rightarrow \text{hydrogen gas}$$
$$\text{(from metal)}$$
$$2H^+ + 2e^- \rightarrow H_2(g)$$

How would you test the gas to show that it was hydrogen? What would be the name and formula of the compound produced during the test you suggested?

$$\text{magnesium} + \text{nitric acid} \rightarrow \text{magnesium nitrate} + \text{hydrogen}$$
$$Mg(s) + 2HNO_3(aq) \rightarrow Mg(NO_3)_2(aq) + H_2(g)$$

The excess magnesium is removed by filtration (Figure 7.11).

Figure 7.11 The excess magnesium is filtered in this way.

The magnesium nitrate solution is evaporated slowly to form a saturated solution of the salt (Figure 7.12).

The hot concentrated magnesium nitrate solution produced is tested by dipping a cold glass rod into it. If salt crystals form at the end of the rod the solution is ready to crystallise and is left to cool. Any crystals produced on cooling are filtered and dried between clean tissues.

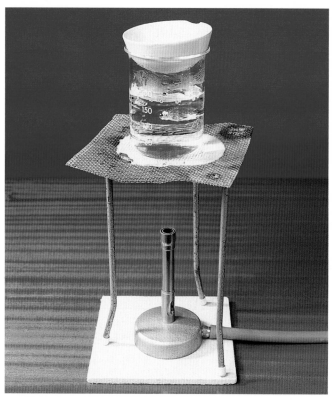

Figure 7.12 The solution of magnesium nitrate is concentrated by slow evaporation.

Acid + carbonate

This method can be used with any metal carbonate and any acid, providing the salt produced is soluble. The typical experimental procedure is similar to that carried out for an acid and a metal. For example, copper(II) carbonate would be added in excess to dilute nitric acid. Effervescence would be observed due to the production of carbon dioxide.

How would you test the gas to show it was carbon dioxide? Write an equation to help you explain what is happening during the test you have chosen.

copper(II) + nitric → copper(II) + carbon + water
carbonate   acid   nitrate   dioxide

$$CuCO_{3(s)} + 2HNO_{3(aq)} \rightarrow Cu(NO_3)_{2(aq)} + CO_{2(g)} + H_2O_{(l)}$$

Metal carbonates contain carbonate ions, CO_3^{2-}. In this reaction the carbonate ions react with the hydrogen ions in the acid.

carbonate + hydrogen → carbon + water
ions   ions   dioxide

$$CO_3^{2-}{}_{(aq)} + 2H^+{}_{(aq)} \rightarrow CO_{2(g)} + H_2O_{(l)}$$

Acid + alkali (soluble base)

This method is generally used for preparing the salts of very reactive metals, such as potassium or sodium. It would certainly be too dangerous to add the metal directly to the acid. In this case, we solve the problem indirectly and use an alkali which contains the particular reactive metal whose salt we wish to prepare.

A **base** is a substance which neutralises an acid, producing a salt and water as the only products. If the base is soluble the term alkali can be used, but there are several bases which are insoluble. It is also a substance which accepts a hydrogen ion (see p. 98). In general, most metal oxides and hydroxides (as well as ammonia solution) are bases. Some examples of soluble and insoluble bases are shown in Table 7.4. Salts can be formed by this method only if the base is soluble.

Table 7.4 Examples of soluble and insoluble bases.

Soluble bases (alkalis)	Insoluble bases
Sodium hydroxide (NaOH)	Iron(III) oxide (Fe_2O_3)
Potassium hydroxide (KOH)	Copper(II) oxide (CuO)
Calcium hydroxide ($Ca(OH)_2$)	Lead(II) oxide (PbO)
Ammonia solution ($NH_{3(aq)}$)	Magnesium oxide (MgO)

Because in this neutralisation reaction both reactants are in solution, a special technique called **titration** is required. Acid is slowly and carefully added to a measured volume of alkali using a burette (Figure 7.13) until the indicator, usually phenolphthalein, changes colour.

Figure 7.13 The acid is added to the alkali until the indicator just changes colour.

An indicator is used to show when the alkali has been neutralised completely by the acid. This is called the **end-point**. Once you know where the end-point is, you can add the same volume of acid to the measured volume of alkali but this time without the indicator.

The solution which is produced can then be evaporated slowly to obtain the salt. For example,

hydrochloric + sodium → sodium + water
acid   hydroxide   chloride

$$HCl(aq) + NaOH(aq) \rightarrow NaCl(aq) + H_2O(l)$$

As previously discussed on p. 98, this reaction can best be described by the ionic equation:

$$H^+(aq) + OH^-(aq) \rightarrow H_2O(l)$$

Acid + insoluble base

This method can be used to prepare a salt of an unreactive metal, such as lead or copper. In these cases it is not possible to use a direct reaction of the metal with an acid so the acid is neutralised using the particular metal oxide (Figure 7.14).

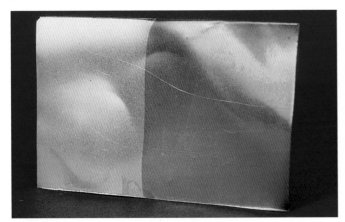

Figure 7.14 The citric acid in the lemon juice reacts with oxide on the surface of the copper and the tarnish dissolves away.

The method is generally the same as that for a metal carbonate and an acid, though some warming of the reactants may be necessary. An example of such a reaction is the neutralisation of sulfuric acid by copper(II) oxide to produce copper(II) sulfate (Figure 7.15).

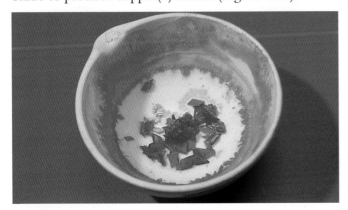

Figure 7.15 After slow evaporation to concentrate the solution, the solution is left to crystallise. Crystals of copper(II) sulfate are produced.

sulfuric   + copper(II) → copper(II) + water
acid   oxide   sulfate

$$H_2SO_4(aq) + CuO(s) \rightarrow CuSO_4(aq) + H_2O(l)$$

Metal oxides contain the oxide ion, O^{2-}. The ionic equation for this reaction is therefore:

$$2H^+(aq) + O^{2-}(s) \rightarrow H_2O(l)$$

or

$$CuO(s) + 2H^+(aq) \rightarrow Cu^{2+}(aq) + H_2O(l)$$

Methods of preparing insoluble salts

An insoluble salt, such as barium sulfate, can be made by precipitation. In this case, solutions of the two chosen soluble salts are mixed (Figure 7.16). To produce barium sulfate, barium chloride and sodium sulfate can be used. The barium sulfate precipitate can be filtered off, washed with distilled water and dried. The reaction that has occurred is:

barium   + sodium → barium + sodium
chloride   sulfate   sulfate   chloride

$$BaCl_2(aq) + Na_2SO_4(aq) \rightarrow BaSO_4(s) + 2NaCl(aq)$$

Figure 7.16 When barium chloride solution is added to sodium sulfate a white precipitate of barium sulfate forms.

The ionic equation for this reaction is:

$$Ba^{2+}(aq) + SO_4^{2-}(aq) \rightarrow BaSO_4(s)$$

This method is sometimes known as **double decomposition** and may be summarised as follows

soluble salt + soluble salt → insoluble salt + soluble salt
(AX)   (BY)   (BX)   (AY)

It should be noted that even salts like barium sulfate dissolve to a very small extent. For example, 1 litre of water will dissolve 2.2×10^{-3} g of barium sulfate at 25 °C. This substance and substances like it are said to be **sparingly soluble**.

More about salts

You have already seen on p. 100 in Table 7.2 that salts are useful substances. Some of the salts shown in that table occur naturally and are mined, for example calcium sulfate (gypsum) and calcium carbonate (limestone). Many of the others must be made by the chemical industry, for example ammonium nitrate, iron(II) sulfate and silver bromide.

With acids such as sulfuric acid, which has two replaceable hydrogen ions per molecule, it is possible to replace only *one* of these with a metal ion. The salt produced is called an **acid salt**. An acid salt is one in which not all of the replaceable hydrogen ions of the acid have been replaced by metal ions or the ammonium ion. Some examples of acid salts are shown in Table 7.5.

Table 7.5 Examples of acid salts.

Acid	Type of acid salt	Example
Carbonic acid (H_2CO_3)	Hydrogencarbonate	Sodium hydrogencarbonate ($NaHCO_3$)
Sulfuric acid (H_2SO_4)	Hydrogensulfate	Potassium hydrogensulfate ($KHSO_4$)

Sodium hydrogencarbonate is the acid salt used as the raising agent in the baking of cakes and some bread, and is often called baking soda (Figure 7.17).

Figure 7.17 Sodium hydrogencarbonate is used as a raising agent in bread and muffins.

Testing for different salts

Sometimes we want to analyse a salt and find out what is in it. There are simple chemical tests which allow us to identify the anion part of the salt. These are often called **spot tests**.

Testing for a sulfate (SO_4^{2-})

You have seen that barium sulfate is an insoluble salt (p. 103). Therefore, if you take a solution of a suspected sulfate and add it to a solution of a soluble barium salt (such as barium chloride) then a white precipitate of barium sulfate will be produced.

$$\text{barium ion} + \text{sulfate ion} \rightarrow \text{barium sulfate}$$
$$Ba^{2+}(aq) + SO_4^{2-}(aq) \rightarrow BaSO_4(s)$$

A few drops of dilute hydrochloric acid are also added to this mixture. If the precipitate does not dissolve, then it is barium sulfate and the unknown salt was in fact a sulfate. If the precipitate does dissolve, then the unknown salt may have been a sulfite (containing the SO_3^{2-} ion).

Testing for a chloride (Cl^-), a bromide (Br^-) or an iodide (I^-)

Earlier in this chapter you saw that silver chloride is an insoluble salt (p. 100). Therefore, if you take a solution of a suspected chloride and add to it a small volume of dilute nitric acid followed by a small amount of a solution of a soluble silver salt (such as silver nitrate) a white precipitate of silver chloride will be produced.

$$\text{chloride ion} + \text{silver ion} \rightarrow \text{silver chloride}$$
$$Cl^-(aq) + Ag^+(aq) \rightarrow AgCl(s)$$

If left to stand, the precipitate goes grey (Figure 7.18).

Figure 7.18 If left to stand the white precipitate of silver chloride goes grey. This photochemical change plays an essential part in black and white photography.

In a similar way, a bromide and an iodide will react to produce either a cream precipitate of silver bromide (AgBr) or a yellow precipitate of silver iodide (AgI) (Figure 7.19).

Figure 7.19 AgCl, a white precipitate, AgBr, a cream precipitate, and AgI, a yellow precipitate.

An alternative test for iodide ions is the addition of lead nitrate solution to the iodide which results in a bright yellow precipitate of lead iodide, PbI_2.

Testing for a carbonate

If a small amount of an acid is added to some of the suspected carbonate (either solid or in solution) then effervescence occurs. If it is a carbonate then carbon dioxide gas is produced, which will turn limewater 'milky' (a cloudy white precipitate of calcium carbonate forms).

$$\begin{array}{c} \text{carbonate} + \text{hydrogen} \rightarrow \text{carbon} + \text{water} \\ \text{ions} \qquad \text{ions} \qquad \text{dioxide} \end{array}$$

$$CO_3^{2-}{}_{(aq)} + 2H^+{}_{(aq)} \rightarrow CO_2{}_{(g)} + H_2O{}_{(l)}$$

Testing for a nitrate

By using Devarda's alloy (45% Al, 5% Zn, 50% Cu) in alkaline solution, nitrates are reduced to ammonia. The ammonia can be identified using damp indicator paper, which turns blue.

Questions

1 Complete the word equations and write balanced chemical equations for the following soluble salt preparations:
   magnesium + sulfuric acid →
   calcium carbonate + hydrochloric acid →
   zinc oxide + hydrochloric acid →
   potassium hydroxide + nitric acid →
   Also write ionic equations for each of the reactions.

2 Lead carbonate and lead iodide are insoluble. Which two soluble salts could you use in the preparation of each substance? Write
   a a word equation
   b a symbol equation
   c an ionic equation
   to represent the reactions taking place.

3 An analytical chemist working for an environmental health organisation has been given a sample of water which is thought to have been contaminated by a sulfate, a carbonate and a chloride.
   a Describe how she could confirm the presence of these three types of salt by simple chemical tests.
   b Write ionic equations to help you explain what is happening during the testing process.

Crystal hydrates

Some salts, such as sodium chloride, copper carbonate and sodium nitrate, crystallise in their anhydrous forms (without water). However, many salts produce **hydrates** when they crystallise from solution. A hydrate is a salt which incorporates water into its crystal structure. This water is referred to as **water of crystallisation**. The shape of the crystal hydrate is very much dependent on the presence of water of crystallisation. Some examples of crystal hydrates are given in Table 7.6 and shown in Figure 7.20.

Table 7.6 Examples of crystal hydrates.

Salt hydrate	Formula
Cobalt(II) chloride hexahydrate	$CoCl_2.6H_2O$
Copper(II) sulfate pentahydrate	$CuSO_4.5H_2O$
Iron(II) sulfate heptahydrate	$FeSO_4.7H_2O$
Magnesium sulfate heptahydrate	$MgSO_4.7H_2O$
Sodium carbonate decahydrate	$Na_2CO_3.10H_2O$
Sodium hydrogensulfate monohydrate	$NaHSO_4.H_2O$
Sodium sulfate decahydrate	$Na_2SO_4.10H_2O$

Figure 7.20 Hydrate crystals (left to right): cobalt nitrate, calcium nitrate and nickel sulfate (top) and manganese sulfate, copper sulfate and chromium potassium sulfate (bottom).

When many hydrates are heated the water of crystallisation is driven away. For example, if crystals of copper(II) sulfate hydrate (blue) are heated strongly, they lose their water of crystallisation. Anhydrous copper(II) sulfate remains as a white powder:

$$\begin{array}{c} \text{copper(II) sulfate} \rightarrow \text{anhydrous copper(II)} + \text{water} \\ \text{pentahydrate} \qquad \qquad \text{sulfate} \end{array}$$

$$CuSO_4.5H_2O{}_{(s)} \rightarrow CuSO_4{}_{(s)} + 5H_2O{}_{(g)}$$

When water is added to anhydrous copper(II) sulfate the reverse process occurs. It turns blue and the pentahydrate is produced (Figure 7.21). This is an extremely exothermic process.

$$CuSO_4(s) + 5H_2O(l) \rightarrow CuSO_4.5H_2O(s)$$

Because the colour change only takes place in the presence of water, the reaction is used to test for the presence of water.

These processes give a simple example of a reversible reaction

$$CuSO_4(s) + 5H_2O(l) \rightleftharpoons CuSO_4.5H_2O(s)$$

Figure 7.21 Anhydrous copper(II) sulfate is a white powder which turns blue when water is added to it.

Some crystal hydrates **effloresce**, that is they lose some or all of their water of crystallisation to the atmosphere. For example, when colourless sodium carbonate decahydrate crystals are left out in the air they become coated with a white powder, which is the monohydrate (Figure 7.22). The process is called **efflorescence**.

$$Na_2CO_3.10H_2O(s) \rightarrow Na_2CO_3.H_2O(s) + 9H_2O(g)$$

Figure 7.22 A white powder forms on the surface of sodium carbonate decahydrate when it is left in the air.

With some substances, not necessarily salt hydrates, the reverse of efflorescence occurs. For example, if anhydrous calcium chloride is left in the air, it absorbs water vapour and eventually forms a very concentrated solution. This process is called **deliquescence**, and substances which behave like this are said to be **deliquescent**. Solid sodium hydroxide will deliquesce.

There are some substances which, if left out in the atmosphere, absorb moisture but do not change their state. For example, concentrated sulfuric acid, a colourless, viscous liquid, absorbs water vapour from the air and becomes a solution. Substances which do this are said to be **hygroscopic**.

Calculation of water of crystallisation

Sometimes it is necessary to work out the percentage, by mass, of water of crystallisation in a hydrated salt. The method is the same as that used in Chapter 5, p. 73, but this time the 'H₂O' is treated as an element in the calculation.

Example

Calculate the percentage by mass of water in the salt hydrate $MgSO_4.7H_2O$. (A_r: H = 1, O = 16, Mg = 24, S = 32)

$$M_r \text{ for } MgSO_4.7H_2O = 24 + 32 + (4 \times 16) + (7 \times 18)$$
$$= 246$$

The mass of water as a fraction of the total mass of hydrate

$$= \frac{126}{246}$$

The percentage of water present

$$= \frac{126}{246} \times 100 = 51.2\%$$

Questions

1 Calculate the percentage by mass of water in the following salt hydrates:
   a $CuSO_4.5H_2O$
   b $Na_2CO_3.10H_2O$
   c $Na_2S_2O_3.5H_2O$.
   (A_r: H = 1, O = 16, Na = 23, S = 32, Cu = 63.5)

2 Devise an experiment to determine the percentage of water of crystallisation present in a salt hydrate of your choice.

Solubility of salts in water

Water is a very good solvent and will dissolve a whole range of solutes, including sodium chloride and copper(II) sulfate, as well as other substances such as sugar. You can dissolve more sugar than sodium chloride in 100 cm³ of water at the same temperature. The sugar is said to be more **soluble** than the sodium chloride at the same temperature. We say that the sugar has a greater **solubility** than the sodium chloride. The solubility of a solute in water at a given temperature is the number of grams of that solute which can be dissolved in 100 g of water to produce a saturated solution at that temperature.

Calculating solubility

Example

21.5 g of sodium chloride dissolve in 60 g of water at 25 °C. Calculate the solubility of sodium chloride in water at that temperature.

If 60 g of water dissolves 21.5 g of sodium chloride, then 1 g of water will dissolve:

$$\frac{21.5}{60} \text{ g of sodium chloride}$$

Therefore, 100 g of water will dissolve:

$$\frac{21.5}{60} \times 100 = 35.8 \text{ g}$$

The solubility of sodium chloride at 25 °C = 35.8 g per 100 g of water.

Solubility curves

It is a well-known fact that usually the amount of solute that a solvent will dissolve increases with temperature. We say that the solubility increases with increasing temperature. Figure 7.23 shows how the solubilities of copper(II) sulfate and potassium nitrate vary with temperature. These graphs of solubility against temperature are known as **solubility curves**. By using curves of this type you can find the solubility of the solute at any temperature.

From a solubility curve such as Figure 7.23, it can be seen that the solubility of potassium nitrate at 50 °C is 80 g per 100 g of water, whereas that of copper(II) sulfate is 35 g per 100 g of water.

When a saturated solution is cooled, some of the solute crystallises out of solution. Thus, by using solubility curves it is possible to determine the amount of solute that will crystallise out of solution at different temperatures. For example, if potassium nitrate is cooled from 70 °C to 40 °C then the difference between the two solubilities at these temperatures tells you the amount that crystallises out.

Solubility at 70 °C = 135 g
Solubility at 40 °C = 62 g
Difference = 73 g

Therefore the amount of potassium nitrate which crystallises out during cooling from 70 °C to 40 °C is 73 g.

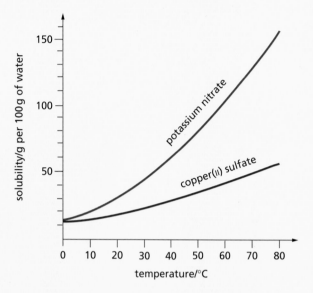

Figure 7.23 Solubility curves for copper(II) sulfate and potassium nitrate.

Questions

1 Calculate the solubility, at the temperature given, of the following salts in water:
   a 12.1 g of copper(II) sulfate dissolved in 50 g of water at 30 °C
   b 11.1 g of potassium chlorate(V) dissolved in 60 g of water at 50 °C
   c 9.72 g of potassium chloride dissolved in 30 g of water at 70 °C.

2 Use the data given below to plot a solubility curve for lead nitrate.

Temperature/°C	10	30	50	70	90
Solubility/g per 100 g of water	44.0	60.7	78.6	97.7	117.4

3 Using your solubility curve from question **2**, find the solubility of lead nitrate at the following temperatures:
   a 25 °C
   b 45 °C
   c 75 °C.

4 Using your answers to question **3**, calculate the amount of lead nitrate that will crystallise out of solution when cooled from:
   a 45 °C to 25 °C
   b 75 °C to 25 °C
   c 75 °C to 45 °C.

5 25 g of a saturated solution containing lead nitrate is cooled from 54 °C to 27 °C. Calculate the amount of lead nitrate that would crystallise out of the solution.

Titration

On p. 102 you saw that it was possible to prepare a soluble salt by reacting an acid with a soluble base (alkali). The method used was that of **titration**. Titration can also be used to find the concentration of the alkali used. In the laboratory, the titration of hydrochloric acid with sodium hydroxide is carried out in the following way.

1 25 cm^3 of sodium hydroxide solution is pipetted into a conical flask to which a few drops of phenolphthalein indicator have been added (Figure 7.24). Phenolphthalein is pink in alkaline conditions but colourless in acid.

dilute hydrochloric acid

burette

Note: remove filter funnel before titrating

Figure 7.25 The burette is filled up to the zero mark with a 0.10 mol dm^{-3} solution of hydrochloric acid.

4 The hydrochloric acid is added to the sodium hydroxide solution in small quantities – usually no more than 0.5 cm^3 at a time (Figure 7.26). The contents of the flask must be swirled after each addition of acid for thorough mixing.

Figure 7.24 Exactly 25.0 cm^3 of sodium hydroxide solution is pipetted into a conical flask.

2 A 0.10 mol dm^{-3} solution of hydrochloric acid is placed in the burette using a filter funnel until it is filled up exactly to the zero mark (Figure 7.25).

3 The filter funnel is now removed.

Figure 7.26 The titration is carried out accurately.

5 The acid is added until the alkali has been neutralised completely. This is shown by the pink colour of the indicator just disappearing.

6 The final reading on the burette at the end-point is recorded and further titrations carried out until consistent results are obtained (within 0.1 cm³ of each other). Some sample data are shown below.

Volume of sodium hydroxide solution
= 25.0 cm³
Average volume of 0.10 mol dm⁻³ solution of hydrochloric acid added
= 21.0 cm³

The neutralisation reaction which has taken place is:

$$\text{hydrochloric} + \text{sodium} \rightarrow \text{sodium} + \text{water}$$
$$\text{acid} \qquad \text{hydroxide} \quad \text{chloride}$$
$$HCl_{(aq)} \quad + NaOH_{(aq)} \rightarrow NaCl_{(aq)} + H_2O_{(l)}$$

From this equation it can be seen that 1 mole of hydrochloric acid neutralises 1 mole of sodium hydroxide.

Now you can work out the number of moles of the acid using the formula given in Chapter 5, p. 71.

$$\text{moles} = \frac{\text{volume}}{1000} \times \text{concentration}$$

$$= \frac{21.0 \times 0.10}{1000}$$

$$= 2.1 \times 10^{-3}$$

$$\frac{\text{number of moles}}{\text{of hydrochloric acid}} = \frac{\text{number of moles}}{\text{of sodium hydroxide}}$$

Therefore, the number of moles of sodium hydroxide
= 2.1×10^{-3}
2.1×10^{-3} moles of sodium hydroxide is present in 25.0 cm³ of solution.
Therefore, in 1 cm³ of sodium hydroxide solution we have

$$\frac{2.1 \times 10^{-3}}{25.0} \text{ moles}$$

Therefore, in 1 litre of sodium hydroxide solution we have

$$\frac{2.1 \times 10^{-3}}{25.0} \times 1000 = 0.084 \text{ mole}$$

The concentration of sodium hydroxide solution is 0.084 mol dm⁻³.

You can simplify the calculation by substituting in the following mathematical equation:

$$\frac{M_1 V_1}{M_{acid}} = \frac{M_2 V_2}{M_{alkali}}$$

where:

M_1 = concentration of the acid used
V_1 = volume of acid used (cm³)
M_{acid} = number of moles of acid shown in the chemical equation
M_2 = concentration of the alkali used
V_2 = volume of the alkali used (cm³)
M_{alkali} = number of moles of alkali shown in the chemical equation

In the example:

$$M_1 = 0.10 \text{ mol dm}^{-3}$$
$$V_1 = 21.0 \text{ cm}^3$$
$$M_{acid} = 1 \text{ mole}$$
$$M_2 = \text{unknown}$$
$$V_2 = 25.0 \text{ cm}^3$$
$$M_{alkali} = 1 \text{ mole}$$

Substituting in the equation:

$$\frac{0.10 \times 21.0}{1} = \frac{M_2 \times 25.0}{1}$$

Rearranging:

$$M_2 = \frac{0.10 \times 21.0 \times 1}{1 \times 25.0}$$

$$M_2 = 0.084$$

The concentration of the sodium hydroxide solution is 0.084 mol dm⁻³.

Questions

1 24.2 cm³ of a solution containing 0.20 mol dm⁻³ of hydrochloric acid just neutralised 25.0 cm³ of a potassium hydroxide solution. What is the concentration of this potassium hydroxide solution?

2 22.4 cm³ of a solution containing 0.10 mol dm⁻³ of sulfuric acid just neutralised 25.0 cm³ of a sodium hydroxide solution. What is the concentration of this sodium hydroxide solution?

Checklist

After studying Chapter 7 you should know and understand the following terms.

- **Acid** A substance which dissolves in water, producing $H^+(aq)$ ions as the only positive ions. An acid is a hydrogen ion (H^+) donor.

- **Acid salt** A substance formed when only some of the replaceable hydrogen of an acid is replaced by metal ions or the ammonium ion (NH_4^+).

- **Alkali** A soluble base which produces $OH^-(aq)$ ions in water.

- **Base** A substance which neutralises an acid, producing a salt and water as the only products. A base is a hydrogen ion (H^+) acceptor.

- **Double decomposition** The process by which an insoluble salt is prepared from solutions of two suitable soluble salts.

- **Efflorescence** The process during which a substance loses water of crystallisation to the atmosphere.

- **Hygroscopic** The ability to absorb water vapour from the atmosphere without forming solutions or changing state, for example, concentrated sulfuric acid.

- **Indicator** A substance used to show whether a substance is acidic or alkaline (basic), for example phenolphthalein.

- **Ionic equation** The simplified equation of a reaction which we can write if the chemicals involved are ionic substances.

- **Neutralisation** The process in which the acidity or alkalinity of a substance is destroyed. Destroying acidity means removing $H^+(aq)$ by reaction with a base, carbonate or metal. Destroying alkalinity means removing the $OH^-(aq)$ by reaction with an acid.

$$H^+(aq) + OH^-(aq) \rightarrow H_2O(l)$$

- **Normal salt** A substance formed when all the replaceable hydrogen of an acid is completely replaced by metal ions or the ammonium ion (NH_4^+).

- **pH scale** A scale running from 0 to 14, used for expressing the acidity or alkalinity of a solution.

- **Salt hydrates** Salts containing water of crystallisation.

- **Solubility** The solubility of a solute in a solvent at a given temperature is the number of grams of that solute which can dissolve in 100 g of solvent to produce a saturated solution at that temperature.

- **Solubility curve** This is a graph of solubility against temperature.

- **Strong acid** One which produces a high concentration of $H^+(aq)$ ions in water solution, for example hydrochloric acid.

- **Strong alkali** One which produces a high concentration of $OH^-(aq)$ ions in water solution, for example sodium hydroxide.

- **Testing for a carbonate** If effervescence occurs when an acid is added to the suspected carbonate and the gas produced tests positively for carbon dioxide, the substance is a carbonate.

- **Testing for a chloride** If a white precipitate is produced when dilute nitric acid and silver nitrate solution are added to the suspected chloride, the solution contains a chloride.

- **Testing for a sulfate** If a white precipitate is produced when dilute hydrochloric acid and barium chloride solution are added to the suspected sulfate, the solution contains a sulfate.

- **Titration** A method of volumetric analysis in which a volume of one reagent (for example an acid) is added to a known volume of another reagent (for example, an alkali) slowly from a burette until an end-point is reached. If an acid and alkali are used, then an indicator is used to show that the end-point has been reached.

- **Water of crystallisation** Water incorporated into the structure of substances as they crystallise, for example in copper(II) sulfate pentahydrate ($CuSO_4.5H_2O$).

- **Weak acid** One which produces a low concentration of $H^+(aq)$ in water solution, for example ethanoic acid. It is only partially ionised.

- **Weak alkali** One which produces a low concentration of $OH^-(aq)$ in water solution, for example, ammonia solution. it is only partially ionised.

Acids, bases and salts
Additional questions

1 Explain, with the aid of examples, what you understand by the following terms:
   a strong acid
   b weak acid
   c strong alkali
   d weak alkali
   e concentrated acid.

2 a Copy out and complete the table, which covers the different methods of preparing salts.

Method of preparation	Name of salt prepared	Two substances used in the preparation
Acid + alkali	Potassium sulfate	. . . . and . . . .
Acid + metal	. . . .	. . . . and dilute hydrochloric acid
Acid + insoluble base	Magnesium sulfate	. . . . and . . . .
Acid + carbonate	Copper	. . . . and . . . .
Precipitation	Lead iodide	. . . . and . . . .

   b Write word and balanced chemical equations for each reaction shown in your table. Also write ionic equations where appropriate.

3 Study the following scheme.

   a Give the names and formulae of substances **A** to **E**.
   b Describe a test which could be used to identify the presence of water.
   c Which indicator is suitable for the initial reaction between the hydroxides and the dilute acids shown?
   d Write balanced chemical equations for the reactions taking place in the scheme.
   e Write an ionic equation for the production of the white precipitate **D**.

4 In a titration involving 24.0 cm³ potassium hydroxide solution against a solution containing 1 mol dm⁻³ of sulfuric acid, 28.0 cm³ of the acid was found to just neutralise the alkali completely.
   a Write a word and balanced chemical equation for the reaction.
   b Name a suitable indicator for the titration and state the colour change you would observe.
   c Calculate the concentration of the alkali in mol dm⁻³.
   d Describe a chemical test which you could use to identify the type of salt produced during the reaction.

5 Explain the following, with the aid of examples:
   a neutralisation
   b titration
   c soluble salt
   d insoluble salt.

6 Ammonium nitrate is used in fertilisers. The table below shows the solubility of this substance in water at various temperatures.

Temperature/°C	0	10	20	30	40	50	60	70
Solubility/g per 100 g H₂O	28	36	44	52	60	70	82	96

   a Plot these results as a solubility curve for ammonium nitrate.
   b Use your graph to answer the following questions.
   (i) What is the solubility of ammonium nitrate at 8 °C, 27 °C, 45 °C and 66 °C?
   (ii) Using your answers to (i), calculate the amount of ammonium nitrate that will crystallise out of solution when it is cooled from 45 °C to 8 °C.
   (iii) Calculate the amount of ammonium nitrate which would crystallise out of solution if 20 g of saturated solution was cooled from 63 °C to 28 °C.

7 Read the following passage and then answer the questions which follow.
   Sodium carbonate decahydrate *effloresces* quite readily. With some substances, such as solid sodium hydroxide, the reverse of efflorescence occurs—they *deliquesce*. There are some substances, such as concentrated sulphuric acid, which when left open to the atmosphere are diluted—they are *hygroscopic*.
   a What are the meanings of the terms in italics?
   b What precautions should be taken to ensure that these substances are not involved in the processes described above?
   c Which of the other salts shown in Table 7.6 on p. 105 are likely to effloresce? Give a reason for your answer.

continued

8 Copper(II) sulfate crystals exist as the *pentahydrate*, $CuSO_4.5H_2O$. It is a salt *hydrate*. If it is heated quite strongly, the *water of crystallisation* is driven off and the *anhydrous* salt remains.

a Explain the meaning of the terms shown in italics.

b Describe the experiment you would carry out to collect a sample of the water given off when the salt hydrate was heated strongly. Your description should include a diagram of the apparatus used and a chemical equation to represent the process taking place.

c Describe a chemical test you could carry out to show that the colourless liquid given off was water.

d Describe one other test you could carry out to show that the colourless liquid obtained in this experiment was **pure** water.

e Sometimes it is necessary to work out the percentage by mass of water of crystallisation as well as the number of moles of water present in a hydrated crystal.

(i) Use the information given below to calculate the percentage, by mass, of water of crystallisation in a sample of hydrated magnesium sulfate.

Mass of crucible	$= 14.20\,g$
Mass of crucible + hydrated $MgSO_4$	$= 16.66\,g$
Mass after heating	$= 15.40\,g$

(ii) Calculate the number of moles of water of crystallisation driven off during the experiment as well as the number of moles of anhydrous salt remaining. (A_r: H = 1; O = 16; Mg = 24; S = 32)

(iii) Using the information you have obtained in (ii), write down, in the form $MgSO_4.xH_2O$, the formula of hydrated magnesium sulfate.

8 Inorganic carbon chemistry

Limestone
Direct uses of limestone
Indirect uses of limestone

Carbonates and hydrogencarbonates
Properties of carbonates
Sodium carbonate – an important industrial chemical
Hydrogencarbonates
Uses of water
Hardness in water

Carbon dioxide
Uses of carbon dioxide
Laboratory preparation of carbon dioxide gas
Properties of carbon dioxide gas

Checklist

Additional questions

■ Limestone

Environmental Problems in the Peak District National Park

Local residents have been protesting against proposals to site a new limestone quarry in the beautiful Derbyshire Peak District in the UK. However, limestone is such a useful and sought-after mineral that demand has encouraged mining in National Park areas.

As the newspaper article on the left says, limestone is found in the Peak District, an example being this gorge on the River Wye in the UK (Figure 8.1). As well as giving rise to beautiful and varied countryside, limestone is a very useful raw material.

Limestone is composed of calcium carbonate ($CaCO_3$) in the form of the mineral calcite (Figure 8.2, overleaf). Chalk and marble are also made of calcite which is the second most abundant mineral in the Earth's crust after the different types of silicates (which include clay, granite and sandstone).

Chalk is made of the 'shells' of marine algae (that is, plants). It is a form of limestone. Most other limestones are formed from the debris of animal structures, for example brachiopods and crinoids.

Marble is a metamorphic rock made of calcium carbonate. It is formed when limestone is subjected to high pressures or high temperatures, or sometimes both acting together, to create crystals of calcium carbonate in the rock.

In a typical year, in excess of 500 million tonnes of limestone are quarried worldwide. Although it is cheap to quarry, as it is found near the surface, there are some environmental costs in its extraction.

Question

1 List the environmental issues which could arise through the quarrying of limestone.

Figure 8.1 This gorge in the Peak District National Park in the UK, is made from limestone.

Figure 8.2 Chalk, calcite and marble are all forms of calcium carbonate.

Direct uses of limestone

Limestone has a variety of uses in, for example, the making of cement, road building, glass making and the extraction of iron (Figure 8.3).

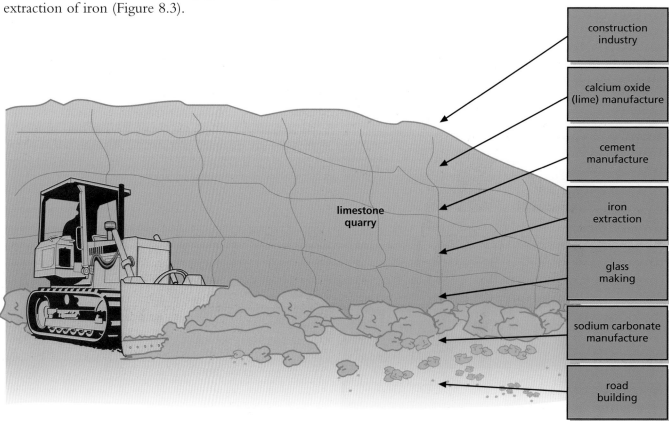

construction industry

calcium oxide (lime) manufacture

cement manufacture

iron extraction

glass making

sodium carbonate manufacture

road building

limestone quarry

Figure 8.3 Uses of limestone.

Neutralisation of acid soil

Powdered limestone is most often used to neutralise acid soil (Figure 8.4) because it is cheaper than any form of lime (calcium oxide), which has to be produced by heating limestone (see p. 116), and because it is slow acting and an excess does not make the soil alkaline. The reaction of limestone with acidic soil can be shown by the following ionic equation.

carbonate ion + hydrogen ion → carbon + water
(from CaCO₃)   (from acid soils)   dioxide

$$CO_3^{2-}(s) + 2H^+(aq) \rightarrow CO_2(g) + H_2O(l)$$

Figure 8.4 Spreading limestone on to soil.

Manufacture of iron and steel

In the blast furnace, limestone is used to remove earthy and sandy materials found in the iron ore. A liquid slag is formed, which is mainly calcium silicate. More details of the extraction of iron and its conversion into steel are given in Chapter 9.

Manufacture of cement and concrete

Limestone (or chalk) is mixed with clay (or shale) in a heated rotary kiln, using coal or oil as the fuel (Figure 8.5). The material produced is called cement. It contains a mixture of calcium aluminate ($Ca(AlO_2)_2$) and calcium silicate ($CaSiO_3$).

The dry product is ground to a powder and then a little calcium sulfate ($CaSO_4$) is added to slow down the setting rate of the cement. When water is added to the mixture slow complex chemical changes occur, resulting in the formation of a hard interlocking mass of crystals of hydrated calcium aluminate and silicate.

Figure 8.5 A modern rotary kiln.

Concrete is a mixture of cement with stone chippings and sand, which help to give it body. After the ingredients have been mixed with water they are poured into wooden moulds and allowed to set hard. Reinforced concrete is made by allowing concrete to set around steel rods or mesh to give it greater tensile strength, which is required for the construction of large bridges (Figure 8.6) and tall buildings.

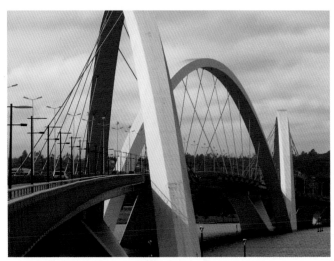

Figure 8.6 This bridge in Brazil uses reinforced concrete.

Manufacture of sodium carbonate – the Solvay process

Sodium carbonate (Na_2CO_3) is one of the world's most important industrial chemicals. It is used in the manufacture of soaps, detergents, dyes, drugs and other chemicals. The world production of sodium carbonate now exceeds 30 million tonnes per year. Sodium carbonate is manufactured ingeniously using calcium carbonate, sodium chloride, carbon dioxide and ammonia in a continuous process, with the Solvay tower the centre of reactions. Sodium carbonate is made as economically as possible because both the carbon dioxide and the ammonia components are recycled continuously.

Indirect uses of limestone

Lime manufacture

When calcium carbonate is heated strongly it thermally dissociates (breaks up reversibly) to form calcium oxide (lime) and carbon dioxide.

$$\text{calcium carbonate} \rightleftharpoons \text{calcium oxide} + \text{carbon dioxide}$$
$$CaCO_3(s) \rightleftharpoons CaO(s) + CO_2(g)$$

This reaction can go in either direction, depending on the temperature and pressure used. This reaction is an important industrial process and takes place in a lime kiln (Figure 8.7). In excess of 60 million tonnes of calcium oxide are produced worldwide every year.

Figure 8.7 The calcium oxide produced from this process is known as quicklime or lime and is used in large quantities in the manufacture of soda glass.

Calcium oxide (CaO) is a base and is still used by some farmers to spread on fields to neutralise soil acidity and to improve drainage of water through soils that contain large amounts of clay. It also has a use as a drying agent in industry. Soda glass is made by heating sand with soda (sodium carbonate, Na_2CO_3) and lime (Figure 8.8). For further discussion of glasses see Chapter 4, p. 64.

Large amounts of calcium oxide are also converted into calcium hydroxide ($Ca(OH)_2$) which is called **slaked lime**.

Figure 8.8 The composition of glass.

Manufacture of calcium hydroxide – slaked lime

Calcium hydroxide is a cheap industrial alkali (Figure 8.9). It is used in large quantities to make bleaching powder, by some farmers to reduce soil acidity, neutralising acidic industrial waste products, in the manufacture of whitewash, in glass manufacture and in water purification. Calcium hydroxide, in its white powder form, is produced by adding an equal amount of water to calcium oxide in a carefully controlled reaction. The control is needed because it is a very exothermic reaction.

Figure 8.9 Large-scale production of calcium hydroxide.

a At the start of the test.

b The limewater turns cloudy.

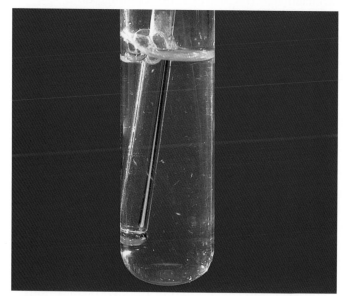

c If the CO_2 is bubbled for longer the white precipitate dissolves.

Figure 8.10 Testing for carbon dioxide.

This process can be shown on the small scale in the laboratory by heating a lump of limestone very strongly to convert it to calcium oxide. Water can then be carefully added dropwise to the calcium oxide. An exothermic reaction takes place as the water and calcium oxide react together in this slaking process to form calcium hydroxide.

$$\text{calcium oxide} + \text{water} \rightarrow \text{calcium hydroxide}$$
$$CaO_{(s)} + H_2O_{(l)} \rightarrow Ca(OH)_{2(s)}$$

A weak solution of calcium hydroxide in water is called **limewater**. It is used to test for carbon dioxide gas, as a white solid of calcium carbonate is formed if carbon dioxide gas is mixed with it (Figure 8.10):

$$\begin{array}{cccc} \text{calcium} & + \text{carbon} & \rightarrow \text{calcium} & + \text{water} \\ \text{hydroxide} & \text{dioxide} & \text{carbonate} & \end{array}$$
$$Ca(OH)_{2(aq)} + CO_{2(g)} \rightarrow CaCO_{3(s)} + H_2O_{(l)}$$

If carbon dioxide is bubbled for a further length of time then the white precipitate of calcium carbonate dissolves and a solution of calcium hydrogencarbonate is produced:

$$\begin{array}{cccc} \text{calcium} & + \text{carbon} & + \text{water} \rightarrow & \text{calcium} \\ \text{carbonate} & \text{dioxide} & & \text{hydrogen-} \\ & & & \text{carbonate} \end{array}$$
$$CaCO_{3(s)} + CO_{2(g)} + H_2O_{(l)} \rightarrow Ca(HCO_3)_{2(aq)}$$

Calcium hydroxide (slaked lime) is mixed with sand to give mortar. When it is mixed with water and allowed to set a strongly bonded material is formed, which is used to hold bricks together. The hardening of mortar takes place as the following reaction occurs.

$$\begin{array}{cccc} \text{calcium} & + \text{carbon} & \rightarrow \text{calcium} & + \text{water} \\ \text{hydroxide} & \text{dioxide} & \text{carbonate} & \end{array}$$
$$Ca(OH)_{2(aq)} + CO_{2(g)} \rightarrow CaCO_{3(s)} + H_2O_{(l)}$$

Questions

1 Some suggested building alternatives to limestone are sandstone and granite. What would be the benefits of using either of these two materials instead of limestone? For further information, consult the Internet.

2 Devise an experiment that you could carry out in the laboratory to establish whether or not powdered limestone is better at curing soil acidity than calcium oxide.

3 Why is calcium oxide described as a 'base' in the reaction which occurs in question 2?

Carbonates and hydrogencarbonates

Carbonates form an important range of compounds. They are all salts of carbonic acid (H_2CO_3) and contain the carbonate ion (CO_3^{2-}). Many of them occur naturally in rock formations. For example, in addition to the calcium carbonate in limestone, chalk and marble, malachite is copper(II) carbonate and dolomite is magnesium carbonate (Figure 8.11).

Figure 8.11 Three naturally occurring carbonates – dolomite (left) with magnesite in front and malachite (right).

Properties of carbonates

- Most metal carbonates thermally decompose when heated to form the metal oxide and carbon dioxide gas. For example,

$$\begin{array}{ccc} \text{copper(II)} & \rightarrow & \text{copper(II)} + \text{carbon} \\ \text{carbonate} & & \text{oxide} \quad\quad \text{dioxide} \end{array}$$
$$CuCO_3(s) \rightarrow CuO(s) + CO_2(g)$$

Group 1 metal carbonates, except for lithium carbonate, do not dissociate on heating. It is generally found that the carbonates of the more reactive metals are much more difficult to dissociate than, for example, copper(II) carbonate.

- Carbonates are generally insoluble in water except for those of sodium, potassium and ammonium.
- Carbonates react with acids to form salts, carbon dioxide and water (Chapter 7, p. 102). For example, calcium carbonate reacts with dilute hydrochloric acid to form calcium chloride, carbon dioxide and water.

$$\begin{array}{ccccc} \text{calcium} & + \text{hydrochloric} & \rightarrow & \text{calcium} & + \text{carbon} + \text{water} \\ \text{carbonate} & \text{acid} & & \text{chloride} & \text{dioxide} \end{array}$$
$$CaCO_3(s) + 2HCl(aq) \rightarrow CaCl_2(aq) + CO_2(g) + H_2O(l)$$

- This reaction is used in the laboratory preparation of carbon dioxide. It is also used as a test for a carbonate because the reaction produces carbon dioxide which causes effervescence and if bubbled through limewater turns it chalky white.

Sodium carbonate – an important industrial chemical

Approximately 33 million tonnes of sodium carbonate are manufactured worldwide annually, mainly by the Solvay process (see p. 115). Sodium carbonate is used for glass manufacture, brine treatment and water purification, as well as in the manufacture of detergents and textiles (Figure 8.12).

Figure 8.12 Sodium carbonate is an important chemical. It has many uses, for example in the manufacture of glass.

Hydrogencarbonates

Almost all the hydrogencarbonates are known only in solution because they are generally too unstable to exist as solids. They contain the hydrogencarbonate ion (HCO_3^-).

Sodium hydrogencarbonate ($NaHCO_3$) is the most common solid hydrogencarbonate. It is found in indigestion remedies because it reacts with the excess stomach acid (HCl) which causes indigestion.

$$\begin{array}{ccccc} \text{sodium} & + \text{hydrochloric} & \rightarrow & \text{sodium} + \text{water} + \text{carbon} \\ \text{hydrogen-} & \text{acid} & & \text{chloride} \quad\quad\quad\quad \text{dioxide} \\ \text{carbonate} & & & \end{array}$$
$$NaHCO_3(s) + HCl(aq) \rightarrow NaCl(aq) + H_2O(l) + CO_2(g)$$

Sodium hydrogencarbonate is the substance which is put into 'plain' flour to make it into 'self-raising' flour. When heated, the sodium hydrogencarbonate in the flour decomposes to give carbon dioxide gas, which makes the cakes rise.

$$\begin{array}{ccccc} \text{sodium} & \rightarrow & \text{sodium} & + \text{water} + \text{carbon} \\ \text{hydrogen-} & & \text{carbonate} & \quad\quad\quad\quad \text{dioxide} \\ \text{carbonate} & & & \end{array}$$
$$2NaHCO_3(s) \rightarrow Na_2CO_3(s) + H_2O(l) + CO_2(g)$$

Calcium hydrogencarbonate and magnesium hydrogencarbonate are two of the hydrogencarbonates found only in solution. They are the substances mainly responsible for the hardness in water.

Uses of water

Water is essential for the existence of life on Earth. It is also used in the manufacture of several important substances such as sulfuric acid (see 229) and aqueous ammonia (see p. 220). It is widely used as a solvent (see Chapter 7). Chlorine dissolves in water, killing bacteria, making the water safe to drink. Water is also used as a coolant in power stations as well as for drinking, washing and cleaning (Figure 8.33). However, there are problems in using water.

Figure 8.13 Examples of the use of water.

Hardness in water

Rainwater dissolves carbon dioxide as it falls through the atmosphere. A small fraction of this dissolved carbon dioxide reacts with the water to produce carbonic acid, which is a weak acid.

$$\text{water} + \text{carbon dioxide} \rightleftharpoons \text{carbonic acid}$$
$$H_2O(l) + CO_2(g) \rightleftharpoons H_2CO_3(aq)$$

Figure 8.14 Evaporating pure water leaves no deposit, while temporary or permanent hard water leaves white deposits behind.

As this solution passes over and through rocks containing limestone and dolomite, the weak acid in the rain attacks these rocks and very slowly dissolves them. The dissolved substances are called calcium and magnesium hydrogencarbonates.

$$\text{calcium} + \text{carbonic} \rightarrow \text{calcium}$$
$$\text{carbonate} \quad \text{acid} \quad \text{hydrogencarbonate}$$
$$CaCO_3(s) + H_2CO_3(aq) \rightarrow Ca(HCO_3)_2(aq)$$

Some of the rock strata may contain gypsum (calcium sulfate, $CaSO_4.2H_2O$), anhydrite ($CaSO_4$) or kieserite ($MgSO_4.H_2O$), which are very sparingly soluble in water. The presence of these dissolved substances causes the water to become **hard**. Hardness in water can be divided into two types – **temporary** and **permanent**. Temporary hardness is so called because it can be removed easily by boiling. Permanent hardness is so called because it is much more difficult to remove and certainly cannot be removed by boiling.

Temporary hardness is caused by the presence of dissolved calcium or magnesium hydrogencarbonates. Permanent hardness is caused by the presence of dissolved calcium or magnesium sulphates. When water containing these substances is evaporated, a white solid deposit is left behind (Figure 8.14). These white deposits are calcium or magnesium sulphates and/or calcium carbonate. Calcium carbonate causes the 'furring' in kettles that occurs in hard water areas (Figure 8.15a). This furring may be removed by the addition of a dilute acid:

$$\text{hydrogen ions} + \text{carbonate ion} \rightarrow \text{carbon dioxide} + \text{water}$$
$$2H^+(aq) + CO_3^{2-}(aq) \rightarrow CO_2(g) + H_2O(l)$$

Also, blockages in hot water pipes (Figure 8.15b) are caused by a similar process to the furring of kettles. A thick deposit of limescale builds up.

a The deposit in kettles ('furring') is caused by calcium carbonate from hard water.

b The thick deposits of limescale can be seen in this hot water pipe.

Figure 8.15

Stalactites and stalagmites are found in underground caverns in limestone areas. They are formed from the slow decomposition of water containing calcium or magnesium hydrogencarbonates (Figure 8.16).

calcium $\rightarrow$ calcium + carbon + water
hydrogencarbonate carbonate dioxide
$$Ca(HCO_3)_{2(aq)} \rightarrow CaCO_{3(s)} + CO_{2(g)} + H_2O_{(l)}$$

Figure 8.16 Stalactites and stalagmites have formed over hundreds of thousands of years.

Effect of hard water on soap

In hard water areas, such as those in the UK shown in Figure 8.17, it is difficult to make the soap lather. Instead, the water becomes cloudy. This cloudiness is caused by the presence of a solid material (a precipitate) formed by the reaction of the dissolved substances in the water with soap (sodium stearate). This white precipitate is known as scum (Figure 8.18).

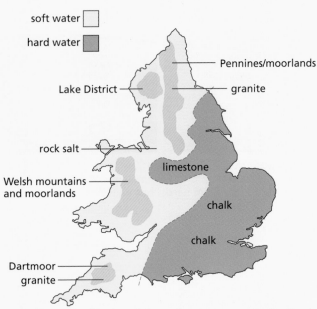

soft water ☐
hard water ▨

Pennines/moorlands
Lake District — granite
rock salt
limestone
Welsh mountains and moorlands
chalk
chalk
Dartmoor granite

Figure 8.17 The hard and the soft water areas in England and Wales.

Figure 8.18 Soap and hard water form scum.

sodium + calcium $\rightarrow$ calcium + sodium
stearate hydrogen- stearate hydrogen-
(soap) carbonate (scum) carbonate
$$2NaSt_{(aq)} + Ca(HCO_3)_{2(aq)} \rightarrow Ca(St)_{2(s)} + 2NaHCO_{3(aq)}$$

St = stearate $NaSt = C_{17}H_{35}COO^-Na^+$

The amount of soap required to just produce a lather can be used to estimate the hardness in water.

To overcome the problem of scum formation, soapless detergents have been produced which do not produce a scum because they do not react with the substances in hard water. For further discussion of soapless detergents see Chapter 14, p. 209.

Removal of hardness

Temporary hardness is easily removed from water by boiling. When heated the calcium hydrogencarbonate decomposes, producing insoluble calcium carbonate.

calcium $\xrightarrow{heat}$ calcium + water + carbon
hydrogen- carbonate dioxide
carbonate
$$Ca(HCO_3)_{2(aq)} \rightarrow CaCO_{3(s)} + H_2O_{(l)} + CO_{2(g)}$$

The substances in permanently hard water are not decomposed when heated and therefore cannot be removed by boiling. Both types of hardness can be removed by any of the following.

- **Addition of washing soda ($Na_2CO_3.10H_2O$) crystals.** In each case the calcium or magnesium ion, which actually causes the hardness, is removed as a precipitate and can, therefore, no longer cause hardness.

calcium ions + carbonate ions $\rightarrow$ calcium
(from hard water) (from washing soda) carbonate
$$Ca^{2+}_{(aq)} + CO_3^{2-}_{(aq)} \rightarrow CaCO_{3(s)}$$

- **Ion exchange.** The water is passed through a container filled with a suitable resin containing sodium ions. The calcium or magnesium ions causing the hardness are exchanged for the sodium ions in the resin (Figure 8.19).

calcium + sodium–resin → calcium–resin + sodium
ion   ion

$$Ca^{2+}(aq) + Na_2-R(s) \rightarrow Ca^{2+}-R(s) + 2Na^+(aq)$$

When all the sodium ions have been removed from the resin, it can be regenerated by pouring a solution of a suitable sodium salt through it.

- **Distillation.** The water is distilled away from the dissolved substances. This method, however, is far too expensive to be used on a large scale.

Advantages and disadvantages of hard water

As we mentioned earlier, there are some problems associated with hard water; however, there are advantages as well. Both are detailed in Table 8.1.

Table 8.1 Disadvantages and advantages of hard water.

Disadvantages	Advantages
Wastes soap	Has a nice taste
Causes kettles to fur	Calcium ions in hard water are required by the body for bones and teeth
Can cause hot water pipes to block	Coats lead pipes with a thin layer of lead(II) sulfate or lead(II) carbonate and cuts down the possibility of lead poisoning
Can spoil the finish of some fabrics	Some industries favour hard water. For example, the leather industry prefers to cure leather in hard water

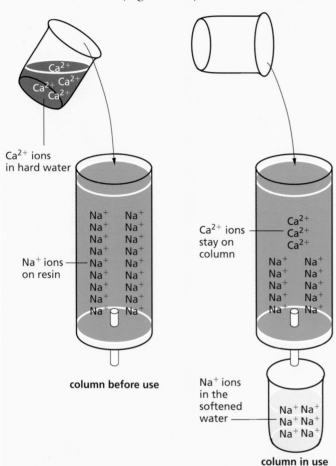

a The ion exchange process.

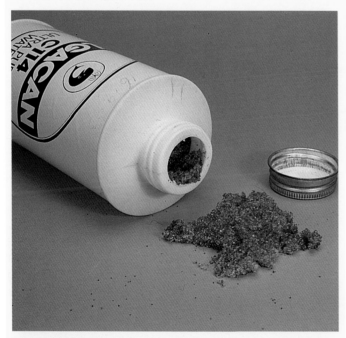

b Ion exchange resin.

Figure 8.19

Questions

1 Write word and balanced chemical equations to show the effect of heat on cobalt carbonate ($CoCO_3$) and nickel carbonate ($NiCO_3$).

2 Make a list of the main methods of softening hard water. In each case write a suitable equation(s) to summarise the chemical reactions involved.

3 One of the substances found in some temporary hard waters is magnesium hydrogencarbonate. Write word and balanced chemical equations to show the effect of heat on this substance in aqueous solution.

4 Write word and balanced chemical equations for the reaction of dilute hydrochloric acid with:
   a copper(II) carbonate
   b sodium carbonate.

Carbon dioxide

Carbon forms two oxides – carbon monoxide (CO) and carbon dioxide (CO_2). Carbon dioxide is the more important of the two, and in industry large amounts of carbon dioxide are obtained from the liquefaction of air. Air contains approximately 0.03% by volume of carbon dioxide. This value has remained almost constant for a long period of time and is maintained via the **carbon cycle** (Figure 8.20). However, scientists have recently detected an increase in the amount of carbon dioxide in the atmosphere.

Carbon dioxide is produced by burning fossil fuels. It is also produced by all living organisms through **aerobic respiration**. Animals take in oxygen and breathe out carbon dioxide.

glucose + oxygen → carbon dioxide + water + energy
$$C_6H_{12}O_{6(aq)} + 6O_{2(g)} \rightarrow 6CO_{2(g)} + 6H_2O_{(l)}$$

Carbon dioxide is taken in by plants through their leaves and used together with water, taken in through their roots, to synthesise sugars. This is the process of **photosynthesis**, and it takes place only in sunlight and only in green leaves, as they contain **chlorophyll** (the green pigment) which catalyses the process.

carbon dioxide + water $\xrightarrow[\text{chlorophyll}]{\text{sunlight}}$ glucose + oxygen

$$6CO_{2(g)} + 6H_2O_{(l)} \xrightarrow{} C_6H_{12}O_{6(aq)} + 6O_{2(g)}$$

The carbon cycle has continued in this manner for millions of years. However, scientists have detected an imbalance in the carbon cycle due to the increase in the amount of carbon dioxide produced through burning fossil fuels and the deforestation of large areas of tropical rain forest. The Earth's climate is affected by the levels of carbon dioxide (and water vapour) in the atmosphere. If the amount of carbon dioxide, in particular, builds up in the air, it is thought that the average temperature of the Earth will rise. This effect is known as the **greenhouse effect** (Figure 8.21).

Some energy from the Sun is absorbed by the Earth and its atmosphere. The remainder is reflected back into space. The energy that is absorbed helps to heat up the Earth. The Earth radiates some heat energy back into space but the 'greenhouse gases', including carbon dioxide, prevent it from escaping. This effect is similar to that observed in a greenhouse where sunlight (visible/ultraviolet radiation) enters through the glass panes but heat (infrared radiation) has difficulty escaping through the glass.

The long-term effect of the higher temperatures will be the gradual melting of ice caps and consequent flooding in low-lying areas of the Earth. There will also be changes in the weather patterns which would affect agriculture worldwide.

These problems have been recognised by nations worldwide. Recent agreements under the Kyoto Accord between nations mean that there will be some reduction in the amount of carbon dioxide (and other greenhouse gases) produced over the next few years. However, there is still a long way to go.

Figure 8.20 The carbon cycle.

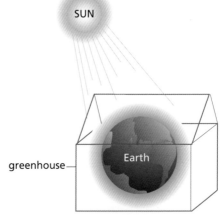

Figure 8.21 The greenhouse effect.

Questions

1 'Burning fossil fuel in an aircraft engine is significantly worse than burning the fuel at ground level'
   Discuss this statement in terms of the greenhouse effect and global warming.

2 Use your research skills to find out about:
   a carbon rationing
   b low carbon homes.

Uses of carbon dioxide

Carbon dioxide has some important uses.

- **Carbonated drinks.** Large quantities are used to make soda and mineral waters. The carbon dioxide gas is bubbled into the liquid under pressure, which increases its solubility.
- **Fire extinguishers.** It is used in extinguishers for use on electrical fires. Carbon dioxide is denser than air and forms a layer around the burning material. It covers the fire and starves it of oxygen. Carbon dioxide does not burn and so the fire is put out (Figure 8.22).

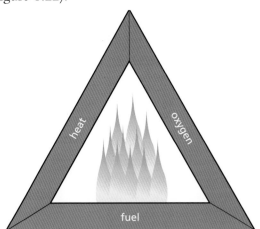

a Any fire needs fuel, oxygen and heat. If any of these is removed, the fire triangle is destroyed and the fire will be extinguished.

b Carbon dioxide fire extinguishers starve the fire of oxygen. They are used mainly for electrical fires.

Figure 8.22

- **Refrigerants.** Solid carbon dioxide (dry ice) is used for refrigerating ice cream, meat and soft fruits. It is used for this purpose because it is colder than ice and it sublimes (p. 5), and so it does not pass through a potentially damaging liquid stage.
- **Special effects.** Carbon dioxide is used to create the 'smoke' effect you may see at pop concerts and on television. Dry ice is placed in boiling water and it forms thick clouds of white 'smoke' (Figure 1.9, p. 5). It stays close to the floor due to the fact that carbon dioxide is denser than air.
- **Heat transfer agents.** Carbon dioxide gas is used for transferring heat in some nuclear power stations.

Laboratory preparation of carbon dioxide gas

In the laboratory the gas is made by pouring dilute hydrochloric acid on to marble chips ($CaCO_3$).

calcium + hydrochloric → calcium + water + carbon
carbonate   acid   chloride   dioxide

$$CaCO_3(s) + 2HCl(aq) \rightarrow CaCl_2(aq) + H_2O(l) + CO_2(g)$$

A suitable apparatus for preparing carbon dioxide is shown in Figure 8.23.

dilute hydrochloric acid

carbon dioxide

water

marble chips

* If dry gas is required, then it is passed through concentrated sulfuric acid (to dry it) and then collected as shown below

carbon * dioxide

CO_2 CO_2

concentrated sulfuric acid

Figure 8.23 Preparation and collection of carbon dioxide gas.

Question

1 List the important uses of carbon dioxide and for each use you have given in your answer explain why carbon dioxide is used.

Properties of carbon dioxide gas

Physical properties
Carbon dioxide is:

- a colourless gas
- sparingly soluble in water
- denser than air.

Chemical properties
- When bubbled into water it dissolves slightly and some of the carbon dioxide reacts, forming a solution of the weak acid carbonic acid which shows a pH of 4 or 5.

$$\text{water} + \text{carbon dioxide} \rightleftharpoons \text{carbonic acid}$$
$$H_2O(l) + CO_2(g) \rightleftharpoons H_2CO_3(aq)$$

- It will support the combustion only of strongly burning substances such as magnesium. This burning reactive metal decomposes the carbon

dioxide to provide oxygen for its continued burning in the gas. This reaction is accompanied by much crackling (Figure 8.24).

$$\text{magnesium} + \text{carbon dioxide} \rightarrow \text{magnesium oxide} + \text{carbon}$$
$$2Mg(s) + CO_2(g) \rightarrow 2MgO(s) + C(s)$$

This reaction is a good example of a **redox** process (Chapter 2, p. 15). Which of these substances has been oxidised and which has been reduced? Which of these substances is the oxidising agent and which is the reducing agent?

- When carbon dioxide is bubbled through limewater (calcium hydroxide solution), a white precipitate is formed. This white solid is calcium carbonate ($CaCO_3$). This reaction is used as a test to show that a gas is carbon dioxide.

$$\text{carbon dioxide} + \text{calcium hydroxide} \rightarrow \text{calcium carbonate} + \text{water}$$
$$CO_2(g) + Ca(OH)_2(aq) \rightarrow CaCO_3(s) + H_2O(l)$$

If carbon dioxide is bubbled through this solution continuously then it will eventually become clear. This is because of the formation of soluble calcium hydrogencarbonate solution.

$$\text{calcium carbonate} + \text{water} + \text{carbon dioxide} \rightarrow \text{calcium hydrogen-carbonate}$$
$$CaCO_3(s) + H_2O(l) + CO_2(g) \rightarrow Ca(HCO_3)_2(aq)$$

- Carbon dioxide reacts with strong alkalis, such as sodium hydroxide, to form carbonates. A solution of sodium hydroxide can be used to absorb carbon dioxide from the air. If excess carbon dioxide is bubbled through a solution of an alkali then a white precipitate may be obtained. When using sodium hydroxide this is due to the formation and precipitation of the sodium hydrogencarbonate:

$$\text{sodium carbonate} + \text{water} + \text{carbon dioxide} \rightarrow \text{sodium hydrogen-carbonate}$$
$$Na_2CO_3(aq) + H_2O(l) + CO_2(g) \rightarrow 2Na(HCO_3)_2(s)$$

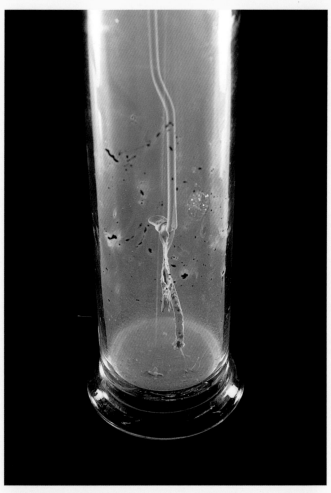

Figure 8.24 When magnesium burns in carbon dioxide gas, magnesium oxide (white) and carbon (black) are produced.

Questions

1 Describe an experiment that you could carry out to show that carbonic acid is a weak acid.

2 Why is it not possible to use dilute sulfuric acid to make carbon dioxide from limestone in the laboratory?

3 When carbon dioxide is 'poured' from a gas jar on to a burning candle the candle goes out. What properties of carbon dioxide does this experiment show?

Checklist

After studying Chapter 8 you should know and understand the following terms.

- **Carbonate**   A salt of carbonic acid containing the carbonate ion, CO_3^{2-}, for example $CuCO_3$.

- **Carbon dioxide**   A colourless, odourless gas, soluble in water, producing a weak acid called carbonic acid. It makes up 0.03% of air. It is produced by respiration in all living things and by the burning of fossil fuels. It is consumed by plants in photosynthesis.

- **Greenhouse effect**   The absorption of reflected infrared radiation from the Earth by gases in the atmosphere such as carbon dioxide (a greenhouse gas) leading to atmospheric warming.

- **Hardness of water**   This is caused by the presence of calcium (or magnesium) ions in water, which form a 'scum' with soap and prevent the formation of a lather. There are two types of hardness:

  - temporary hardness – caused by the presence of dissolved calcium (or magnesium) hydrogencarbonate
  - permanent hardness – this results mainly from dissolved calcium (or magnesium) sulfate.

- **Lime**   A white solid known chemically as calcium oxide (CaO). It is produced by heating limestone. It is used to counteract soil acidity and to manufacture calcium hydroxide (slaked lime). It is also used as a drying agent in industry.

- **Limestone**   A form of calcium carbonate ($CaCO_3$). Other forms include chalk, calcite and marble. It is used directly to neutralise soil acidity and in the manufacture of iron and steel, glass, cement, concrete, sodium carbonate and lime.

- **Photosynthesis**   The chemical process by which green plants synthesise their carbon compounds from atmospheric carbon dioxide using light as the energy source and chlorophyll as the catalyst.

- **Removal of hardness**   Temporary hardness is removed by boiling. Both temporary and permanent hardness are removed by:

  - addition of washing soda (sodium carbonate)
  - ion exchange
  - distillation.

- **Thermal decomposition**   The breakdown of a substance under the influence of heat.

Inorganic carbon chemistry
Additional questions

1 This question is about the limestone cycle.

Limestone $\xrightarrow{\text{heat}}$ Compound **A** + Carbon dioxide

CO$_2$(g) ↑   water ↓

Limewater ← water ← Compound **B** ('slaked lime')

a Name and give the formula of:
(i) compound **A**
(ii) compound **B**.
b Write balanced chemical equations for the formation of both compounds **A** and **B**.
c Name and give the symbol/formula for the ions present in limewater.
d Describe with the aid of a balanced chemical equation what happens if carbon dioxide is bubbled through limewater until there is no further change.

2 The diagram shown below is a simplified version of the carbon cycle.

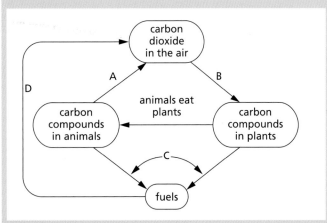

a Name the processes **A**, **B**, **C** and **D**.
b Write balanced chemical equations to represent the processes taking place in **A** and **B**.
c There are a number of fuels which could be placed in the 'fuels' box. Name three such fuels.
d 'The air contains approximately 0.03% of carbon dioxide and this amount is almost constant.' Explain why this is a true statement.
e (i) What is the 'greenhouse effect'?
(ii) Discuss the possible consequences of the greenhouse effect if nothing is done to counteract it.

3 Limestone is an important raw material used in many different industries.
a One of the properties of limestone is that it reacts with acids.
(i) Why do farmers spread powdered limestone on their fields?
(ii) How can buildings made of limestone be affected by 'acid rain'?
(iii) Write an ionic equation which would represent the reactions taking place in both (i) and (ii).
b Limestone is used in the manufacture of iron (Chapter 9, p. 135).
(i) Why is it added to the blast furnace along with coke and haematite?
(ii) Write chemical equations for the reactions it is involved in, in this process.
c (i) Name a building material made by heating a mixture of limestone and clay in a rotary kiln.
(ii) What substance is the dry product produced in the rotary kiln added to?
Explain why this substance is added.

4

Carbon dioxide $\underset{\text{dissolves in water}}{\rightleftharpoons}$ Acid **A**

Carbon dioxide → bubbled through calcium hydroxide solution → White precipitate **D** → further bubbling of carbon dioxide → Colourless solution **E**

Acid **A** → + burning magnesium → White solid **B** + carbon → + water → Alkaline solution **C**

a Name and give the formulae of substances **A** to **E**.
b Write balanced chemical equations for the reactions in which compounds **B**, **C** and **E** were formed.
c Where would you expect to find acid **A**?
d Universal indicator solution was added to solution **C**. What colour did it go?
e Upon addition of dilute hydrochloric acid to solution **C**, a neutralisation reaction took place.
(i) Write a balanced chemical equation for the reaction taking place.
(ii) Name the salt produced in this reaction.

5 The following question is about carbon dioxide.

a Name and give the formula of each of the substances **A**, **B** and **C**.

b Identify by name the different pieces of apparatus **D**, **E**, **F** and **G**.

c Draw and label the apparatus that should be used if a dry sample of carbon dioxide gas was required.

d When a gas jar containing carbon dioxide is held over a burning wooden splint and the cover removed, the flame goes out. State two properties of carbon dioxide illustrated by this observation.

e Carbon dioxide is also produced when zinc carbonate is heated strongly.

(i) Write a balanced chemical equation for the reaction taking place.

(ii) Name the process which is taking place as the zinc carbonate is heated.

(iii) Calculate the volume of carbon dioxide that would be produced (measured at room temperature and pressure (rtp)) if 12.5 g of zinc carbonate were heated strongly.

(One mole of any gas occupies 24 dm³ at rtp. A_r: C = 12; O = 16; Zn = 65)

6 Explain the following.

a Industry normally requires water which has been softened.

b Hard water is good for the promotion of healthy bones and teeth.

c Hard water causes kettles to fur. This 'fur' can be removed by using a dilute acid

d Hard water wastes soap.

e Hard water can coat lead pipes and reduce the possibility of lead poisoning.

7 The results of testing five samples of water from different areas are shown in the table below. The soap solution was gradually added to 25 cm³ of each sample of water with shaking until a permanent lather (one which lasts for 20 seconds) was obtained.

Water sample/25 cm³	Volume of soap solution added/cm³	
	Before boiling	After boiling
A	12	1
B	13	6
C	11	11
D	14	3
E	16	16

a (i) Which samples are permanently hard?

(ii) Which samples are temporarily hard?

(iii) Which sample contains both temporary and permanent hardness?

b Name a compound which could be present in sample **D** but not in sample **E**.

c Name a compound which could be present in sample **E** but not in sample **D**.

d Explain how the compound you have named in **c** gets into the water.

e Sample **E** was distilled. The water collected was tested with soap solution. What volume of soap solution might you expect to be added to produce a permanent lather? Comment on your answer.

8 Lime (calcium oxide) is produced in very large quantities in a lime kiln. The equation for the reaction is:

$$CaCO_3(s) \rightarrow CaO(s) + CO_2(g)$$

a How much limestone would be needed to produce 61.60 tonnes of lime?

(A_r: C = 12; O = 16; Ca = 40)

b Why is the carbon dioxide swept out of the lime kiln?

c Give three uses of lime.

d (i) What problems are associated with the large-scale quarrying of limestone?

(ii) What steps have been taken to overcome or reduce the problems you have outlined in (i)?

9 Metal extraction and chemical reactivity

Metal reactions
With acid
With air/oxygen
With water/steam
Reactivity of metals and their uses
Using the reactivity series
Competition reactions in the solid state
Competition reactions in aqueous solutions

Identifying metal ions
Amphoteric hydroxides

Discovery of metals

Extraction of metals from their ores
Extraction of reactive metals
Extraction of fairly reactive metals
Extraction of unreactive metals

Recycling metals

Rusting of iron
Rust prevention
Corrosion

Alloys
Production of steel
Alloys to order

Checklist

Additional questions

You have already seen in Chapter 2, p. 12, that metals usually have similar physical properties. However, they differ in other ways. Look closely at the three photographs in Figure 9.1.

Sodium is soft and reacts violently with both air and water. Iron also reacts with air and water but much more slowly, forming rust. Gold, however, remains totally unchanged after many hundreds of years. Sodium is said to be more reactive than iron and, in turn, iron is said to be more reactive than gold.

b Iron rusts when left unprotected.

c Gold is used in leaf form on this giant Buddha as it is unreactive.

a Sodium burning in air/oxygen.

Figure 9.1

Metal reactions

By carrying out reactions in the laboratory with other metals and air, water and dilute acid, it is possible to produce an order of reactivity of the metals.

With acid

Look closely at the photograph in Figure 9.2 showing magnesium metal reacting with dilute hydrochloric acid. You will notice effervescence, which is caused by bubbles of hydrogen gas being formed as the reaction between the two substances proceeds. The other product of this reaction is the salt, magnesium chloride.

magnesium + hydrochloric → magnesium + hydrogen
acid chloride

$$Mg_{(s)} + 2HCl_{(aq)} \rightarrow MgCl_{2(aq)} + H_{2(g)}$$

If a metal reacts with dilute hydrochloric acid then hydrogen and the metal chloride are produced.

If similar reactions are carried out using other metals with acid, an order of reactivity can be produced by measuring the rate of evolution of hydrogen. This is known as a **reactivity series**. An order of reactivity, giving the most reactive metal first, using results from experiments with dilute acid, is shown in Table 9.1. The table also shows how the metals react with air/oxygen and water/steam, and, in addition, the ease of extraction of the metal.

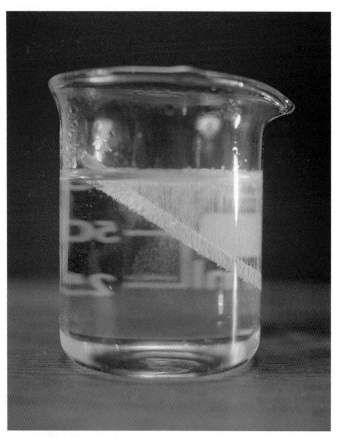

Figure 9.2 Effervescence occurs when magnesium is put into acid.

Table 9.1 Order of reactivity.

Reactivity series	Reaction with dilute acid	Reaction with air/oxygen	Reaction with water	Ease of extraction
Potassium (K) Sodium (Na)	Produce H_2 with decreasing vigour	Burn very brightly and vigorously	Produce H_2 with decreasing vigour with cold water	Difficult to extract
Calcium (Ca) Magnesium (Mg)		Burn to form an oxide with decreasing vigour	React with steam with decreasing vigour	Easier to extract
Aluminium (Al*)				
[Carbon (C)]				
Zinc (Zn)				
Iron (Fe)				
Lead (Pb) [Hydrogen (H)]		React slowly to form the oxide	Do not react with cold water or steam	
Copper (Cu)	Do not react with dilute acids			
Silver (Ag)		Do not react		Found as the element (native)
Gold (Au)				
Platinum (Pt)				

Increasing reactivity of metal

* Because aluminium reacts so readily with the oxygen in the air, a protective oxide layer is formed on its surface. This often prevents any further reaction and disguises aluminium's true reactivity. This gives us the use of a light and strong metal.

With air/oxygen

Many metals react directly with oxygen to form oxides. For example, magnesium burns brightly in oxygen to form the white powder magnesium oxide.

$$\text{magnesium} + \text{oxygen} \rightarrow \text{magnesium oxide}$$
$$2Mg(s) + O_2(g) \rightarrow 2MgO(s)$$

With water/steam

Reactive metals such as potassium, sodium and calcium react with cold water to produce the metal hydroxide and hydrogen gas. For example, the reaction of sodium with water produces sodium hydroxide and hydrogen.

$$\text{sodium} + \text{water} \rightarrow \text{sodium hydroxide} + \text{hydrogen}$$
$$2Na(s) + 2H_2O(l) \rightarrow 2NaOH(aq) + H_2(g)$$

The moderately reactive metals, magnesium, zinc and iron, react slowly with water. They will, however, react more rapidly with steam (Figure 9.3). In their reaction with steam, the metal oxide and hydrogen are formed. For example, magnesium produces magnesium oxide and hydrogen gas.

$$\text{magnesium} + \text{steam} \rightarrow \text{magnesium oxide} + \text{hydrogen}$$
$$Mg(s) + H_2O(g) \rightarrow MgO(s) + H_2(g)$$

Figure 9.3 Apparatus used to investigate how metals such as magnesium react with steam.

labels: rocksil wool soaked in water; metal sample (magnesium, zinc, iron); rubber bung; hydrogen; water; heat

Generally, it is the unreactive metals that we find the most uses for; for example, the metals iron and copper can be found in many everyday objects (Figure 9.4). However, magnesium is one of the metals used in the construction of the Airbus A380 (Figure 9.5).

Both sodium and potassium are so reactive that they have to be stored under oil to prevent them from coming into contact with water or air. However, because they have low melting points and are good conductors of heat, they are used as coolants for nuclear reactors.

a This wood-burning stove is made of iron.

b Copper pots and pans.

Figure 9.4

Figure 9.5 Planes are made of an alloy which contains magnesium and aluminium.

Questions

1 Write balanced chemical equations for the reactions between:
   a iron and dilute hydrochloric acid
   b zinc and oxygen
   c calcium and water.

2 Make a list of six things you have in your house made from copper or iron.
   Give a use for each of the other unreactive metals shown in the reactivity series.

Reactivity of metals and their uses

Generally, it is the unreactive metals for which we find most uses. For example, the metals iron and copper can be found in everyday objects such as car bodies and coins, respectively. However, the metal aluminium is an exception. Aluminium appears in the reactivity series just below magnesium and is quite reactive. Fortunately, it forms a relatively thick oxide layer on the surface of the metal which prevents further reaction. This gives us the use of a light, strong metal for items such as door handles and cooking foil. There are many reactive metals you would not expect to find many uses for. For example, sodium is so reactive it is stored under oil to prevent it coming into contact with water and the air. However, because it has a low melting point and is a good conductor of heat, it can be used as a coolant in nuclear reactors.

Question

1 Choose a reactive metal (e.g. magnesium) and an unreactive metal (e.g. lead) and compare the uses of these metals with their reactivity and properties.

Using the reactivity series

The reactivity series is useful for predicting how metals will react. It can also be used to predict the reactions of some metal compounds. The tables on page 284 show how some metal nitrates and hydroxides behave when heated. You can see that metal compounds from a similar part of the series behave in a similar manner.

Competition reactions in the solid state

A more reactive metal has a greater tendency to form a metal ion by losing electrons than a less reactive metal does. Therefore, if a more reactive metal is heated with the oxide of a less reactive metal, then it will remove the oxygen from it (as the oxide anion). You can see from the reactivity series that iron is less reactive than aluminium (p. 129). If iron(III) oxide is mixed with aluminium and the mixture is heated using a magnesium fuse (Figure 9.6), a very violent reaction occurs as the competition between the aluminium and the iron for the oxygen takes place.

The aluminium, being the more reactive metal, takes the oxygen from the less reactive iron. It is a very exothermic reaction. When the reaction is over, a solid lump of iron is left along with a lot of white aluminium oxide powder.

$$\text{iron(III) oxide} + \text{aluminium} \xrightarrow{\text{heat}} \text{aluminium} + \text{iron} \atop \text{oxide}$$

$$Fe_2O_3(s) + 2Al(s) \rightarrow Al_2O_3(s) + 2Fe(s)$$

or

$$2Fe^{3+}(s) + 2Al(s) \rightarrow 2Al^{3+}(s) + 2Fe(s)$$

This is a **redox reaction** (see Chapter 2, p. 15 and Chapter 6, p. 82 for a discussion of this type of reaction).

This particular reaction is known as the **Thermit reaction** (Figure 9.7). Since large amounts of heat are given out and the iron is formed in a molten state, this reaction is used to weld together damaged railway lines. It is also used in incendiary bombs.

Some metals, such as the transition metals chromium and titanium, are prepared from their oxides using this type of competition reaction.

Carbon, a non-metal, is included in Table 9.1 (p. 129) just below aluminium. It is able to reduce metal oxides below it in the series.

mixture of aluminium powder and iron(III) oxide powder

magnesium fuse

iron vessel

heatproof mat

Figure 9.6 Thermit reaction mixture apparatus.

Figure 9.7 The Thermit reaction in a laboratory. The same reaction is used to weld damaged railway lines.

Competition reactions in aqueous solutions

In another reaction, metals compete with each other for other anions. This type of reaction is known as a **displacement reaction**. As in the previous type of competitive reaction, the reactivity series can be used to predict which of the metals will 'win'.

In a displacement reaction, a more reactive metal will displace a less reactive metal from a solution of its salt. Zinc is above copper in the reactivity series. Figure 9.8 shows what happens when a piece of zinc metal is left to stand in a solution of copper(II) nitrate. The copper(II) nitrate slowly loses its blue colour as the zinc continues to displace the copper from the solution and eventually becomes colourless zinc nitrate.

$$\text{zinc} + \text{copper(II) nitrate} \rightarrow \text{zinc nitrate} + \text{copper}$$
$$Zn(s) + Cu(NO_3)_2(aq) \rightarrow Zn(NO_3)_2(aq) + Cu(s)$$

The ionic equation for this reaction is:

$$\text{zinc} + \text{copper ions} \rightarrow \text{zinc ions} + \text{copper}$$
$$Zn(s) + Cu^{2+}(aq) \rightarrow Zn^{2+}(aq) + Cu(s)$$

This is also a redox reaction involving the transfer of two electrons from the zinc metal to the copper ions. The zinc is oxidised to zinc ions in aqueous solution, while the copper ions are reduced. (See Chapter 6, p. 82, for a discussion of oxidation and reduction in terms of electron transfer.) It is possible to confirm the reactivity series for metals using competition reactions of the types discussed in this section.

Figure 9.8 Zinc displaces copper.

Questions

1 Predict whether or not the following reactions will take place:
 a magnesium + copper(II) oxide
 b iron + aluminium oxide
 c calcium + magnesium oxide.
 Complete the word equations, and write balanced chemical and ionic equations for those reactions which do take place.

2 Predict whether or not the following reactions will take place:
 a magnesium + calcium nitrate solution
 b iron + copper(II) nitrate solution
 c copper + silver nitrate solution.
 Complete the word equations, and write balanced chemical and ionic equations for those reactions which do take place.

3 How could you carry out a series of reactions between metals and solutions of their nitrates to establish a reactivity series?

■ *Identifying metal ions*

When an alkali dissolves in water, it produces hydroxide ions. It is known that most metal hydroxides are insoluble. So if hydroxide ions from a solution of an alkali are added to a solution of a metal salt, an insoluble, often coloured, metal hydroxide is precipitated from solution (Figure 9.9).

Let's take the example of iron(III) chloride with sodium hydroxide solution:

$$\begin{array}{c}\text{iron(III)} + \text{sodium} \rightarrow \text{iron(III)} + \text{sodium}\\ \text{chloride} \quad \text{hydroxide} \quad \text{hydroxide} \quad \text{chloride}\end{array}$$
$$FeCl_3(aq) + 3NaOH(aq) \rightarrow Fe(OH)_3(s) + 3NaCl(aq)$$

The ionic equation for this reaction is:

$$\text{iron(III) ions} + \text{hydroxide ions} \rightarrow \text{iron(III) hydroxide}$$
$$Fe^{3+}(aq) + 3OH^-(aq) \rightarrow Fe(OH)_3(s)$$

Figure 9.9 a Iron(III) hydroxide is precipitated. **b** Copper(II) hydroxide is precipitated.

Table 9.2 shows the effects of adding a few drops of sodium hydroxide solution to solutions containing various metal ions, and of adding an excess. The colours of the insoluble metal hydroxides can be used to identify the metal cations present in solution. In some cases the precipitate dissolves in excess hydroxide, owing to the amphoteric nature of the metal hydroxide. This amphoteric nature can also be used to help identify metals such as aluminium and zinc.

Table 9.2 The effect of adding sodium hydroxide solution to solutions containing various metal ions.

Metal ion present in solution	Effect of adding sodium hydroxide solution	
	A few drops	An excess
Aluminium	White precipitate of aluminium hydroxide	Precipitate dissolves
Calcium	White precipitate of calcium hydroxide	Precipitate does not dissolve
Copper(II)	Blue precipitate of copper(II) hydroxide	Precipitate does not dissolve
Iron(II)	Green precipitate of iron(II) hydroxide	Precipitate does not dissolve
Iron(III)	Brown precipitate of iron(III) hydroxide	Precipitate does not dissolve
Zinc	White precipitate of zinc hydroxide	Precipitate dissolves

Amphoteric hydroxides

The hydroxides of metals are basic and they react with acids to form salts (Chapter 7, p. 100). The hydroxides of some metals, however, will also react with strong bases, such as sodium hydroxide, to form soluble salts. Hydroxides of this type are said to be **amphoteric**. For example,

zinc + hydrochloric → zinc + water
hydroxide   acid   chloride
$$Zn(OH)_2(aq) + 2HCl(aq) \rightarrow ZnCl_2(aq) + 2H_2O(l)$$

and

zinc hydroxide + sodium hydroxide → sodium zincate
$$Zn(OH)_2(aq) + 2NaOH(aq) \rightarrow Na_2Zn(OH)_4(aq)$$

Other amphoteric hydroxides are lead hydroxide ($Pb(OH)_2$) and aluminium hydroxide ($Al(OH)_3$). We can use this sort of behaviour to help identify metal cations, as their hydroxides are soluble in strong bases.

Both aluminium and zinc metals will also react readily with moderately concentrated acids and alkalis.

zinc + hydrochloric → zinc chloride + hydrogen
acid
$$Zn(s) + 2HCl(aq) \rightarrow ZnCl_2(aq) + H_2(g)$$

zinc + sodium + water → sodium + hydrogen
   hydroxide   zincate
$$Zn(s) + 2NaOH(aq) + 2H_2O(l) \rightarrow Na_2Zn(OH)_4(aq) + H_2(g)$$

It should be noted that the oxides of the metals used as examples above are also amphoteric. Aluminium oxide and zinc oxide will react with both acids and alkalis.

zinc oxide + hydrochloric acid → zinc chloride + water
$$ZnO(s) + 2HCl(aq) \rightarrow ZnCl_2(aq) + H_2O(l)$$

zinc oxide + sodium hydroxide + water → sodium zincate
$$ZnO(s) + 2NaOH(aq) + H_2O(l) \rightarrow Na_2Zn(OH)_4(aq)$$

Questions

1 Write equations for the reaction between:
   a aluminium and moderately concentrated hydrochloric acid
   b aluminium and moderately concentrated sodium hydroxide (producing sodium aluminate, $NaAl(OH)_4$).

2 Write ionic equations for the reactions which take place to produce the metal hydroxides shown in Table 9.2.

3 Describe what you would see when sodium hydroxide is added slowly to a solution containing iron(II) nitrate.

Discovery of metals

Metals have been used since prehistoric times. Many primitive iron tools have been excavated. These were probably made from small amounts of native iron found in rock from meteorites. It was not until about 2500 BC that iron became more widely used. This date marks the dawn of the iron age, when people learned how to get iron from its **ores** in larger quantities by reduction using charcoal. An ore is a naturally occurring mineral from which a metal can be extracted.

Over the centuries other metals, which like iron are also relatively low in the reactivity series, were isolated in a similar manner. These included copper, lead, tin and zinc. However, due to the relatively low abundance of the ores containing these metals, they were not extracted and used in large amounts.

Metals high in the reactivity series have proved very difficult to isolate. It was not until more recent times, through Sir Humphry Davy's work on electrolysis, that potassium (1807), sodium (1807), calcium (1808) and magnesium (1808) were isolated. Aluminium, the most plentiful reactive metal in the Earth's crust, was not extracted from its ore until 1827, by Friedrich Wöhler (p. 81), and the extremely reactive metal rubidium was not isolated until 1861 by Robert Bunsen and Gustav Kirchhoff.

Extraction of metals from their ores

The majority of metals are too reactive to exist on their own in the Earth's crust, and they occur naturally in rocks as compounds in ores (Figure 9.10). These ores are usually carbonates, oxides or sulfides of the metal, mixed with impurities.

Figure 9.10 Metal ores – chalcopyrite (top) and galena.

Figure 9.11 Gold crystals.

Some metals, such as gold and silver, occur in a native form as the free metal (Figure 9.11). They are very unreactive and have withstood the action of water and the atmosphere for many thousands of years without reacting to become compounds.

Some of the common ores are shown in Table 9.3.

Large lumps of the ore are first crushed and ground up by very heavy machinery. Some ores are already fairly concentrated when mined. For example, in some parts of the world, haematite contains over 80% Fe_2O_3. However, other ores, such as copper pyrites, are often found to be less concentrated, with only 1% or less of the copper compound, and so they have to be concentrated before the metal can be extracted. The method used to extract the metal from its ore depends on the position of the metal in the reactivity series.

Extraction of reactive metals

Because reactive metals, such as sodium, hold on to the element(s) they have combined with, they are usually difficult to extract. For example, sodium chloride (as rock salt) is an ionic compound with the Na^+ and Cl^- ions strongly bonded to one another. The separation of these ions and the subsequent isolation of the sodium metal is therefore difficult.

Electrolysis of the molten, purified ore is the method used in these cases. During this process, the metal is produced at the cathode while a non-metal is produced at the anode. As you might expect, extraction of metal by electrolysis is expensive. In order to keep costs low, many metal smelters using electrolysis are situated in regions where there is hydroelectric power. Hydroelectric power is discussed further in Chapter 13, p. 188.

For further discussion of the extraction of sodium and aluminium, see Chapter 6, pp. 81–85.

Table 9.3 Some common ores.

Metal	Name of ore	Chemical name of compound in ore	Formula	Usual method of extraction
Aluminium	Bauxite	Aluminium oxide	$Al_2O_3.2H_2O$	Electrolysis of oxide dissolved in molten cryolite
Copper	Copper pyrites	Copper iron sulfide	$CuFeS_2$	The sulfide ore is roasted in air
Iron	Haematite	Iron(III) oxide	Fe_2O_3	Heat oxide with carbon
Sodium	Rock salt	Sodium chloride	$NaCl$	Electrolysis of molten sodium chloride
Zinc	Zinc blende	Zinc sulfide	ZnS	Sulfide is roasted in air and the oxide produced is heated with carbon

Extraction of fairly reactive metals

Metals towards the middle of the reactivity series, such as iron and zinc, may be extracted by reducing the metal oxide with the non-metal carbon.

Iron

Iron is extracted mainly from its oxides, haematite (Fe_2O_3) and magnetite (Fe_3O_4), in a blast furnace (Figures 9.12 and 9.13). These ores contain at least 60% iron. The iron ores used are a blend of those extracted in Australia, Canada, Sweden, Venezuela and Brazil. The blast furnace is a steel tower approximately 50 m high lined with heat-resistant bricks. It is loaded with the 'charge' of iron ore (usually haematite), coke (made by heating coal) and limestone (calcium carbonate).

Figure 9.12 A blast furnace.

hot waste gases – these are used to heat air going into the blast furnace

skip dumps charge iron ore, coke and limestone into hopper

hopper

height approximately 90 m

heat-resistant brick

50 m

650°C

1000°C

1100°C

50 rows of water-cooled copper coolers in the lining of the furnace

hot air blown through tuyère

slag notch

molten iron

molten slag

tap hole

firebrick pad

Figure 9.13 Cross-section of a blast furnace.

A blast of hot air is sent in near the bottom of the furnace through holes (tuyères) which makes the 'charge' glow, as the coke burns in the preheated air.

$$\text{carbon} + \text{oxygen} \rightarrow \text{carbon dioxide}$$
$$C(s) + O_2(g) \rightarrow CO_2(g)$$

A number of chemical reactions then follow.

- The limestone begins to decompose:

$$\begin{array}{ccc} \text{calcium} & \rightarrow \text{calcium} + & \text{carbon} \\ \text{carbonate} & \text{oxide} & \text{dioxide} \end{array}$$
$$CaCO_3(s) \rightarrow CaO(s) + CO_2(g)$$

- The carbon dioxide gas produced reacts with more hot coke higher up in the furnace, producing carbon monoxide in an endothermic reaction.

$$\text{carbon dioxide} + \text{coke} \rightarrow \text{carbon monoxide}$$
$$CO_2(g) + C(s) \rightarrow 2CO(g)$$

- Carbon monoxide is a reducing agent. It rises up the furnace and reduces the iron(III) oxide ore. This takes place at a temperature of around 700 °C:

$$\begin{array}{cccc} \text{iron(III)} + & \text{carbon} & \rightarrow \text{iron} + & \text{carbon} \\ \text{oxide} & \text{monoxide} & & \text{dioxide} \end{array}$$
$$Fe_2O_3(s) + 3CO(g) \rightarrow 2Fe(l) + 3CO_2(g)$$

The molten iron trickles to the bottom of the furnace.

- Calcium oxide is a base and this reacts with acidic impurities such as silicon(IV) oxide in the iron, to form a slag which is mainly calcium silicate.

$$\begin{array}{ccc} \text{calcium} + & \text{silicon(IV)} & \rightarrow \text{calcium} \\ \text{oxide} & \text{oxide} & \text{silicate} \end{array}$$
$$CaO(s) + SiO_2(s) \rightarrow CaSiO_3(l)$$

The slag trickles to the bottom of the furnace, but because it is less dense than the molten iron, it floats on top of it.

Generally, metallic oxides, such as calcium oxide (CaO), are basic and non-metallic oxides, such as silicon (IV) oxide (SiO_2), are acidic.

Certain oxides, such as carbon monoxide (CO), are neutral and others, such as zinc oxide (ZnO), are amphoteric (p. 133).

The molten iron, as well as the molten slag, may be **tapped off** (run off) at regular intervals.

The waste gases, mainly nitrogen and oxides of carbon, escape from the top of the furnace. They are used in a heat exchange process to heat incoming air and so help to reduce the energy costs of the process. Slag is the other waste material. It is used by builders and road makers (Figure 9.14, overleaf) for foundations.

The extraction of iron is a continuous process and is much cheaper to run than an electrolytic method.

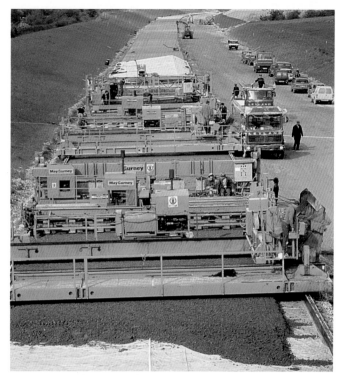

Figure 9.14 Slag is used in road foundations.

The iron obtained by this process is known as 'pig' or cast iron and contains about 4% carbon (as well as some other impurities). The name pig iron arises from the fact that if it is not subsequently converted into steel it is poured into moulds called pigs. Because of its brittle and hard nature, the iron produced by this process has limited use. Gas cylinders are sometimes made of cast iron, since they are unlikely to get deformed during their use.

The majority of the iron produced in the blast furnace is converted into different steel alloys (p. 141) such as manganese and tungsten steels as well as the well-known example of stainless steel (p. 142).

The annual production of iron worldwide is 560 million tonnes. The larger blast furnaces are capable of producing 10 000 tonnes of iron per day (Figure 9.15).

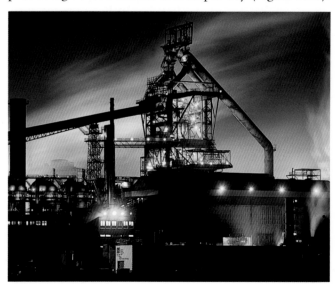

Figure 9.15 Redcar steelworks at Teesside, UK.

Zinc extraction

The principal source of zinc is zinc sulfide or zinc blende. This ore occurs mainly in Australia, Canada and the US. The zinc ore is first concentrated by a process called **froth flotation**. The crushed ore is fed into tanks of water containing a chemical frothing agent. Air is blown through so that the whole mixture froths up. The rock particles become soaked with water and sink to the bottom of the tank. However, the zinc sulfide particles, into which the water cannot soak, are carried to the top of the tank by the air bubbles and are skimmed off and dried. This ore now contains 55–75% of zinc sulfide. The zinc sulfide is then heated very strongly in a current of air in a furnace (Figure 9.16) to convert it to the oxide:

$$\text{zinc sulfide} + \text{oxygen} \rightarrow \text{zinc oxide} + \text{sulfur dioxide}$$
$$2ZnS(s) + 3O_2(g) \rightarrow 2ZnO(s) + 2SO_2(g)$$

The sulfur dioxide is a useful co-product and is used in the manufacture of sulfuric acid. The zinc oxide is mixed with powdered coke in a furnace and heated very strongly to a temperature of approximately 1400 °C. The zinc oxide is reduced by the coke to zinc:

$$\text{zinc oxide} + \text{coke (carbon)} \rightarrow \text{zinc} + \text{carbon monoxide}$$
$$ZnO(s) + C(s) \rightarrow Zn(g) + CO(g)$$

The mixture of zinc vapour and carbon monoxide passes through an outlet near the top of the furnace and the zinc metal cools and condenses. The heating costs of the furnace are reduced by burning the carbon monoxide which is produced.

Zinc is used in alloys such as brass. It is also used to galvanise steel and for electrodes in batteries.

Figure 9.16 Zinc is extracted in a furnace like this one.

Extraction of unreactive metals

Copper

Copper is quite a long way down the reactivity series. It can be found as the free metal element or 'native' in the US. It is principally extracted, however, from copper pyrites, $CuFeS_2$. The crushed ore is concentrated by froth flotation. A chemical known as a collector is added to an ore/water mixture and sticks to the surface of the copper pyrites particles, giving them a water-repellent coating. Detergent is added and air is blown into the mixture to make it froth. The copper pyrites particles are concentrated in the froth and can be removed easily. They are then roasted in a limited supply of air to ensure conversion of copper pyrites to copper(I) sulfide:

copper + oxygen → copper(I) + sulfur + iron(II)
pyrites   sulfide   dioxide   oxide
$$2CuFeS_2(s) + 4O_2(g) \rightarrow Cu_2S(s) + 3SO_2(g) + 2FeO(s)$$

Silicon is then added and the mixture is heated in the absence of air. The iron(II) oxide is converted into iron(II) silicate ($FeSiO_3$), which is run off. The remaining copper(I) sulfide is then reduced to copper by heating in a controlled amount of air.

copper(I) + oxygen → copper + sulfur
sulfide   dioxide
$$Cu_2S(s) + O_2(g) \rightarrow 2Cu(s) + SO_2(g)$$

Copper is then refined by electrolysis (Chapter 6, p. 88) to give a product which is at least 99.92% pure. The purified copper is easily drawn into wires (it is highly ductile), which makes it useful for electrical wiring. It is also used in alloys, such as bronze and brass (Figure 9.17). Copper is also used to make cooking utensils (Figure 9.4b, on p. 130), as well as water and central heating pipes and steam boilers (it is a good conductor of heat).

The annual production of copper 13.8 million tonnes is worldwide; 1.8 million tonnes of this total is from recycling copper.

Silver and gold

These are very unreactive metals. Silver exists mainly as silver sulfide, Ag_2S (silver glance). The extraction involves treatment of the pulverised ore with sodium cyanide. Zinc is then added to displace the silver from solution. The pure metal is obtained by electrolysis. Silver also exists to a small extent native in the Earth's crust. Gold is nearly always found in its native form (Figure 9.18). It is also obtained in significant amounts during both the electrolytic refining of copper and the extraction of lead.

Silver and gold, because of their resistance to corrosion, are used to make jewellery. Both of these metals are also used in the electronics industry because of their high electrical conductivity.

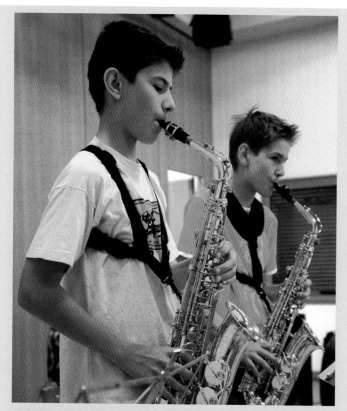

Figure 9.17 Copper is alloyed to produce brass.

Figure 9.18 These people are panning for gold.

Questions

1 How does the method used for extracting a metal from its ore depend on the metal's position in the reactivity series?

2 'It is true to say that almost all the reactions by which a metal is extracted from its ore are reduction reactions.' Discuss this statement with respect to the extraction of iron, aluminium and zinc.

3 Titanium is extracted from rutile (TiO_2). Use your research skills, including the Internet, to suggest the detail of the extraction process.

4 Write a word and symbol equation for the reduction of Cu(I) sulfide to copper by heating in air.

Recycling metals

Recycling has become commonplace in recent years (Figure 9.19). Why should we really want to recycle metals? Certainly, if we extract fewer metals from the Earth then the existing reserves will last that much longer. Also, recycling metals prevents the creation of a huge environmental problem (Figure 9.20). However, one of the main considerations is that it saves money.

The main metals which are recycled include aluminium and iron. Aluminium is saved by many households as drinks cans and milk bottle tops, to be melted down and recast. Iron is collected at community tips in the form of discarded household goods and it also forms a large part of the materials collected by scrap metal dealers. Iron is recycled to steel. Many steel-making furnaces run mainly on scrap iron.

Aluminium is especially easy to recycle at low cost. Recycling uses only 5% of the energy needed to extract the metal by electrolysis from bauxite. Approximately 60% of the European need for aluminium is obtained by recycling.

Figure 9.19 Aluminium can recycling.

Figure 9.20 If we did not recycle metals, then this sight would be commonplace.

Rusting of iron

After a period of time, objects made of iron or steel will become coated with rust. The rusting of iron is a serious problem and wastes enormous amounts of money each year. Estimates are difficult to make, but it is thought that upwards of £1 billion a year is spent worldwide on replacing iron and steel structures.

Rust is an orange–red powder consisting mainly of hydrated iron(III) oxide ($Fe_2O_3.xH_2O$). Both water and oxygen are essential for iron to rust, and if one of these two substances is not present then rusting will not take place. The rusting of iron is encouraged by salt. Figure 9.21 shows an experiment to show that oxygen (from the air) and water are needed for iron to rust.

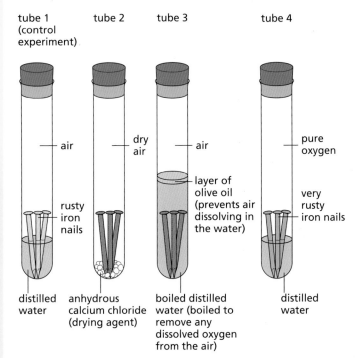

Figure 9.21 Rusting experiment with nails.

Rust prevention

To prevent iron rusting, it is necessary to stop oxygen (from the air) and water coming into contact with it. There are several ways of doing this.

Painting

Ships, lorries, cars, bridges and many other iron and steel structures are painted to prevent rusting (Figure 9.22). However, if the paint is scratched, the iron beneath it will start to rust (Figure 9.23) and corrosion can then spread under the paintwork which is still sound. This is why it is essential that the paint is kept in good condition and checked regularly.

Oiling/greasing

The iron and steel in the moving parts of machinery are coated with oil to prevent them from coming into contact with air or moisture. This is the most common way of protecting moving parts of machinery, but the protective film must be renewed.

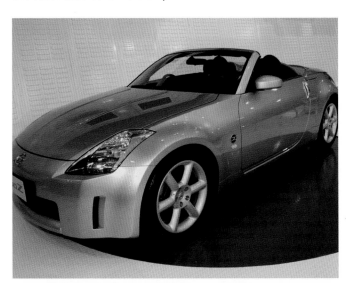

Figure 9.22 Painting keeps the air and water away from the steel used to build a ship.

Figure 9.23 A brand new car is protected against corrosion (top). However, if the paintwork is damaged, then rusting will result.

Coating with plastic

The exteriors of refrigerators, freezers and many other items are coated with plastic, such as PVC, to prevent the steel structure rusting (Figure 9.24).

Figure 9.24 A coating of plastic stops metal objects coming into contact with oxygen or water.

Plating

Cans for food can be made from steel coated with tin. The tin is deposited on to the steel used to make food cans by dipping the steel into molten tin. Some car bumpers, as well as bicycle handlebars, are electroplated with chromium to prevent rusting. The chromium gives a decorative finish as well as protecting the steel beneath.

Galvanising

Some steel girders, used in the construction of bridges and buildings, are **galvanised**. Coal bunkers and steel dustbins are also galvanised. This involves dipping the object into molten zinc (Figure 9.25). The thin layer of the more reactive zinc metal coating the steel object slowly corrodes and loses electrons to the iron, thereby protecting it. This process continues even when much of the layer of zinc has been scratched away, so the iron continues to be protected.

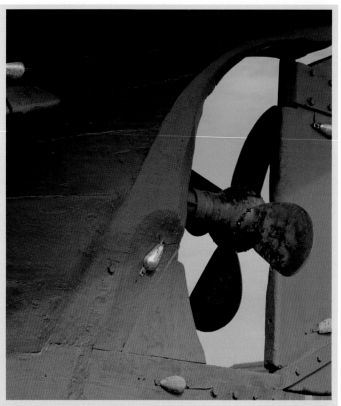

a Bars of zinc on the hull of a ship.

Figure 9.25 The Burnley Singing Ringing Tree is a sculpture made from galvanized tubes.

Sacrificial protection

Bars of zinc are attached to the hulls of ships and to oil rigs (as shown in Figure 9.26a). Zinc is above iron in the reactivity series and will react in preference to it and so is corroded. It forms positive ions more readily than the iron:

$$Zn(s) + Fe^{2+}(aq) \rightarrow Zn^{2+}(aq) + Fe(s)$$

As long as some of the zinc bars remain in contact with the iron structure, the structure will be protected from rusting. When the zinc runs out, it must be renewed. Gas and water pipes made of iron and steel are connected by a wire to blocks of magnesium to obtain the same result. In both cases, as the more reactive metal corrodes it loses electrons to the iron and so protects it (Figure 9.26b).

b The zinc is sacrificed to protect the steel. Electrons released from the dissolving zinc cause reduction to occur at the surface of the hull.

Figure 9.26 Sacrificial protection.

Corrosion

Rusting is the most common form of corrosion but this term is confined to iron and steel. **Corrosion** is the general name given to the process which takes place when metals and alloys are chemically attacked by oxygen, water or any other substances found in their immediate environment. The metals in the reactivity series will corrode to a greater or lesser extent. Generally, the higher the metal is in the reactivity series, the more rapidly it will corrode. If sodium and potassium were not stored under oil they would corrode very rapidly indeed. Magnesium, calcium and aluminium are usually covered by a thin coating of oxide after initial reaction with oxygen in the air. Freshly produced copper is pink in colour (Figure 6.19 on p. 88). However, on exposure to air, it soon turns brown due to the formation of copper(II) oxide on the surface of the metal.

In more exposed environments, copper roofs and pipes quickly become covered in verdigris. Verdigris is green in colour (Figure 9.27) and is composed of copper salts formed on copper. The composition of verdigris varies depending on the atmospheric conditions, but includes mixed copper(II) carbonate and copper(II) hydroxide ($CuCO_3.Cu(OH)_2$).

Gold and platinum are unreactive and do not corrode, even after thousands of years.

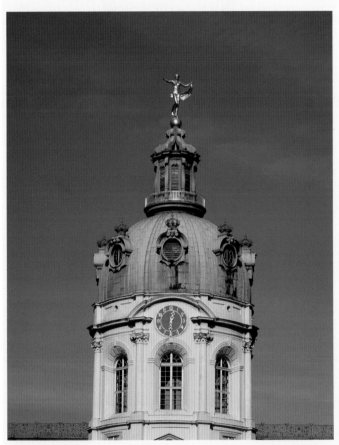

Figure 9.27 Verdigris soon covers copper roofs in exposed environments.

Questions

1 What is rust? Explain how rust forms on structures made of iron or steel.

2 Rusting is a redox reaction. Explain the process of rusting in terms of oxidation and reduction (Chapter 6, p. 82).

3 Design an experiment to help you decide whether steel rusts faster than iron.

4 Why do car exhausts rust faster than other structures made of steel?

Alloys

The majority of the metallic substances used today are **alloys**. Alloys are mixtures of two or more metals and are formed by mixing molten metals thoroughly. It is generally found that alloying produces a metallic substance that has more useful properties than the original pure metals it was made from. Steel, which is a mixture of the metal iron and the non-metal carbon, is also considered to be an alloy.

Of all the alloys we use, steel is perhaps the most important. Many steels have been produced; they contain not only iron but also carbon and other metals. For example, nickel and chromium are the added metals when stainless steel is produced (Figure 9.28). The chromium prevents the steel from rusting while the nickel makes it harder.

Figure 9.28 A stainless steel exhaust system. Why do you think more people are buying these exhaust systems?

Production of steel

The 'pig iron' obtained from the blast furnace contains between 3% and 5% of carbon and other impurities, such as sulfur, silicon and phosphorus. These impurities make the iron hard and brittle. In order to improve the quality of the metal, most of the impurities must be removed and in doing this, steel is produced.

The impurities are removed in the **basic oxygen process**, which is the most important of the steel-making processes. In this process:

- Molten pig iron from the blast furnace is poured into the basic oxygen furnace (Figure 9.29).
- A water-cooled 'lance' is introduced into the furnace and oxygen at 5–15 tm pressure is blown onto the surface of the molten metal.
- Carbon is oxidised to carbon monoxide and carbon dioxide, while sulfur is oxidised to sulfur dioxide. These escape as gases.
- Silicon and phosphorus are oxidised to silicon(IV) oxide and phosphorus pentoxide, which are solid oxides.
- Some calcium oxide (lime) is added to remove these solid oxides as slag. The slag may be skimmed or poured off the surface.
- Samples are continuously taken and checked for carbon content. When the required amount of carbon has been reached, the blast of oxygen is turned off.

The basic oxygen furnace can convert up to 300 tonnes of pig iron to steel per hour. Worldwide production by this process is 430 million tonnes.

There are various types of steel that differ only in their carbon content. The differing amounts of carbon present confer different properties on the steel and they are used for different purposes (Table 9.4). If other types of steel are required then up to 30% scrap steel is added, along with other metals (such as tungsten), and the carbon is burned off.

Steel recycling

The recycling of scrap steel contributes 310 million tonnes to the world supply of the alloy – 750 million tonnes. It has been calculated that the energy savings are equivalent to 160 million tonnes of coal. Also, it has been calculated that the raw materials conserved are equivalent to 200 million tonnes of iron ore.

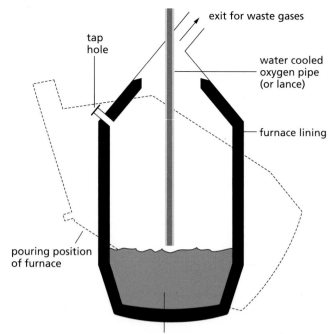

a A section through the basic oxygen furnace.

b Basic oxygen furnace.
Figure 9.29

Table 9.4 Different types of steel.

Steel	Typical composition	Properties	Uses
Mild steel	99.5% iron, 0.5% carbon	Easily worked Lost most of brittleness	Car bodies, large structures, machinery
Hard steel	99% iron, 1% carbon	Tough and brittle	Cutting tools, chisels, razor blades
Manganese steel	87% iron, 13% manganese	Tough, springy	Drill bits, springs
Stainless steel	74% iron, 18% chromium, 8% nickel	Tough, does not corrode	Cutlery, kitchen sinks, surgical instruments, chemical plant reaction vessels
Tungsten steel	95% iron, 5% tungsten	Tough, hard, even at high temperatures	Edges of high-speed cutting tools

Alloys to order

Just as the properties of iron can be changed by alloying, so the same can be done with other useful metals. Metallurgists have designed alloys to suit a wide variety of different uses. Many thousands of alloys are now made, with the majority being 'tailor-made' to do a particular job (Figure 9.30).

Table 9.5 shows some of the more common alloys, together with some of their uses.

b A polarised light micrograph of brass showing the distinct grain structure of this alloy.

a Bronze is often used in sculptures.

Figure 9.30

c Alloy structure. The dark circles represent atoms of a metal; the pale circles are the larger atoms of a different metal added to make the alloy. The different size of these atoms gives the alloy different physical properties from those of the pure metal.

Table 9.5 Uses of common alloys.

Alloy	Composition	Use
Brass	65% copper, 35% zinc	Jewellery, machine bearings, electrical connections, door furniture
Bronze	90% copper, 10% tin	Castings, machine parts
Cupro-nickel	30% copper, 70% nickel	Turbine blades
	75% copper, 25% nickel	Coinage metal
Duralumin	95% aluminium, 4% copper, 1% magnesium, manganese and iron	Aircraft construction, bicycle parts
Magnalium	70% aluminium, 30% magnesium	Aircraft construction
Pewter	30% lead, 70% tin, a small amount of antimony	Plates, ornaments and drinking mugs
Solder	70% lead, 30% tin	Connecting electrical wiring

Questions

1 Calcium oxide is a base. It combines with solid, acidic oxides in the basic oxygen furnace.
Write a chemical equation for one of these oxides reacting with the added lime.

2 'Many metals are more useful to us when mixed with some other elements.' Discuss this statement with respect to stainless steel.

■ *Checklist*

After studying Chapter 9 you should know and understand the following terms.

■ **Alloy**   Generally, a mixture of two or more metals (for example, brass is an alloy of zinc and copper) or a metal and a non-metal (for example, steel is an alloy of iron and carbon, sometimes with other metals included). They are formed by mixing the molten substances thoroughly. Generally, it is found that alloying produces a metallic substance which has more useful properties than the original pure metals it was made from.

■ **Amphoteric hydroxide**   A hydroxide which can behave as an acid (react with an alkali) or a base (react with an acid), for example zinc hydroxide.

■ **Blast furnace**   A furnace for smelting iron ores such as haematite (Fe_2O_3) and magnetite (Fe_3O_4) to produce pig (or cast) iron. In a modified form it can be used to extract metals such as zinc.

■ **Competition reactions**   Reactions in which metals compete for oxygen or anions. The more reactive metal:

- takes the oxygen from the oxide of a less reactive metal
- displaces the less reactive metal from a solution of that metal salt – this type of competition reaction is known as a displacement reaction.

■ **Corrosion**   The name given to the process that takes place when metals and alloys are chemically attacked by oxygen, water or any other substances found in their immediate environment.

■ **Metal extraction**   The method used to extract a metal from its ore depends on the position of the metal in the reactivity series.

- Reactive metals are usually difficult to extract. The preferred method is by electrolysis of the molten ore (electrolytic reduction); for example, sodium from molten sodium chloride.

- Moderately reactive metals (those near the middle of the reactivity series) are extracted using a chemical reducing agent (for example carbon) in a furnace; for example, iron from haematite in the blast furnace.
- Unreactive metals, for example gold and silver, occur in an uncombined (native) state as the free metal.

■ **Metal ion precipitation**   These are reactions in which certain metal cations form insoluble hydroxides. The colours of these insoluble hydroxides can be used to identify the metal cations which are present; for example, copper(II) hydroxide is a blue precipitate.

■ **Ore**   A naturally occurring mineral from which a metal can be extracted.

■ **Reactivity series of metals**   An order of reactivity, giving the most reactive metal first, based on results from experiments with oxygen, water and dilute hydrochloric acid.

■ **Recycling metals**   Metal drink cans such as those made of aluminium are collected in large 'banks' for the sole purpose of recycling them. Reusing the metal in this way saves money.

■ **Rust**   A loose, orange–brown, flaky layer of hydrated iron(III) oxide found on the surface of iron or steel. The conditions necessary for rusting to take place are the presence of oxygen and water. The rusting process is encouraged by other substances such as salt. It is an oxidation process.

■ **Rust prevention**   To prevent iron rusting it is necessary to stop oxygen and water coming into contact with it. The methods employed include painting, oiling/greasing, coating with plastic, plating, galvanising and sacrificial protection.

Metal extraction and chemical reactivity

Additional questions

1 Use the following list of metals to answer the questions **a** to **i**: iron, calcium, potassium, gold, aluminium, magnesium, sodium, zinc, platinum, titanium.
   a Which of the metals is found native?
   b Which of the metals is found in nature as the ore:
      (i) rock salt?
      (ii) rutile?
   c Which metal has a carbonate found in nature called chalk?
   d Which of the metals will not react with oxygen to form an oxide?
   e Which of the metals will react violently with cold water?
   f Choose one of the metals in your answer to **e** and write a balanced chemical equation for the reaction which takes place.
   g Which of the metals has a protective oxide coating on its surface?
   h Which of the metals reacts very slowly with cold water but extremely vigorously with steam?
   i Which of the metals is used to galvanise iron?

2

Copper
↓
heat in air (oxygen)
↓
Black solid **A**
↓
heat with powdered magnesium
↓
Brown–pink solid **B** +
white powder **C**
↓
dilute hydrochloric acid
↓
Brown–pink solid **B** +
colourless solution **D** + water
↓
filter and evaporate filtrate to dryness
↓
Solid **D**
↓
electrolysis of molten **D**
↓
Silvery metal **E** +
green gas **F**

   a Name and give the formulae of the substances **A** to **F**.

   b Write balanced chemical equations for the reactions in which:
      (i) black solid **A** was formed
      (ii) white powder **C** and brown–pink solid **B** were formed
      (iii) colourless solution **D** was formed.
   c The reaction between black solid **A** and magnesium is a redox reaction. With reference to this reaction, explain what you understand by this statement.
   d Write anode and cathode reactions for the processes which take place during the electrolysis of molten **D**.
   e Suggest a use for:
      (i) brown–pink solid **B**
      (ii) silvery metal **E**
      (iii) green gas **F**.

3 Explain the following:
   a metals such as gold and silver occur native in the Earth's crust
   b the parts of shipwrecks made of iron rust more slowly in deep water
   c zinc bars are attached to the structure of oil rigs to prevent them from rusting
   d copper roofs quickly become covered with a green coating when exposed to the atmosphere
   e recycling metals can save money.

4 Iron is extracted from its ores haematite and magnetite. Usually it is extracted from haematite (iron(III) oxide). The ore is mixed with limestone and coke and reduced to the metal in a blast furnace. The following is a brief outline of the reactions involved.

$$coke + oxygen \rightarrow gas\ X$$
$$gas\ X + coke \rightarrow gas\ Y$$
$$iron(III)\ oxide + gas\ Y \rightarrow iron + gas\ X$$

   a Name the gases **X** and **Y**.
   b Give a chemical test to identify gas **X**.
   c Write balanced chemical equations for the reactions shown above.
   d The added limestone is involved in the following reactions:

$$limestone \rightarrow calcium\ oxide + gas\ X$$
$$calcium\ oxide + silicon(IV)\ oxide \rightarrow slag$$

   (i) Give the chemical names for limestone and slag.
   (ii) Write balanced chemical equations for the reactions shown above.
   (iii) Why is the reaction between calcium oxide and silicon(IV) oxide called an acid–base reaction?
   (iv) Describe what happens to the liquid iron and slag when they reach the bottom of the furnace.
   e Why do you think that the furnace used in the extraction of iron is called a blast furnace?

continued

5 The iron obtained from the blast furnace is known as pig or cast iron. Because of the presence of impurities, such as carbon, it has a hard and brittle nature. Most of this type of iron is therefore converted into steel in the basic oxygen process. During this process either all or some of the carbon is removed. Calculated quantities of other elements are then added to produce the required type of steel.

a Explain the meaning of the term alloy as applied to steel.

b Name two impurities, other than carbon, which are present in cast iron and which are removed completely during the steel manufacture.

c Describe the method of steel manufacture used which removes the impurities referred to in b.

d Name two metallic elements which may be added to the basic oxygen furnace to produce different varieties of steel.

e Give two uses of stainless steel.

f Give two advantages of stainless steel compared with cast iron.

6 The metal titanium is extracted from its ore rutile (TiO_2).

a Describe the processes involved in the extraction of titanium from rutile.

b Name the area of the periodic table in which titanium is found.

c The final part of the process involves the following reaction:

$$TiCl_4(l) + 4Na(l) \rightarrow 4NaCl(s) + Ti(s)$$

(i) Calculate the quantity of titanium produced from 19 tonnes of titanium(IV) chloride ($TiCl_4$).
(ii) How much sodium would have to be used to extract the titanium from the 19 tonnes of $TiCl_4$ used in (i)?

d Give two uses of titanium metal.

7 Zinc can be reacted with steam using the apparatus shown below. When gas A is collected, mixed with air and ignited it gives a small pop. A white solid B remains in the test tube when the reaction has stopped and the apparatus cooled down.

a Name and give the formula of gas A.

b (i) Name the product formed when gas A burns in air.
(ii) Write a balanced chemical equation for this reaction.

c (i) Name white solid B.
(ii) Write a balanced chemical equation to represent the reaction between magnesium and steam.

d Name two other metals which could be safely used to replace zinc and produce another sample of gas A.

e When zinc reacts with dilute hydrochloric acid, gas A is produced again. Write a balanced chemical equation to represent this reaction and name the other product of this reaction.

8 Copper is extracted in the final stage by roasting copper(I) sulfide (Cu_2S) in a current of air. This converts the sulfide to the metal.

$$Cu_2S(s) + O_2(g) \rightarrow 2Cu(s) + SO_2(g)$$

a (i) What is the name of the chemical process by which copper(I) sulfide is converted to copper?
(ii) At what temperature does this process take place?

b It is a very expensive business to maintain the high temperature needed to efficiently convert copper(I) sulfide to copper. How are the heating costs offset?

c Calculate the quantity of copper(I) sulfide required to produce 1270 tonnes of copper. (A_r: O = 16; S = 32, Cu = 63.5)

d For use in electrical wiring, the copper must be 99.99% pure. How is the copper purified to obtain this high purity?

10 Atmosphere and oceans

The developing atmosphere

The structure of the atmosphere
The composition of the atmosphere

Fractional distillation of liquid air

Uses of the gases
Oxygen
Nitrogen
Noble gases

Resources in the oceans
Bromine
Magnesium
Sodium chloride

The water cycle

Pollution
Water pollution
Atmospheric pollution

Checklist

Additional questions

The developing atmosphere

About 4500 million years ago the planets in the solar system were formed. Each planet had a thick layer of gases, mainly hydrogen and helium, surrounding its core. This layer of gas is known as the **primary atmosphere** (Figure 10.1). Over a period of time intense solar activity caused these lighter gases to be removed from this primary atmosphere of planets nearest to the Sun. During this time the Earth was cooling to become a molten mass upon which a thin crust formed.

b Aurora borealis.

a Lightning flashes.

c A tornado or 'twister'.

Figure 10.1 The Earth's atmosphere formed over many millions of years and involved various atmospheric phenomena.

Volcanic activity through the crust pushed out huge quantities of gases, such as ammonia, nitrogen, methane, carbon monoxide, carbon dioxide and a small amount of sulfur dioxide, which formed a secondary atmosphere around the Earth (Figure 10.2a).

About 3800 million years ago, when the Earth had cooled below 100 °C, the water vapour in this secondary atmosphere condensed and fell as rain. This caused the formation of the first oceans, lakes and seas on the now rapidly cooling Earth. The structure of the surface of the Earth, as we know it today, has evolved as a result of the presence of these large expanses of water (Figure 10.2b). The oceans are even today an important reservoir for carbon dioxide gas, removing it from the atmosphere.

Eventually, early forms of life developed in these oceans, lakes and seas at depths which prevented potentially harmful ultraviolet light from the Sun affecting them. About 3000 million years ago the first forms of bacteria appeared. These were the predecessors of algae-like organisms which used the light from the Sun, via photosynthesis, to produce their own food, and oxygen was released into the atmosphere as a waste product. This process also acted to reduce the amount of CO_2 in the atmosphere.

The process of photosynthesis can be described by the following equation:

$$\text{carbon dioxide} + \text{water} \xrightarrow[\text{chlorophyll}]{\text{sunlight}} \text{glucose} + \text{oxygen}$$

$$6CO_2(g) + 6H_2O(l) \longrightarrow C_6H_{12}O_6(aq) + 6O_2(g)$$

The ultraviolet radiation broke down some of the oxygen molecules, which had been released, into oxygen atoms.

$$\text{oxygen molecules} \xrightarrow{\text{UV light}} \text{oxygen atoms}$$

$$O_2(g) \longrightarrow 2O(g)$$

Some of the highly reactive oxygen atoms reacted with molecules of oxygen to form ozone molecules, $O_3(g)$.

$$\text{oxygen atom} + \text{oxygen molecule} \rightarrow \text{ozone}$$

$$O(g) + O_2(g) \rightarrow O_3(g)$$

Ozone is an unstable molecule which readily decomposes, under the action of ultraviolet radiation, into single oxygen atoms and oxygen molecules.

$$\text{ozone} \xrightarrow{\text{UV light}} \text{oxygen atom} + \text{oxygen molecule}$$

$$O_3(g) \longrightarrow O(g) + O_2(g)$$

However, the single oxygen atoms would react quickly, re-forming ozone molecules.

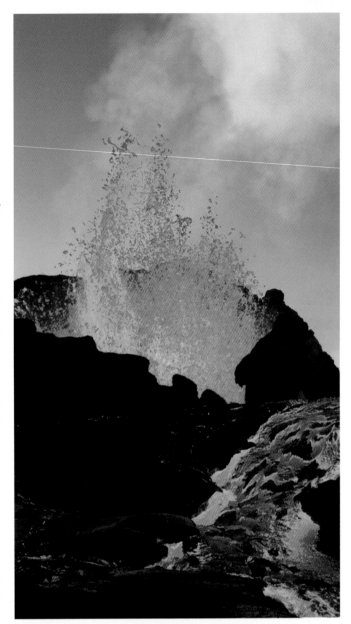

a Volcanic activity like this expelled gases through the crust to form a secondary atmosphere.

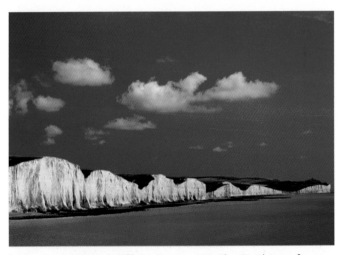

b The Seven Sisters cliffs in Sussex, UK. The Earth's surface has evolved as a result of lakes and other large expanses of water.

Figure 10.2

Ozone is an important gas in the atmosphere (stratosphere). It prevents harmful ultraviolet radiation from reaching the Earth. Over many millions of years the amount of ultraviolet radiation was reduced significantly. About 400 million years ago the first land plants appeared on the Earth, and so the amount of oxygen and hence ozone increased.

Oxygen is a reactive gas, and over millions of years organisms have adapted to make use of it. The oxygen from the atmosphere was used, along with the carbon they obtained from their food, to produce energy in a process known as **respiration**. The process of respiration can be represented as:

$$glucose \ + oxygen \rightarrow \begin{array}{c}carbon \\ dioxide\end{array} + \ water \ + energy$$

$$C_6H_{12}O_6(aq) \ + \ 6O_2(g) \rightarrow 6CO_2(g) + 6H_2O(l) \ + energy$$

Over recent years, scientists have become aware of a reduction in the amount of ozone in our atmosphere and of the formation of ozone holes (Figure 10.3). The reduction of ozone in our atmosphere will lead to an increased risk of skin cancer as more harmful ultraviolet radiation reaches the surface of the Earth.

EP/TOMS Total Ozone for Oct 11, 2000

GSFC/916

NASA

Dobson Units
Dark Gray < 100, Red > 500 DU

Figure 10.3 This diagram comes from NASA's ozone monitoring programme TOMS (Total Ozone Mapping Spectrometer). You can view diagrams similar to this at http://toms.gsfc.nasa.gov/. The ozone hole over the Antarctic (shown in purple and pink on the diagram) is largest in the Antarctic spring. Note: Dobson Units are a measure of the total amount of ozone in a vertical column from the ground to the top of the atmosphere.

Question

1 What precautions are necessary to prevent an increase in skin cancers?

The structure of the atmosphere

The gases in the atmosphere are held in an envelope around the Earth by its gravity. The atmosphere is 80 km thick (Figure 10.4) and it is divided into four layers:

* troposphere
* stratosphere
* mesosphere
* thermosphere.

About 75% of the mass of the atmosphere is found in the layer nearest the Earth called the troposphere. Beyond this layer the atmosphere reaches into space but becomes extremely thin beyond the mesosphere.

Figure 10.4 The Earth's atmosphere.

The composition of the atmosphere

If a sample of dry, unpolluted air was taken from any location in the troposphere and analysed, the composition by volume of the sample would be similar to that shown in Table 10.1.

Table 10.1 Composition of the atmosphere.

Component	%
Nitrogen	78.08
Oxygen	20.95
Argon	0.93
Carbon dioxide	0.04
Neon	0.002
Helium	0.000 5
Krypton	0.000 1
Xenon plus minute amounts of other gases	0.000 01

149

Measuring the percentage of oxygen in the air

When $100\,cm^3$ of air is passed backwards and forwards over heated copper turnings it is found that the amount of gas decreases (Figure 10.5). This is because the reactive part of the air, the oxygen gas, is reacting with the copper to form black copper(II) oxide (Figure 10.6). In such an experiment the volume of gas in the syringe decreases from $100\,cm^3$ to about $79\,cm^3$, showing that the air contained $21\,cm^3$ of oxygen gas. The percentage of oxygen gas in the air is:

$$\frac{21}{100} \times 100 = 21\%$$

The composition of the atmosphere is affected by the following factors:

- respiration
- photosynthesis
- volcanic activity
- radioactive decay, in which helium is formed
- human activity, involving burning of fossil fuels, in which carbon dioxide and water vapour are produced as well as other gases (p. 186).

Compare the components of our atmosphere with those of the other planets in the solar system (as shown in Table 10.2).

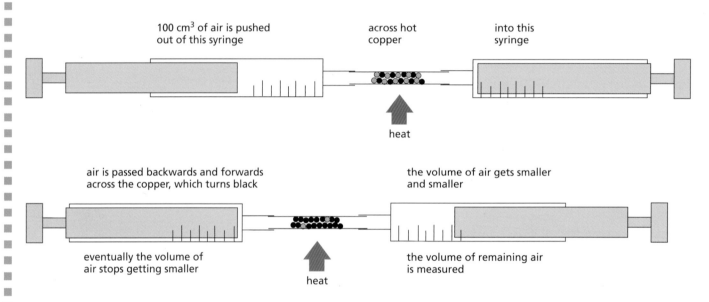

Figure 10.5 This apparatus can be used to find out the volume of oxygen gas in the air.

Figure 10.6 Copper turnings before and after reaction.

Table 10.2 Atmospheres of the other planets* in the solar system.

Planet	Atmosphere
Mercury	No atmosphere – the gases were burned off by the heat of the Sun
Venus	Carbon dioxide and sulfur dioxide
Mars	Mainly carbon dioxide
Jupiter	Ammonia, helium, hydrogen, methane
Saturn	Ammonia, helium, hydrogen, methane
Uranus	Ammonia, helium, hydrogen, methane
Neptune	Helium, hydrogen, methane

*Pluto is no longer recognised as a planet.

Questions

1 Draw a pie chart to show the data given in Table 10.1.

2 Is air a compound or a mixture? Explain your answer.

3 Design an experiment to find out how much oxygen there is in exhaled air.

Fractional distillation of liquid air

Air is the major source of oxygen, nitrogen and the noble gases. The gases are obtained by fractional distillation of liquid air but it is a complex process, involving several different steps (Figure 10.7).

Figure 10.7 Large amounts of gases are obtained from the fractional distillation of liquid air in industrial plants such as this.

- The air is passed through fine filters to remove dust.
- The air is cooled to about $-80\,°C$ to remove water vapour and carbon dioxide as solids. If these are not removed, then serious blockages of pipes can result.
- Next, the cold air is compressed to about $100\,atm$ of pressure. This warms up the air, so it is passed into a heat exchanger to cool it down again.
- The cold, compressed air is allowed to expand rapidly and that cools it still further.
- The process of compression followed by expansion is repeated until the air reaches a temperature below $-200\,°C$. At this temperature the majority of the air liquefies (Table 10.3).
- The liquid air is passed into a fractionating column and it is fractionally distilled. The gases can be separated because they have different boiling points.
- The gases are then stored separately in large tanks and cylinders.

Table 10.3 Boiling points of atmospheric gases.

Gas	Boiling point/°C
Helium	−269
Neon	−246
Nitrogen	−196
Argon	−186
Oxygen	−183
Krypton	−157
Xenon	−108

It should be noted that the noble gases neon, argon, krypton and xenon are obtained by this method; however, helium is more profitably obtained from natural gas.

Question

1 Use information given in the text to construct a flow chart to show the processes involved in the extraction of gases from air.

Uses of the gases

Oxygen

- Large quantities are used in industry to convert pig iron into steel (Chapter 9, p. 142) and for producing very hot flames for welding by mixing with gases such as ethyne (acetylene).
- It is used in hospitals to help patients with breathing difficulties (Figure 10.8).
- People such as mountaineers and divers use oxygen.
- It is carried in space rockets so that the hydrogen and kerosene fuels can burn.
- Space shuttles use oxygen gas in fuel cells which convert chemical energy into electrical energy.
- Astronauts must carry their own supply of oxygen, as do fire-fighters.
- It is used to restore life to polluted lakes and rivers and in the treatment of sewage.

Figure 10.8 Incubators have their own oxygen supply for babies.

Nitrogen

- Nitrogen is used in large quantities in the production of ammonia gas (Chapter 15, p. 219), which is used to produce nitric acid. Nitric acid is used in the manufacture of dyes, explosives and fertilisers.
- Liquid nitrogen is used as a refrigerant. Its low temperature makes it useful for freezing food quickly.
- Because of its unreactive nature, nitrogen is used as an inert atmosphere for some processes and chemical reactions. For example, empty oil tankers are filled with nitrogen to prevent fires.
- It is used in food packaging, for example in crisp packets, to keep the food fresh and in this case to prevent the crisps being compressed (Figure 10.9).

Figure 10.9 Inert nitrogen gas is used in food packaging.

Noble gases

Argon is used:

- to fill ordinary and long-life light bulbs to prevent the tungsten filament from reacting with oxygen in the air and forming the oxide (Figure 10.10)
- to provide an inert atmosphere in arc welding and in the production of titanium metal.

Neon is used:

- in advertising signs, because it glows red when electricity is passed through it
- in the helium–neon gas laser (Figure 10.11)
- in Geiger–Müller tubes, which are used for the detection of radioactivity.

Helium is used:

- to provide an inert atmosphere for welding
- as a coolant in nuclear reactors
- with 20% oxygen as a breathing gas for deep-sea divers
- to inflate the tyres of large aircraft
- to fill airships and weather balloons (Figure 10.12)
- in the helium–neon laser
- in low-temperature research, because of its low boiling point.

Krypton and xenon are used:

- in lamps used in photographic flash units, in stroboscopic lamps and in lamps used in lighthouses.

Figure 10.10 This long-life bulb contains argon.

Figure 10.11 A helium–neon laser used in eye surgery.

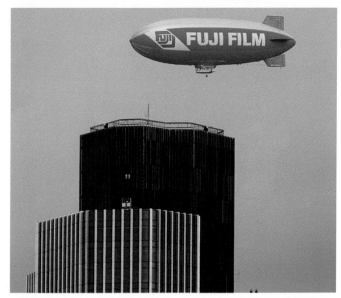

Figure 10.12 Helium is used to fill this airship as it has a low density and is unreactive.

Questions

1. How does oxygen help to restore life to polluted lakes?
2. Why is it important to have nitrogen in fertilisers?
3. Why is helium needed to produce an inert atmosphere for welding?

Resources in the oceans

Water covers at least 70% of the Earth's surface. In some hot, arid countries the sea is the main source of pure water for drinking. The water is obtained by a process known as desalination (Chapter 2, p. 21).

Sea water contains about 35 g of dissolved substances in each kilogram. A typical analysis of the varied elements present in sea water is shown in Table 10.4. Some elements, such as bromine, are extracted from sea water on a commercial basis.

Table 10.4 Elements present in sea water.

Element	Concentration/g dm^{-3} of sea water
Bromine	0.07
Calcium	0.4
Chlorine	19.2
Magnesium	1.3
Potassium	0.4
Sodium	10.7
Sulfur	0.9
Other elements	1.4

Bromine

Eighty per cent of the world's supply of bromine is extracted from sea water. One of the biggest factories for extracting bromine is to be found on the south-west shores of the Dead Sea in Israel (Figure 10.13). The Dead Sea is a particularly rich source of bromine as it contains 5.2 g of the element per dm^3 of sea water.

Worldwide 55 000 tonnes of bromine are harvested from the sea each year.

The extraction of bromine from sea water begins with the evaporation of the sea water using energy from the Sun, forming a more concentrated solution. The bromine is present in solution as bromide ions ($Br^-_{(aq)}$), which can be converted to bromine via a displacement reaction with chlorine gas. The process is as follows.

- Chlorine gas is bubbled through the warm, partially evaporated sea water and displacement takes place.

$$\text{chlorine} + \text{bromide} \rightarrow \text{bromine} + \text{chloride}$$
$$\text{gas} \qquad \text{ions} \qquad \text{gas} \qquad \text{ions}$$
$$Cl_2(g) + 2Br^-_{(aq)} \rightarrow Br_2(aq) + 2Cl^-_{(aq)}$$

- Steam is blown through the resulting solution and the volatile bromine is given off as bromine vapour mixed with steam.
- Bromine is not very soluble in water and therefore two separate layers form when the bromine and steam condense. The less dense water floats on top of the denser bromine layer. The bromine layer is then run off.
- The impure bromine is then purified by distillation and dried.

Half a tonne of chlorine is required for each tonne of bromine manufactured. Bromine is used in the manufacture of the fuel additive tetraethyllead(IV) (TEL), although to a decreasing extent. Bromine compounds are also used in photography (AgBr), medicines (KBr) and herbicides as well as in the manufacture of flame-retardant materials.

Figure 10.13 The Dead Sea is a particularly rich source of salts, including bromides from which bromine is extracted.

Magnesium

Magnesium can also be extracted from sea water. The sea water is first treated with calcium hydroxide, and magnesium hydroxide is precipitated and filtered off.

$$\text{magnesium} + \text{hydroxide} \rightarrow \text{magnesium}$$
$$\text{ions} \qquad \text{ions} \qquad \text{hydroxide}$$
$$Mg^{2+}(aq) + 2OH^-(aq) \rightarrow Mg(OH)_2(s)$$

a Magnesium (as the element) is used in alloys to build space shuttles and racing cycles.

b As a compound it is used in the manufacture of indigestion remedies and toothpaste.

Figure 10.14

154

The magnesium hydroxide is then dissolved in hydrochloric acid and the resulting solution is evaporated to produce magnesium chloride. This is then fused, at about 700 °C, with additives to lower its melting point and is electrolysed with a graphite anode and a steel cathode. The magnesium produced is liberated at the cathode and is 99.9% pure. Around 250 000 tonnes are produced worldwide each year.

$$\text{magnesium ions} + \text{electrons} \rightarrow \text{magnesium}$$
$$Mg^{2+}(aq) \qquad + \qquad 2e^- \qquad \rightarrow \qquad Mg(s)$$

Magnesium is used in alloys which are used in the space and aviation industries, and in racing cycles (Figure 10.14a). Compounds of magnesium ($Mg(OH)_2$ and $MgSO_4.7H_2O$) are also used in medicines, for example indigestion remedies (Figure 10.14b), and in toothpastes.

Sodium chloride

As you might guess, salt is the most abundant resource in sea water. There is about 25 g of sodium chloride per dm^3 of sea water. It is extracted in several areas of the world by evaporation, for example in France, Saudi Arabia and Australia.

The sea water is kept in shallow ponds until all the water has been evaporated by the heat of the Sun. The salt is then harvested.

Sodium chloride is used to flavour food, in the manufacture of sodium carbonate and sodium hydrogencarbonate and as the raw material for the chlor-alkali industry (Chapter 6, p. 86).

Questions

1 Use the information in the text and any other information you can obtain from other sources to construct a flow diagram to show the processes involved in the extraction of bromine from sea water.

2 In the extraction of magnesium from sea water, the magnesium hydroxide precipitate is treated with dilute hydrochloric acid. Write both a word and a balanced chemical equation to describe this reaction. In addition, write an electrode equation to show the production of chlorine gas at the anode.

The water cycle

Figure 10.15 illustrates the water cycle, which shows how water circulates around the Earth. The driving force behind the water cycle is the heat of the Sun.

- Heat from the Sun causes evaporation from oceans, seas and lakes. Water vapour is also formed from the evaporation of water from leaves (transpiration), through respiration and through combustion. The water vapour rises and cools, and condenses to form tiny droplets of water. These droplets form clouds.
- The clouds are moved along by air currents. As they cool, the tiny droplets join to form larger droplets, which fall as rain when they reach a certain size.
- The water that falls as rain runs into streams and rivers and then on into lakes, seas and oceans.

Question

1 Construct a simplified version of the water cycle using 'key words' in boxes and the 'processes involved' over linking arrows.

Pollution

The two major resources considered in this chapter, water and air, are essential to our way of life and our very existence. However, we are continually guilty of polluting these resources. We now look at the effects of the various sources of pollution and the methods used to control or eliminate them.

Water pollution

Water is very good at dissolving substances. It is, therefore, very unusual to find really pure water on this planet. As water falls through the atmosphere and down on to and through the surface of the Earth, it dissolves a tremendous variety of substances. Chemical fertilisers washed off surrounding land will add nitrate ions (NO_3^-) and phosphate ions (PO_4^{3-}) to the water, owing to the use of artificial fertilisers such as ammonium nitrate and ammonium phosphate. The nitrates encourage the growth of algae which eventually die and decay, removing oxygen from the water. It may also contain human waste as well as insoluble impurities such as grit and bacteria, and oil and lead 'dust' (to a decreasing extent) from the exhaust fumes of lorries and cars (Figure 10.16, overleaf).

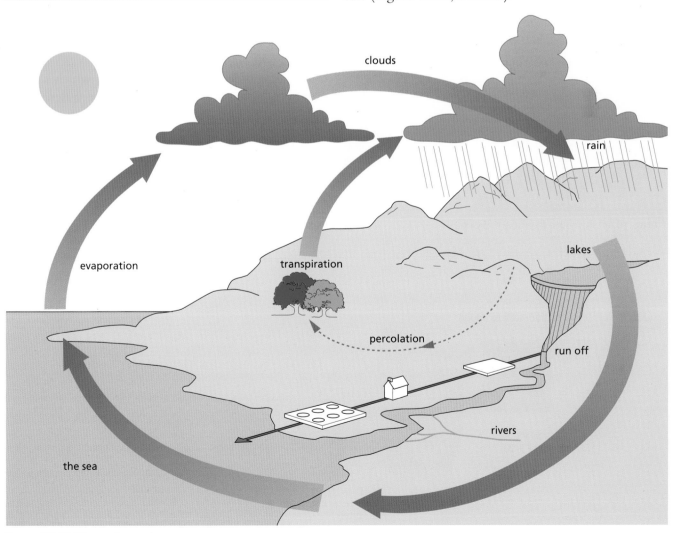

Figure 10.15 The water cycle.

Figure 10.16 A badly polluted river.

All these artificial as well as natural impurities must be removed from the water before it can be used (Figure 10.17). Recent European Union (EU) regulations have imposed strict guidelines on the amounts of various substances allowed in drinking water.

Figure 10.17 A water treatment works.

Most drinking water in the West is obtained from lakes and rivers where the pollution levels are low. Undesirable materials removed from water include:

- colloidal clay (clay particles in the water)
- bacteria
- chemicals which cause the water to be coloured and foul tasting
- acids, which are neutralised.

The process of water treatment involves both filtration and chlorination and is summarised in Figure 10.18.

1 Impure water is first passed through screens to filter out floating debris.
2 Aluminium sulfate is added to coagulate small particles of clay so that they form larger clumps, which settle more rapidly.
3 Filtration through coarse sand traps larger, insoluble particles. The sand also contains specially grown microbes which remove some of the bacteria.
4 A sedimentation tank has chemicals known as flocculants, for example aluminium sulfate, added to it to make the smaller particles (which remain in the water as colloidal clay) stick together and sink to the bottom of the tank.
5 These particles are removed by further filtration through fine sand. Sometimes a carbon slurry is used to remove unwanted tastes and odours and a lime slurry is used to adjust the acidity.
6 Finally, a little chlorine gas is added, which kills any remaining bacteria. This sterilises the water. Excess chlorine can be removed by the addition of sulphur dioxide gas. The addition of chlorine gas makes the water more acidic and so appropriate amounts of sodium hydroxide solution are added.

Fluoride is sometimes added to water if there is insufficient occurring naturally, as it helps to prevent tooth decay.

Figure 10.18 The processes involved in water treatment.

The 'iron problem'

If the acidity level of the treated water is not controlled then problems occur due to the precipitation of iron(III) hydroxide. These include:

- vegetables turning brown
- tea having an inky appearance and a bitter taste
- clothes showing rusty stains after washing.

Sewage treatment

After we have used water it must be treated again before it can be returned to rivers, lakes and seas. This multi-stage process known as sewage treatment is shown in Figure 10.19.

Figure 10.19 The processes involved in sewage treatment.

Used water, sewage, contains waste products such as human waste and washing-up debris as well as everything else that we put down a drain or sink. The processes that are involved are as follows.

1. Large screens remove large pieces of rubbish.
2. Sand and grit are separated in large sedimentation tanks. The process is speeded up by adding aluminium sulphate, which helps the solids to coagulate into larger particles that separate more rapidly. The sand and grit often contain large amounts of useful chemicals which, by the action of selected microbes, can be used as fertilisers.
3. The impure water is then removed and sent to a trickling filter, where it is allowed to drain through gravel on which microbes have been deposited. These kill off any remaining bacteria in the water by aerobic processes. This stage is known as biological filtration.
4. The treated water is then chlorinated and returned to a river, after checking.
5. Anaerobic bacteria digest what remains from the other stages. Methane gas is produced, which can be used as a fuel.

For a discussion of the uses of water see Chapter 8, p. 119.

Atmospheric pollution

Air pollution is all around us. Concentrations of gases in the atmosphere such as carbon monoxide, sulfur dioxide and nitrogen oxides are increasing with the increasing population. As the population rises there is a consequent increase in the need for energy, industries and motor vehicles. These gases are produced primarily from the combustion of the fossil fuels coal, oil and gas, but they are also produced by the smoking of cigarettes.

Motor vehicles are responsible for much of the air pollution in large towns and cities. They produce four particularly harmful pollutants:

- carbon monoxide
- sulfur dioxide
- hydrocarbons
- oxides of nitrogen.

Nitrogen monoxide is formed by the reaction of nitrogen and oxygen inside the internal combustion engine.

$$\text{nitrogen} + \text{oxygen} \rightarrow \text{nitrogen monoxide}$$
$$N_2(g) + O_2(g) \rightarrow 2NO(g)$$

The nitrogen monoxide, emitted as an exhaust gas, reacts with oxygen from the air to form the brown acidic gas nitrogen(IV) oxide (nitrogen dioxide).

$$\text{nitrogen monoxide} + \text{oxygen} \rightarrow \text{nitrogen(IV) oxide}$$
$$2NO(g) + O_2(g) \rightarrow 2NO_2(g)$$

Concern about pollution due to cars has led to the introduction of strict regulations by the EU and now all new cars must have a device known as a catalytic converter fitted to eliminate the production of some of these gases.

The catalytic converter acts as a device to speed up reactions which involve the pollutant gases, converting them to less harmful products, such as nitrogen and carbon dioxide. It should be noted that catalytic converters can only be used with unleaded petrol as the lead 'poisons' the catalyst, preventing it from catalysing the reactions. For a further discussion of catalytic converters see Chapter 11, p. 167.

Another method that has been introduced to reduce the amount of pollutants is that of the 'lean burn' engine. Although this type of engine reduces the amounts of carbon monoxide and oxides of nitrogen produced, it actually increases the amount of hydrocarbons in the exhaust gases.

A further method of regulating pollutant gases is to convert petrol-burning engines to LPG (liquid petroleum gas) engines, whilst retaining the ability to burn petrol. These cars are known as dual-fuel cars. Much research is directed towards an efficient electric motor that can be fitted in place of the petrol engine in a car. This development is moving forward at a pace.

Internationally, the use of lead compounds as additives remains an issue because of the continued use of leaded petrol. Perhaps of greater concern globally will be the increase of **particulates** in the air (Figure 10.20). ('Particulates' is a general term used to describe very small particles in the atmosphere, such as certain types of smoke emitted from diesel engines, as well as dust.)

Figure 10.20 Particulates produced by diesel engines are a real health problem worldwide.

Heavy industry (Figure 10.21) and power stations are major sources of sulfur dioxide, formed by the combustion of coal, oil and gas, which contain small amounts of sulfur.

$$\text{sulfur} + \text{oxygen} \rightarrow \text{sulfur dioxide}$$
$$S_{(s)} + O_{2(g)} \rightarrow SO_{2(g)}$$

This sulfur dioxide gas dissolves in rainwater to form the weak acid, sulfurous acid (H_2SO_3).

$$\text{sulfur dioxide} + \text{water} \rightleftharpoons \text{sulfurous acid}$$
$$SO_{2(g)} + H_2O_{(l)} \rightleftharpoons H_2SO_{3(aq)}$$

A further reaction occurs in which the sulfurous acid is oxidised to sulfuric acid.

Solutions of these acids are the principal contributors to acid rain. For a further discussion of acid rain, see Chapter 16, p. 228.

Units called flue gas desulfurisation (FGD) units are being fitted to some power stations to prevent the emission of sulfur dioxide gas. Here, the sulfur dioxide gas is removed from the waste gases by passing it through calcium hydroxide slurry. This not only removes the sulfur dioxide but also creates calcium sulfate, which can be sold to produce plasterboard. The FGD units are very expensive and therefore the sale of the calcium sulfate is an important economic part of the process.

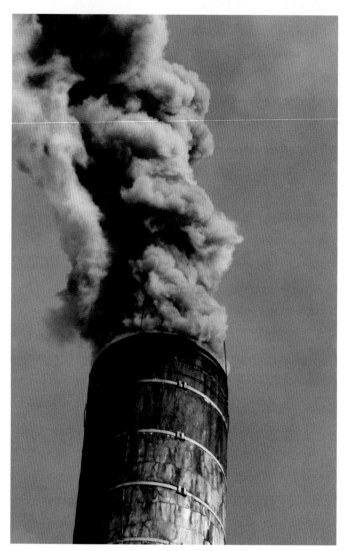

Figure 10.21 Sulfur dioxide is a major pollutant produced by industry.

Questions

1 Make a list of four major water pollutants and explain where they come from. What damage can these pollutants do?

2 Write a balanced chemical equation to represent the reaction which takes place between sulfur dioxide and calcium hydroxide slurry in the FGD unit of a power station.

3 In the treatment of water for public use, state the purpose of the addition of:
 a aluminium sulfate
 b chlorine
 c sodium hydroxide
 d sulfur dioxide.

4 Many industries use water as a coolant. Suggest the sorts of problems that may be created by this 'thermal pollution'.

5 It has been found that air pollution in cities damages lung development in children. Use the information in the text and any other information to make a list of the pollutants responsible for this problem.

Checklist

After studying Chapter 10 you should know and understand the following terms.

■ **Fractional distillation of air**   The process to extract individual gases from the air. Air is a major raw material. The mixture of gases is separated by first liquefying the mixture at low temperature and high pressure. The temperature is then allowed to rise and the gases collected as they boil off. The gases so produced have many and varied uses.

■ **Mesosphere**   A section of the atmosphere above the stratosphere in which the temperature falls with increasing height.

■ **Ozone (trioxygen)**   A colourless gas produced in the stratosphere by the action of high-energy ultraviolet radiation on oxygen gas, producing oxygen atoms. These oxygen atoms then react with further oxygen molecules to produce ozone. Its presence in the stratosphere acts as a screen (ozone layer) against dangerous ultraviolet radiation.

■ **Particulates**   Very small particles found in the atmosphere.

■ **Pollution**   The modification of the environment caused by human influence. It often renders the environment harmful and unpleasant to life. Water pollution is caused by many substances, such as those found in fertilisers and in industrial effluent. Atmospheric pollution is caused by gases such as sulfur dioxide, carbon monoxide and nitrogen oxides being released into the atmosphere by a variety of industries and also by the burning of fossil fuels.

■ **Primary atmosphere**   The original thick layer of gases, mainly hydrogen and helium, that surrounded the Earth's core soon after the planet was formed 4500 million years ago.

■ **Sea water as a resource**   Sea water is a major resource. By evaporation (desalination) we can produce water for drinking or irrigation. Also, bromine, magnesium and salt are extracted in large quantities from sea water.

■ **Secondary atmosphere**   A layer of a mixture of gases created by early volcanic activity. The mixture that formed this atmosphere included ammonia, nitrogen, methane, carbon monoxide, carbon dioxide and sulfur dioxide gases.

■ **Stratosphere**   A layer of the atmosphere above the troposphere in which the temperature increases with increasing height.

■ **Troposphere**   A layer of the atmosphere closest to the Earth which contains about 75% of the mass of the atmosphere. The composition of dry air is relatively constant in this layer of the atmosphere.

■ **Water cycle**   This cycle shows how water circulates around the Earth. The driving force behind the water cycle is the Sun.

Atmosphere and oceans
Additional questions

1 The apparatus shown on p. 150, Figure 10.5, was used to estimate the proportion of oxygen in the atmosphere. A volume of dry air (200 cm³) was passed backwards and forwards over heated copper until no further change in volume took place. The apparatus was then allowed to cool down to room temperature and the final volume reading was then taken. Some typical results are shown below.

Volume of gas before = 200 cm³
Volume of gas after = 157 cm³

During the experiment the copper slowly turned black.

a Why was the apparatus allowed to cool back to room temperature before the final volume reading was taken?

b (i) Using the information given above, calculate the volume reduction which has taken place.
(ii) Calculate the percentage reduction in volume.

c Explain briefly why there is a change in volume.

d What observation given above supports your explanation in **c**? Write a balanced chemical equation for any reaction which has occurred.

e Give the name of the main residual gas at the end of the experiment.

f Would you expect the copper to have increased or decreased in mass during the experiment? Explain your answer.

2 Explain the following.
a Air is a mixture of elements and compounds.
b The percentage of carbon dioxide in the atmosphere does not significantly vary from 0.03%.
c When liquid air has its temperature slowly raised from −270°C, helium is the first gas to boil off.
d Power stations are thought to be a major cause of acid rain.

3 Air is a raw material from which several useful substances can be separated. They are separated in the following process.
Dry and 'carbon dioxide free' air is cooled under pressure. Most of the gases liquefy as the temperature falls below −200°C. The liquid mixture is separated by fractional distillation. The boiling points of the gases left in the air after removal of water vapour and carbon dioxide are given in the table below:

Gas	Boiling point/°C
Argon	−186
Helium	−269
Krypton	−157
Neon	−246
Nitrogen	−196
Oxygen	−183
Xenon	−108

a Why is the air dried and carbon dioxide removed before it is liquefied?
b Which of the gases will not become liquid at −200°C?
c Which of the substances in the liquid mixture will be the first to change from liquid to gas as the temperature is slowly increased?
d Give a use for each of the gases shown in the table.
e Use the data given in Table 10.1, p. 149, to calculate the volume of each of the gases found in 1 dm³ of air.

4 Explain what is meant by the term 'pollution' with reference to air and water.
a (i) Name an air pollutant produced by the burning of coal.
(ii) Name a different air pollutant produced by the combustion of petrol in a car engine.
b Some of our drinking water is obtained by purifying river water.
(i) Would distillation or filtration produce the purest water from river water? Give a reason for your answer.
(ii) Which process, distillation or filtration, is actually used to produce drinking water from river water? Comment on your answer in comparison to your answer in **c** (i).
c Power stations produce warm water. This causes thermal pollution as this warm water is pumped into nearby rivers.
(i) Why do power stations produce such large quantities of warm water?
(ii) What effect does this warm water have on aquatic life?

5 In the final stage of the extraction of magnesium from sea water, molten magnesium chloride is electrolysed. Substances are added to bring the working temperature of the electrolysis cell down to 700°C.
a Write equations to represent the reactions taking place at the cathode and anode. State clearly whether oxidation or reduction is taking place.
b Why are substances added to the molten magnesium chloride before electrolysis takes place?
c The industrial production of magnesium by electrolysis uses a current of 14 000 amperes. Calculate the time required to produce 50 kg of magnesium from molten magnesium chloride.
(1 mole of electrons = 96 500 coulombs; A_r: Mg = 24)
d A large proportion of the magnesium produced (43%) is used to make alloys.
(i) Why has it such a use in the production of magnesium alloys?
(ii) The world production of magnesium is approximately 240 000 tonnes. Calculate the number of tonnes used in the production of magnesium alloys.

6 The developed countries produce about 230 000 tonnes of bromine per year: 80% of this production is from sea water.
   a Calculate the amount of bromine obtained in developed countries from sea water.
   b Sea water contains 0.07 g of bromine per dm^3. Using your answer to a, calculate the volume of sea water required to obtain that amount of bromine.
   c The main reaction in the extraction process is that involving displacement of bromine using chlorine gas.
      (i) Write an ionic equation for this reaction.
      (ii) Explain why you could not use iodine instead of chlorine in the displacement reaction.
   d Chlorine and bromine are hazardous substances. Describe some of the precautions that have to be taken to ensure the safety of members of the workforce who deal with these two substances.

7 France obtains some of its sodium chloride by evaporation of sea water.
   a If France produces 1.1 million tonnes of salt per year by this method and sea water contains 25 g of sodium chloride per dm^3, calculate the volume of sea water required to produce the annual salt production.
   b Give four important uses of sodium chloride.

   c Sodium chloride is an ionic substance.
      (i) Draw a diagram to show the bonding which takes place within sodium chloride.
      (ii) What are the properties of ionic substances such as sodium chloride?

8 In plants, during photosynthesis, carbon dioxide and water are converted into carbohydrates such as glucose ($C_6H_{12}O_6$), and oxygen is released.
   a Write a balanced chemical equation for the reaction taking place during photosynthesis.
   b What conditions are essential for this reaction to take place?
   c Why are animals unable to photosynthesise?
   d Which process occurs if we reverse photosynthesis?
   e Some of the oxygen released during photosynthesis is broken up by ultraviolet radiation. Some of the oxygen atoms produced combine with further oxygen molecules to produce an important allotrope of oxygen.
      (i) Name and give the formula of this allotrope of oxygen.
      (ii) Write a balanced chemical equation for the reaction in which this allotrope is produced.
      (iii) Why is this allotrope of oxygen so important to us?

11 *Rates of reaction*

Factors that affect the rate of a reaction
Surface area
Concentration
Temperature
Light
Catalysts

Enzymes

Checklist

Additional questions

Figure 11.1 shows some slow and fast reactions. The two photographs on the left show examples of slow reactions. The ripening of apples takes place over a number of weeks, and the making and maturing of cheese may take months. The burning of solid fuels, such as coal, can be said to involve chemical reactions taking place at a medium speed or rate. The other example shows a fast reaction. The chemicals inside explosives, such as TNT, react very rapidly in reactions which are over in seconds or fractions of seconds.

As new techniques have been developed, the processes used within the chemical industry have become more complex. Therefore, chemists and chemical engineers have increasingly looked for ways to control the rates at which chemical reactions take place. In doing so, they have discovered that there are five main ways in which you can alter the rate of a chemical reaction. These ideas are not only incredibly useful to industry but can also be applied to reactions which occur in the school laboratory.

Figure 11.1 Some slow (ripening fruit and cheese making), medium (coal fire) and fast (explosion) reactions.

Factors that affect the rate of a reaction

- Surface area of the reactants.
- Concentration of the reactants and gas pressure.
- Temperature at which the reaction is carried out.
- Light.
- Use of a catalyst, including enzymes.

Surface area

In Chapter 8, we discussed the use of limestone (calcium carbonate) as a substance which can be used to neutralise soil acidity. Powdered limestone is used as it neutralises the acidity faster than if lumps of limestone are used. Why do you think this is the case?

In the laboratory, the reaction between acid and limestone in the form of lumps or powder can be observed in a simple test-tube experiment. Figure 11.2 shows the reaction between dilute hydrochloric acid and limestone in lump and powdered form.

$$\begin{array}{ccccccc}
\text{hydrochloric} & + & \text{calcium} & \rightarrow & \text{calcium} & + & \text{carbon} & + & \text{water} \\
\text{acid} & & \text{carbonate} & & \text{chloride} & & \text{dioxide} \\
\end{array}$$

$$2HCl_{(aq)} + CaCO_{3(s)} \rightarrow CaCl_{2(aq)} + CO_{2(g)} + H_2O_{(l)}$$

The rates at which the two reactions occur can be found by measuring either:

- the volume of the carbon dioxide gas which is produced, or
- the loss in mass of the reaction mixture with time.

These two methods are generally used for measuring the rate of reaction for processes involving the formation of a gas as one of the products.

Figure 11.2 The powdered limestone (left) reacts faster with the acid than the limestone in the form of lumps.

The apparatus shown in Figure 11.3 is used to measure the loss in mass of the reaction mixture. The mass of the conical flask plus the reaction mixture is measured at regular intervals. The total loss in mass is calculated for each reading of the balance, and this is plotted against time. Some sample results from experiments of this kind have been plotted in Figure 11.4 (overleaf).

Figure 11.3 After 60 seconds the mass has fallen by 1.24 g.

Figure 11.4 Sample results for the limestone/acid experiment.

The reaction is generally at its fastest in the first minute. This is indicated by the slopes of the curves during this time. The steeper the slope, the faster the rate of reaction. You can see from the two traces in Figure 11.4 that the rate of reaction is greater with the powdered limestone than the lump form.

The surface area has been increased by powdering the limestone (Figure 11.5). The acid particles now have an increased amount of surface of limestone with which to collide. The products of a reaction are formed when collisions occur between reactant particles. Therefore, the increase in surface area of the limestone increases the rate of reaction.

Figure 11.5 A powder has a larger surface area.

In certain industries the large surface area of fine powders and dusts can be a problem. For example, there is a risk of explosion in flourmills and mines, where the large surface area of the flour or coal dust can – and has – resulted in explosions through a reaction with oxygen gas in the air when a spark has been created by machinery or the workforce (Figure 11.6).

Figure 11.6 The dust created by this cement plant in India is a potential hazard.

Questions

1 What apparatus would you use to measure the rate of reaction of limestone with dilute hydrochloric acid by measuring the volume of carbon dioxide produced?

2 The following results were obtained from an experiment of the type you were asked to design in question 1.

Time/min	0	0.5	1.0	1.5	2.0	2.5	3.0	3.5	4.0	4.5	5.0
Total volume of CO_2 gas/cm^3	0	15	24	28	31	33	35	35	35	35	35

a Plot a graph of the total volume of CO_2 against time.
b At which point is the rate of reaction fastest?
c What volume of CO_2 was produced after 1 minute 15 seconds?
d How long did it take to produce 30 cm^3 of CO_2?

Concentration

A yellow precipitate is produced in the reaction between sodium thiosulfate and hydrochloric acid.

sodium + hydrochloric → sodium + sulfur + sulfur + water
thiosulfate   acid   chloride   dioxide

$$Na_2S_2O_3(aq) + 2HCl(aq) \rightarrow 2NaCl(aq) + S(s) + SO_2(g) + H_2O(l)$$

The rate of this reaction can be followed by recording the time taken for a given amount of sulfur to be precipitated. This can be done by placing a conical flask containing the reaction mixture on to a cross on a piece of paper (Figure 11.7). As the precipitate of sulfur forms, the cross is obscured and finally disappears from view. The time taken for this to occur is a measure of the rate of this reaction. To obtain sufficient information about the effect of changing the concentration of the reactants, several experiments of this type must be carried out, using different concentrations of sodium thiosulfate or hydrochloric acid.

Figure 11.7 The precipitate of sulfur obscures the cross.

Some sample results of experiments of this kind have been plotted in Figure 11.8. You will note from the graph that when the most concentrated sodium thiosulfate solution was used, the reaction was at its fastest. This is shown by the shortest time taken for the cross to be obscured.

As discussed earlier, the products of the reaction are formed as a result of the collisions between reactant particles. There are more particles in a more concentrated solution and the collision rate is higher. The more often the particles collide, the greater the chance they have of reacting. This means that the rate of a chemical reaction will increase if the concentration of reactants is increased, because there are more particles per unit volume.

In reactions involving only gases, for example the Haber process (Chapter 15, p. 218), an increase in the overall pressure at which the reaction is carried out increases the rate of the reaction. The increase in pressure results in the gas particles being pushed closer together. This means that they collide more often and so react faster.

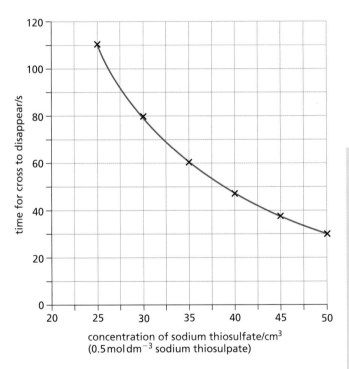

Figure 11.8 Sample data for the sodium thiosulfate/acid experiment at different concentrations of sodium thiosulfate.

Question

1 Devise an experiment to show the effect of changing the concentration of dilute acid on the rate of reaction between magnesium and hydrochloric acid.

Temperature

Why do you think food is stored in a refrigerator? The reason is that the rate of decay is slower at lower temperatures. This is a general feature of the majority of chemical processes.

The reaction between sodium thiosulfate and hydrochloric acid can also be used to study the effect of temperature on the rate of a reaction. Figure 11.9 shows some sample results of experiments with sodium thiosulfate and hydrochloric acid (at fixed concentrations) carried out at different temperatures. You can see from the graph that the rate of the reaction is fastest at high temperatures.

Figure 11.9 Sample data for the sodium thiosulfate/acid experiment at different temperatures.

When the temperature at which the reaction is carried out is increased, the kinetic energy that the particles have also increases – the particles move faster. This increases the collision rate of sodium thiosulfate and hydrochloric acid particles. The collision theory assumes that before two chemicals can react together their particles must collide with sufficient energy to overcome the **activation energy** for the reaction to form products. The activation energy is the energy which reactants must overcome, when their particles collide, to react successfully and form products (Figure 11.10). Therefore, if the temperature at which a reaction takes place is increased then the rate of reaction will increase.

Figure 11.10 Energy level diagram showing activation energy.

Questions

1 Explain why potatoes cooked in oil cook faster than those cooked in water.

2 Devise an experiment to study the effect of temperature on the reaction between magnesium and hydrochloric acid.

3 Explain why food cooks faster in a pressure cooker.

Light

Some chemical reactions are affected by light. Photosynthesis is a very important reaction (Chapter 8, p. 122) which occurs only when sunlight falls on leaves containing the green pigment chlorophyll. The chlorophyll acts as a catalyst and the rate of photosynthesis depends on the intensity of the light. Another chemical reaction that takes place only in light is that which occurs in photographic film. This is a transparent plastic strip coated with emulsion: a layer of gelatin throughout which are spread many millions of tiny crystals of silver halides, in particular, silver bromide (AgBr). The emulsion used is similar for both black-and-white and colour film.

When light hits a silver bromide crystal, silver cations (Ag^+) accept an electron from the bromide ions (Br^-) so silver atoms and bromine atoms are produced in the emulsion.

$$\text{silver ion} + \text{electron} \rightarrow \text{silver atom}$$
$$Ag^+ \quad + \quad e^- \quad \rightarrow \quad Ag$$

The more light that falls on the film, the greater the amount of silver deposited. Places where the most silver is deposited are darkest when the negative is developed and lightest on the photographic print.

Question

1 Devise an experiment to show how sunlight affects the rate of formation of silver from the silver salts, silver chloride and silver bromide.

Catalysts

Over 90% of industrial processes use **catalysts**. A catalyst is a substance which can alter the rate of a reaction without being chemically changed itself. In the laboratory, the effect of a catalyst can be observed using the decomposition of hydrogen peroxide as an example.

$$\text{hydrogen peroxide} \rightarrow \text{water} + \text{oxygen}$$
$$2H_2O_{2(aq)} \rightarrow 2H_2O_{(l)} + O_{2(g)}$$

The rate of decomposition at room temperature is very slow. There are substances, however, which will speed up this reaction, one being manganese(IV) oxide. When black manganese(IV) oxide powder is added to hydrogen peroxide solution, oxygen is produced rapidly. The rate at which this occurs can be followed by measuring the volume of oxygen gas produced with time.

Some sample results from experiments of this type have been plotted in Figure 11.11. At the end of the reaction, the manganese(IV) oxide can be filtered off and used again. The reaction can proceed even faster by increasing the amount and surface area of the catalyst. This is because the activity of a catalyst involves its surface. Note that, in gaseous reactions, if dirt or impurities are present on the surface of the catalyst, it will not act as efficiently; it is said to have been 'poisoned'. Therefore, the gaseous reactants must be pure.

Figure 11.11 Sample data for differing amounts of MnO_2 catalyst.

Chemists have found that:

- a small amount of catalyst will produce a large amount of chemical change
- catalysts remain unchanged chemically after a reaction has taken place, but they can change physically. For example, a finer manganese(IV) oxide powder is left behind after the decomposition of hydrogen peroxide
- catalysts are specific to a particular chemical reaction.

Some examples of chemical processes and the catalysts used are shown in Table 11.1.

Table 11.1 Examples of catalysts.

Process	Catalyst
Haber process – for the manufacture of ammonia	Iron
Contact process – for the manufacture of sulfuric acid	Vanadium(V) oxide
Oxidation of ammonia to give nitric acid	Platinum
Hydrogenation of unsaturated oils to form fats in the manufacture of margarines	Nickel

A catalyst increases the rate by providing an alternative reaction path with a lower activation energy. Those particles with sufficient kinetic energy to overcome the activation energy barrier can react. If the barrier is lowered by using a catalyst then, on collision, more particles will go on to react at a given temperature.

Catalytic converters

In the previous chapter you saw that European regulations state that all new cars have to be fitted with catalytic converters as part of their exhaust system (Figure 11.12). Car exhaust fumes contain pollutant gases such as carbon monoxide (CO) formed from the incomplete combustion of hydrocarbons in the fuel, and nitrogen(II) oxide (NO) formed by the reaction of nitrogen gas and oxygen gas from the air. The following reactions proceed of their own accord but very slowly under the conditions inside an exhaust.

carbon monoxide + oxygen → carbon dioxide
$$2CO(g) + O_2(g) \rightarrow 2CO_2(g)$$

nitrogen(II) + carbon → nitrogen + carbon
oxide   monoxide   dioxide
$$2NO(g) + 2CO(g) \rightarrow N_2(g) + 2CO_2(g)$$

The catalyst in the converter speeds up these reactions considerably. In these reactions, the pollutants are converted to carbon dioxide and nitrogen, which are naturally present in the air. The removal of oxides of nitrogen is important because they cause respiratory disease. They are also involved in the production of photochemical smogs (Figure 11.13) which occur worldwide in major cities, especially in the summer. It should be noted, however, that the catalytic converter can only be used with unleaded petrol and that, due to impurities being deposited on the surface of the catalyst, it becomes poisoned and has to be replaced every five or six years.

Figure 11.13 The haze is due to pollution caused mainly by cars without catalyst exhaust systems.

Questions

1 Using a catalysed reaction of your choice, devise an experiment to follow the progress of the reaction and determine how effective the catalyst is.

2 Why do some people consider catalytic converters not to be as environmentally friendly as suggested in their advertising material?

3 Unreacted hydrocarbons such as octane, C_8H_{18} (from petrol), also form part of the exhaust gases. These gases are oxidised in the converter to carbon dioxide and water vapour. Write an equation for the oxidation of octane.

a Catalytic converter.

b A section through a catalytic converter.

Figure 11.12

Enzymes

Enzymes are protein molecules produced in living cells. They are catalysts which speed up hundreds of different chemical reactions going on inside living cells. These biological catalysts are very specific in that each chemical reaction has a different enzyme catalyst. There are literally hundreds of different kinds of enzyme. Enzymes all have an active site. The active site is a particular shape and locks into a corresponding shape in a reactant molecule. When this has happened, the enzyme can work to break up the reactant (Figure 11.14).

For example, hydrogen peroxide is produced within our bodies. However, it is extremely damaging and must be decomposed very rapidly. Catalase is the enzyme which converts hydrogen peroxide into harmless water and oxygen within our livers:

$$\text{hydrogen peroxide} \xrightarrow{\text{catalase}} \text{water} + \text{oxygen}$$

$$2H_2O_2(aq) \xrightarrow{\text{catalase}} 2H_2O(l) + O_2(g)$$

Although many chemical catalysts can work under various conditions of temperature and pressure as well as alkalinity or acidity, biological catalysts operate only under very particular conditions. For example, they operate over a very narrow temperature range and if the temperature becomes too high, they become inoperative. At temperatures above about 45°C, they denature. This means that the specific shape of the active site of the enzyme molecule changes due to the breaking of bonds. This means that the reactant molecules are no longer able to fit into the active site.

A huge multimillion-pound industry has grown up around the use of enzymes to produce new materials. Biological washing powders (Figure 11.15) contain enzymes to break down stains such as sweat, blood and egg, and they do this at the relatively low temperature of 40°C. This reduces energy costs, because the washing water does not need to be heated as much.

Figure 11.15 These biological washing powders contain enzymes.

There were problems associated with the early biological washing powders. Some customers suffered from skin rashes, because they were allergic to the enzymes (Figure 11.16). This problem has been overcome to a certain extent by advising that extra rinsing is required. Also, many manufacturers have placed warnings on their packets, indicating that the powder contains enzymes which may cause skin rashes.

Figure 11.14 The enzyme molecules (red, pink, green and blue) lock on exactly to a particular reactant molecule (yellow). Once the enzyme is locked on, the reactant molecule breaks up into pieces.

Figure 11.16 An allergic reaction to a biological detergent.

Other industrial processes also make use of enzymes.

- In the manufacture of baby foods, enzymes called proteases are used to 'pre-digest' the protein part of the baby food. This is because young babies are unable to digest protein molecules.
- The enzyme isomerase is used to convert glucose syrup to fructose syrup. Fructose syrup is much sweeter than glucose syrup and can be used as a sweetener in slimming foods as less is needed.
- In the production of yoghurt, milk is initially heated to 90 °C for 15–30 minutes to kill any bacteria in the milk. After cooling to 40 °C, a starter culture of *Lactobacillus* bacteria is added and the mixture incubated at 40 °C for eight hours (Figure 11.17). The bacteria ferment the lactose in the milk to lactic acid, which causes the milk protein to become solid.

- In cheese making, milk is initially heated to kill bacteria and then cooled. A starter culture of *Streptococcus* bacteria is then added, which coagulates the milk into curds and whey (Figure 11.18). The curds are put into steel or wooden drums and pressed and allowed to dry.

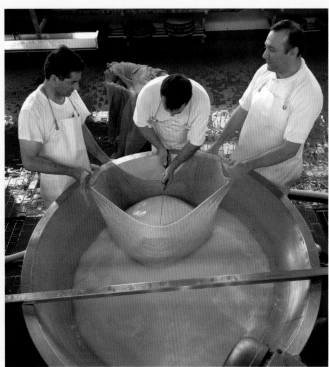

Figure 11.18 Cheese is separated into curds and whey by the addition of bacteria. The liquid, whey, is separated from the curds which are then pressed.

In industry, enzymes are used to bring about reactions at normal temperatures and pressures that would otherwise require expensive conditions and equipment. Successful processes using enzymes need to ensure that:

- the enzyme is able to function for long periods of time by optimising the environment
- the enzyme is kept in place by trapping it on the surface of an inert solid (some processes immobilise the enzymes when the process is complete)
- continuous processes occur rather than batch processes.

Figure 11.17 Yoghurt is incubated in these tanks, and allowed to mature.

Questions

1 When using biological washing powders what factors have to be taken into consideration?

2 Enzymes in yeast are used in the fermentation of glucose. Why, when the temperature is raised to 45 °C, is very little ethanol actually produced compared with the amount formed at room temperature?

■ *Checklist*

After studying Chapter 11 you should know and understand the following terms.

■ **Activation energy**   The excess energy that the reactants must acquire to permit the reaction to occur.

■ **Catalyst**   A substance which alters the rate of a chemical reaction without itself being chemically changed.

■ **Catalytic converter**   A device for converting dangerous exhaust gases from cars into less harmful emissions. For example, carbon monoxide gas is converted to carbon dioxide gas.

■ **Enzymes**   Protein molecules produced in living cells. They act as biological catalysts and are specific to certain reactions. They operate only within narrow temperature and pH ranges.

■ **Reaction rate**   A measure of the change which happens during a reaction in a single unit of time. It may be affected by the following factors:

- surface area of the reactants
- concentration of the reactants
- the temperature at which the reaction is carried out
- light
- use of a catalyst.

Rates of reaction
Additional questions

1 Explain the following statements.
  a A car exhaust pipe will rust much faster if the car is in constant use.
  b Vegetables cook faster when they are chopped up.
  c Industrial processes become more economically viable if a catalyst can be found for the reactions involved.
  d In fireworks it is usual for the ingredients to be powdered.
  e Tomatoes ripen faster in a greenhouse.
  f The reaction between zinc and dilute hydrochloric acid is slower than the reaction between zinc and concentrated hydrochloric acid.

2 A student performed two experiments to establish how effective manganese(IV) oxide was as a catalyst for the decomposition of hydrogen peroxide. The results below were obtained by carrying out these experiments with two different quantities of manganese(IV) oxide. The volume of the gas produced was recorded against time.

Time/s	0	30	60	90	120	150	180	210
Volume for 0.3 g/cm³	0	29	55	79	98	118	133	146
Volume for 0.5 g/cm³	0	45	84	118	145	162	174	182

  a Draw a diagram of the apparatus you could use to carry out these experiments.
  b Plot a graph of the results.
  c Is the manganese(IV) oxide acting as a catalyst in this reaction? Explain your answer.
  d (i) At which stage does the reaction proceed most quickly?
    (ii) How can you tell this from your graph?
    (iii) In terms of particles, explain why the reaction is quickest at the point you have chosen in (i).
  e Why does the slope of the graph become less steep as the reaction proceeds?
  f What volume of gas has been produced when using 0.3 g of manganese(IV) oxide after 50 s?
  g How long did it take for 60 cm³ of gas to be produced when the experiment was carried out using 0.5 g of the manganese(IV) oxide?
  h Write a balanced chemical equation for the decomposition of hydrogen peroxide.

3 a Which of the following reaction mixtures will produce hydrogen more quickly at room temperature:
    (i) zinc granules + dilute nitric acid?
    (ii) zinc powder + dilute nitric acid?
  b Give an explanation of your answer to a.
  c Suggest two other methods by which the speed of this reaction can be altered.

4 A flask containing dilute hydrochloric acid was placed on a digital balance. An excess of limestone chippings was added to this acid, a plug of cotton wool was placed in the neck of the flask and the initial mass was recorded. The mass of the apparatus was recorded every two minutes. At the end of the experiment the loss in mass of the apparatus was calculated and the following results were obtained.

Time/min	0	2	4	6	8	10	12	14	16
Loss in mass/g	0	2.1	3.0	3.1	3.6	3.8	4.0	4.0	4.0

  a Plot the results of the experiment.
  b Which of the results would appear to be incorrect? Explain your answer.
  c Write a balanced chemical equation to represent the reaction taking place.
  d Why did the mass of the flask and its contents decrease?
  e Why was the plug of cotton wool used?
  f How does the rate of reaction change during this reaction? Explain this using particle theory.
  g How long did the reaction last?
  h How long did it take for half of the reaction to occur?

5 a What is a catalyst?
  b List the properties of catalysts.
  c Name the catalyst used in the following processes:
    (i) the Contact process
    (ii) the Haber process
    (iii) the hydrogenation of unsaturated fats.
  d Which series of metallic elements, in the periodic table, do the catalysts you have named in c belong to?
  e What are the conditions used in the industrial processes named in c? The following references will help you: Chapters 12, 15 and 16.

6 This question concerns the reaction of copper(II) carbonate with dilute hydrochloric acid. The equation for the reaction is:

$$CuCO_3(s) + 2HCl(aq) \rightarrow CuCl_2(aq) + CO_2(aq) + H_2O(l)$$

  a Sketch a graph to show the rate of production of carbon dioxide when an excess of dilute hydrochloric acid is added. The reaction lasts 40 s and produces 60 cm³ of gas.
  b Find on your graph the part which shows:
    (i) where the reaction is at its fastest
    (ii) when the reaction has stopped.
  c Calculate the mass of copper(II) carbonate used to produce 60 cm³ of carbon dioxide. (A_r: C = 12; O = 16; Cu = 63.5. One mole of a gas occupies 24 dm³ at room temperature and pressure (rtp).)

continued

d Sketch a further graph using the same axes to show what happens to the rate at which the gas is produced if:
  (i) the concentration of the acid is decreased
  (ii) the temperature is increased.

7 European regulations state that all new cars have to be fitted with catalytic converters as part of their exhaust system.
  a Why are these regulations necessary?
  b Which gases are removed by catalytic converters?
  c Which metals are often used as catalysts in catalytic converters?
  d What does the term 'poisoned' mean with respect to catalysts?
  e The latest converters will also remove unburnt petrol. An equation for this type of reaction is:

$$2C_7H_{14}(g) + 21O_2(g) \rightarrow 14CO_2(g) + 14H_2O(g)$$

  (i) Calculate the mass of carbon dioxide produced by 1.96 g of unburnt fuel.
  (ii) Convert this mass of carbon dioxide into a volume measured at rtp.
  (iii) If the average car produces 7.84 g of unburnt fuel a day, calculate the volume of carbon dioxide produced by the catalytic converter measured at rtp. (A_r: H = 1; C = 12; O = 16. One mole of any gas occupies 24 dm^3 at rtp.)

8 Suggest practical methods by which the rate of reaction can be investigated in each of the following cases:
  a magnesium reacting with hydrochloric acid
  b nitrogen monoxide reacting with oxygen.

12 *The petroleum industry*

Substances from oil
Oil refining

Alkanes
Naming the alkanes
Structural isomerism
The chemical behaviour of alkanes
The ozone hole problem
Methane – another greenhouse gas
Other uses of alkanes

Alkenes
Where do we get alkenes from?
The chemical behaviour of alkenes
A test for unsaturated compounds

Checklist

Additional questions

Substances from oil

What do the modes of transport shown in Figure 12.1 have in common? They all use liquids obtained from **crude oil** as fuels.

Figure 12.1 These modes of transport all use fuels that have been extracted from crude oil.

Oil refining

Crude oil is a complex mixture of compounds known as **hydrocarbons** (Figure 12.2a). Hydrocarbons are molecules which contain only the elements carbon and hydrogen bonded together covalently (Chapter 4, p. 57). These carbon compounds form the basis of a group called **organic compounds**. All living things are made from organic compounds based on chains of carbon atoms similar to those found in crude oil. Crude oil is not only a major source of fuel but is also a raw material of enormous importance. It supplies a large and diverse chemical industry to make dozens of products (Figure 12.2b).

Crude oil is not very useful to us until it has been processed. The process, known as **refining**, is carried out at an oil refinery (Figure 12.3).

a Crude oil is a mixture of hydrocarbons.

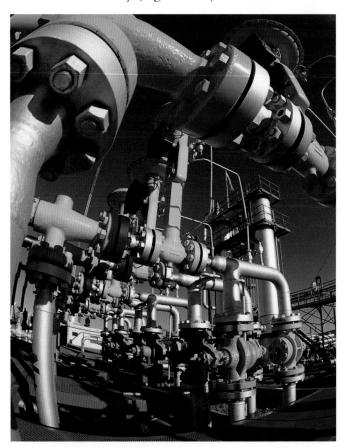

Figure 12.3 An oil refinery.

b The objects above are made from substances obtained from oil.

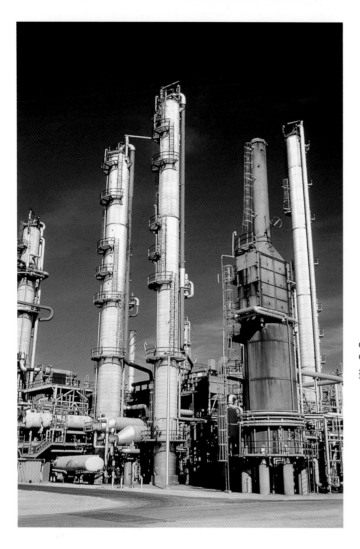

Figure 12.4a Fractional distillation of crude oil in a refinery.

refinery gas used as a fuel

30°C
gasoline
used as fuel in cars (petrol)

110°C
naphtha
used to make chemicals

fractionating tower

180°C
kerosene/paraffin
used as a fuel in jet engines
and oil stoves

260°C
diesel oil or gas oil
used as a fuel in diesel
engines

crude oil

fuel oil
used as a fuel for ships and
for home heating systems

lubricating oil

heater

340°C
residue
used to make bitumen for
surfacing roads

b Uses of the different fractions obtained from crude oil.

Refining involves separating crude oil into various batches or **fractions**. Chemists use a technique called **fractional distillation** to separate the different fractions. This process works in a similar way as that discussed in Chapter 2, p. 22, for separating ethanol (alcohol) and water. The different components (fractions) separate because they have different boiling points. The crude oil is heated to about 400°C to vaporise all the different parts of the mixture. The mixture of vapours passes into the fractionating column near the bottom (Figure 12.4). Each fraction is obtained by collecting hydrocarbon molecules which have a boiling point in a given range of temperatures (Figure 12.4). For example, the fraction we know as petrol contains molecules which have boiling points between 30°C and 110°C. The molecules in this fraction contain between five and ten carbon atoms. These smaller molecules with lower boiling points condense higher up the tower. The bigger hydrocarbon molecules which have the higher boiling points condense in the lower half of the tower.

The liquids condensing at different levels are collected on **trays**. In this way the crude oil is separated into different fractions. These fractions usually contain a number of different hydrocarbons. The individual single hydrocarbons can then be obtained, again by refining the fraction by further distillation.

It is important to realise that the uses of the fractions depend on their properties. For example, one of the lower fractions, which boils in the range 250–350°C, is quite thick and sticky and makes a good lubricant. However, the petrol fraction burns very easily and this therefore makes it a good fuel for use in engines.

Questions

1 What do you understand by the term hydrocarbon?

2 All organisms are composed of compounds which contain carbon. Why do you think carbon chemistry is often called 'organic chemistry'?

3 List the main fractions obtained by separating the crude oil mixture and explain how they are obtained in a refinery.

■ *Alkanes*

Most of the hydrocarbons in crude oil belong to the family of compounds called **alkanes**. The molecules within the alkane family contain carbon atoms covalently bonded to four other atoms by single bonds. Because these molecules possess only single bonds they are said to be **saturated**, as no further atoms can be added (Figure 12.5). For a discussion of the covalent bonding scheme for methane see Chapter 4, p. 57. The physical properties of the first six members of the alkane family are shown in Table 12.1.

Table 12.1 Some alkanes and their physical properties.

Alkane	Formula	Melting point/°C	Boiling point/°C	Physical state at room temperature
Methane	CH_4	−182	−162	Gas
Ethane	C_2H_6	−183	−89	Gas
Propane	C_3H_8	−188	−42	Gas
Butane	C_4H_{10}	−138	0	Gas
Pentane	C_5H_{12}	−130	36	Liquid
Hexane	C_6H_{14}	−95	69	Liquid

You will notice from Figure 12.5 and Table 12.1 that the compounds have a similar structure and similar name endings. They also behave chemically in a similar way and the family of compounds can be represented by a general formula. In the case of the alkanes the general formula is:

$$C_nH_{(2n\ +\ 2)}$$

where *n* is the number of carbon atoms present.

A family with the above factors in common is called a **homologous series**. As you go up a homologous series, in order of increasing number of carbon atoms, the physical properties of the compounds gradually change. For example, the melting and boiling points of the alkanes shown in Table 12.1 gradually increase. This is due to an increase in the intermolecular forces (van der Waals' forces) as the size and mass of the molecule increases (Chapter 4, p. 59).

Under normal conditions molecules with up to four carbon atoms are gases, those with between five and 16 carbon atoms are liquids, while those with more than 16 carbon atoms are solids.

Questions

1 Estimate the boiling points for the alkanes with formulae:
   a C_7H_{16}
   b C_8H_{18}.

2 Name the alkanes which have the following formulae:
   a C_7H_{16}
   b $C_{10}H_{22}$.

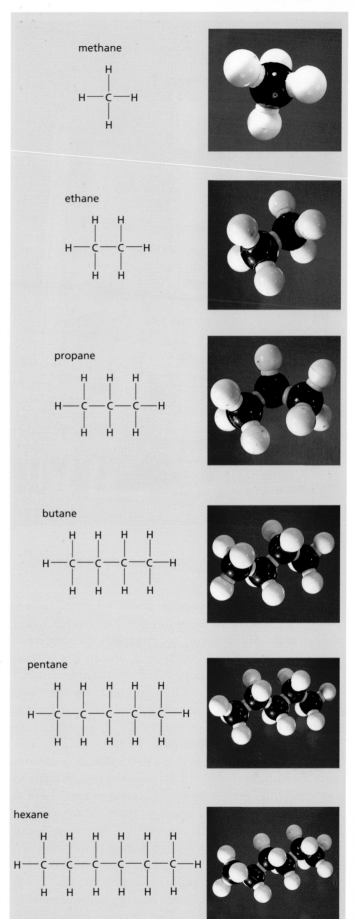

Figure 12.5 The alkane molecules look like the models in the photographs.

Naming the alkanes

All the alkanes have names ending in *-ane*. The rest of the name tells you the number of carbon atoms present in the molecule. For example, the compound whose name begins with:

- *meth-* has one carbon atom
- *eth-* has two carbon atoms
- *prop-* has three carbon atoms
- *but-* has four carbon atoms
- *pent-* has five carbon atoms

and so on.

Structural isomerism

Sometimes it is possible to write more than one structural formula to represent a molecular formula. The structural formula of a compound shows how the atoms are joined together by the covalent bonds. For example, there are two different compounds with the molecular formula C_4H_{10}. The structural formulae of these two substances along with their names and physical properties are shown in Figure 12.6.

Compounds such as those in Figure 12.6 are known as **isomers**. Isomers are substances which have the same molecular formula but different structural formulae. The different structures of the compounds shown in Figure 12.6 have different melting and boiling points. Molecule (b) contains a branched chain and has a lower melting point than molecule (a), which has no branched chain. All the alkane molecules with four or more carbon atoms possess isomers. Perhaps now you can see why there are so many different organic compounds!

a butane

melting point
−138°C

boiling point
0°C

b 2-methylpropane

melting point
−159°C

boiling point
−12°C

Figure 12.6 The isomers of C_4H_{10}.

Question

1 Draw the structural formulae for the isomers of:
   a C_5H_{12}
   b C_6H_{14}.

The chemical behaviour of alkanes

Alkanes are rather unreactive compounds. For example, they are generally not affected by alkalis, acids or many other substances. Their most important property is that they burn easily.

Gaseous alkanes, such as methane, will burn in a good supply of air, forming carbon dioxide and water as well as plenty of heat energy.

methane + oxygen → carbon dioxide + water + energy

$$CH_4(g) + 2O_2(g) \rightarrow CO_2(g) + 2H_2O(g)$$

The gaseous alkanes are some of the most useful fuels. Methane, better known as natural gas, is used for cooking as well as for heating offices, schools and homes (Figure 12.7a). Propane and butane burn with very hot flames and they are sold as liquefied petroleum gas (LPG). In rural areas where there is no supply of natural gas, central heating systems can be run on propane gas (Figure 12.7b). Butane, sometimes mixed with propane, is used in portable blowlamps and in gas lighters.

a This is burning methane.

b Central heating systems can be run on propane.

Figure 12.7

Another useful reaction worth noting is that between the alkanes and the halogens. For example, methane and chlorine react in the presence of sunlight (or ultraviolet light). The ultraviolet light splits the chlorine molecules into atoms. When this type of reaction takes place, these atoms are called **free radicals** and they are very reactive.

chlorine gas $\xrightarrow{\text{sunlight}}$ chlorine atoms (free radicals)

$$Cl_2(g) \longrightarrow 2Cl(g)$$

The chlorine atoms then react further with methane molecules, and a hydrogen chloride molecule is produced along with a methyl free radical.

$$\text{chlorine} + \text{methane} \rightarrow \text{methyl} + \text{hydrogen}$$
$$\text{atom} \qquad\qquad \text{radical} \quad \text{chloride}$$
$$Cl(g) \;+\; CH_4(g) \rightarrow CH_3(g) + HCl(g)$$

The methyl free radical reacts further.

$$\text{methyl} + \text{chlorine} \rightarrow \text{chloromethane} + \text{chlorine}$$
$$\text{radical} \quad\;\; \text{gas} \qquad\qquad\qquad\qquad\quad \text{atom}$$
$$CH_3(g) \;+\; Cl_2(g) \;\rightarrow\; CH_3Cl(g) \;+\; Cl(g)$$

This chlorine free radical, in turn, reacts further and the process continues until all the chlorine and the methane have been used up. This type of process is known as a **chain reaction** and it is very fast. The overall chemical equation for this process is:

$$\text{methane} + \text{chlorine} \rightarrow \text{chloromethane} + \text{hydrogen}$$
$$\text{chloride}$$
$$CH_4(g) \;+\; Cl_2(g) \;\rightarrow\; CH_3Cl(g) \;+\; HCl(g)$$

We can see that one hydrogen atom of the methane molecule is **substituted** by a chlorine atom. This type of reaction is known as a **substitution reaction**.

Because we cannot control the chlorine free radicals produced in this reaction, we also obtain small amounts of other 'substituted' products – CH_2Cl_2 (dichloromethane), $CHCl_3$ (trichloromethane or chloroform) and CCl_4 (tetrachloromethane) – by further reactions such as those shown below.

$$\text{chloromethane} + \text{chlorine} \rightarrow \text{chloromethyl} + \text{hydrogen}$$
$$\text{atom} \qquad\quad \text{radical} \qquad \text{chloride}$$
$$CH_3Cl(g) \;+\; Cl(g) \;\rightarrow\; CH_2Cl(g) \;+\; HCl(g)$$

$$\text{chloromethyl} + \text{chlorine} \rightarrow \text{dichloromethane} + \text{chlorine}$$
$$\text{radical} \qquad\quad \text{gas} \qquad\qquad\qquad\qquad\qquad \text{atom}$$
$$CH_2Cl(g) \;+\; Cl_2(g) \;\rightarrow\; CH_2Cl_2(g) \;+\; Cl(g)$$

Many of these so-called **halogenoalkanes** are used as solvents. For example, dichloromethane is used as a solvent in paint stripper (Figure 12.8).

Figure 12.8 Dichloromethane is used as a solvent in paint stripper.

Early anaesthetics relied upon trichloromethane, $CHCl_3$, or chloroform. Unfortunately, this anaesthetic had a severe problem since the lethal dose was only slightly higher than that required to anaesthetise the patient. In 1956, halothane was discovered by chemists working at ICI. This is a compound containing chlorine, bromine and fluorine. Its formula is $CF_3CHBrCl$. However, even this is not the perfect anaesthetic since evidence suggests that prolonged exposure to this substance may cause liver damage. The search continues for even better anaesthetics.

A group of compounds were discovered in the 1930s and were called the chlorofluorocarbons (CFCs). Because of their inertness they found many uses, especially as a propellant in aerosol cans. CFC-12 or dichlorodifluoromethane, CF_2Cl_2, was one of the most popular CFCs in use in aerosols. Scientists now believe that CFCs released from aerosols are destroying the ozone layer.

The ozone hole problem

Our atmosphere protects us from harmful ultraviolet radiation from the Sun. This damaging radiation is absorbed by the relatively thin ozone layer found in the stratosphere (Figure 12.9).

Figure 12.9 The ozone layer is between 25 km and 50 km above sea level.

Large holes have recently been discovered in the ozone layer over Antarctica, Australasia and Europe (Figure 10.3, p. 149). Scientists think that these holes have been produced by CFCs such as CFC-12. CFCs escape into the atmosphere and, because of their inertness, remain without further reaction until they reach the stratosphere and the ozone layer. In the stratosphere the high-energy ultraviolet radiation causes a chlorine atom to split off from the CFC molecule. This chlorine atom, or free radical, then reacts with the ozone.

$$Cl(g) + O_3(g) \rightarrow OCl(g) + O_2(g)$$

This is not the only problem with CFCs. They are also significant 'greenhouse gases' (Chapter 8, p. 122).

The ozone depletion and greenhouse effects have become such serious problems that an international agreement known as the *Montreal Protocol on Substances that Deplete the Ozone Layer* was agreed in 1987. The proposed controls were tightened in 1990 by the second meeting of the parties to the *Montreal Protocol*. Modifications have since been made at meetings held in 1992, 1995 and 1996.

Research is now going ahead, with some success, to produce safer alternatives to CFCs. At present, better alternatives called hydrochlorofluorocarbons (HCFCs) have been developed – these have lower ozone-depletion effects and are not effective greenhouse gases.

Methane – another greenhouse gas!

Methane, the first member of the alkanes, occurs naturally. Cows produce it in huge quantities when digesting their food and it is also formed by growing rice. Like carbon dioxide, it is a greenhouse gas (Chapter 8, p. 122) because it acts like the glass in a greenhouse – it will let in heat from the Sun but will not let all of the heat back out again. It is thought that the greenhouse effect may contribute to climate change, which could have disasterous effects for life on this planet.

As world population increases, so does the need for food, such as meat from cattle, and rice. Increased food production causes more methane to be added to the atmosphere. Methane is also added to the atmosphere from landfill sites – waste vegetation in landfill decays, and produces methane.

Question

1 Use your research skills to find out:
   a any other sources of methane found in nature
   b how climate change might affect your particular environment.

Other uses of alkanes

Besides their major use as fuels (p. 177), some of the heavier alkanes are used as waxes in candles, as lubricating oils and in the manufacture of another family of hydrocarbons – the alkenes.

Questions

1 Write a balanced chemical equation to represent the combustion of propane.

2 In what mole proportions should chlorine and methane be mixed to produce:
   a mainly chloromethane?
   b mainly tetrachloromethane?

3 Describe a method you would use to separate chloromethane from the other possible reaction products when methane reacts with chlorine.

4 Why do you think that CFCs release chlorine atoms into the stratosphere and not fluorine atoms?

■ *Alkenes*

Alkenes form another homologous series of hydrocarbons of the general formula:

$$C_nH_{2n}$$

where n is the number of carbon atoms. The alkenes are more reactive than the alkanes because they each contain a double covalent bond between the carbon atoms (Figure 12.10). Molecules that possess a double covalent bond of this kind are said to be **unsaturated**, because it is possible to break one of the two bonds to add extra atoms to the molecule.

Figure 12.10 The bonding in ethene, the simplest alkene.

All alkenes have names ending in *-ene*. Alkenes, especially ethene, are very important industrial chemicals. They are used extensively in the plastics industry and in the production of alcohols such as ethanol and propanol. Table 12.2 gives the names, formulae and some physical properties of the first three members of the alkene family. Figure 12.11 shows the structure of these members as well as models of their shape.

Table 12.2 The first three alkenes and their physical properties.

Alkene	Formula	Melting point/°C	Boiling point/°C	Physical state at room temperature
Ethene	C_2H_4	−169	−104	Gas
Propene	C_3H_6	−185	−47	Gas
Butene	C_4H_8	−184	−6	Gas

Figure 12.11 Structure and shape of the first three alkenes.

Where do we get alkenes from?

Very few alkenes are found in nature. Most of the alkenes used by the petrochemical industry are obtained by breaking up larger, less useful alkane molecules obtained from the fractional distillation of crude oil. This is usually done by a process called **catalytic cracking**. In this process the alkane molecules to be 'cracked' (split up) are passed over a mixture of aluminium and chromium oxides heated to about 500°C.

$$\text{dodecane} \longrightarrow \text{decane} + \text{ethene}$$
$$C_{12}H_{26}(g) \longrightarrow C_{10}H_{22}(g) + C_2H_4(g)$$
$$\text{(found in kerosene)} \quad \text{shorter alkane} \quad \text{alkene}$$

Another possibility is:

$$C_{12}H_{26}(g) \rightarrow C_8H_{16}(g) + C_4H_8(g) + H_2(g)$$

Note that in this reaction hydrogen may also be formed during cracking. The amount of hydrogen produced depends on the conditions used. Since smaller hydrocarbons are generally in greater demand than the larger ones, cracking is used to match demand.

Fraction	Approx % in crude oil	Approx % demand
Refinery gas	2	5
Gasoline	21	28
Kerosene	13	8
Diesel oil	17	25
Fuel oil and bitumen	47	34

This means that oil companies are not left with large surpluses of fractions containing the larger molecules.

Figure 12.12 shows the simple apparatus that can be used to carry out cracking reactions in the laboratory. You will notice that in the laboratory we may use a catalyst of broken, unglazed pottery.

Figure 12.12 The cracking of an alkane in the laboratory.

Questions

1 Using the information in Table 12.2 (p. 179), make an estimate of the boiling point of pentene.

2 Write a balanced chemical equation to represent the process that takes place when decane is cracked.

The chemical behaviour of alkenes

The double bond makes alkenes more reactive than alkanes in chemical reactions. For example, hydrogen adds across the double bond of ethene, under suitable conditions, forming ethane (Figure 12.13).

Figure 12.13 The addition of hydrogen to ethene using molecular models.

This reaction is called **hydrogenation**. Hydrogenation reactions like the one shown with ethene are used in the manufacture of margarines from vegetable oils. Vegetable oils contain fatty acids, such as linoleic acid ($C_{18}H_{32}O_2$). These are unsaturated molecules, containing several double bonds. These double bonds make the molecule less flexible. Hydrogenation can convert these molecules into more saturated ones. Now the molecules are less rigid and can flex and twist more easily, and hence pack more closely together. This in turn causes an increase in the intermolecular forces and so raises the melting point. The now solid margarines can be spread on bread more easily than liquid oils.

There is another side to this process. Many doctors now believe that unsaturated fats are more healthy than saturated ones. Because of this, many margarines are left partially unsaturated. They do not have all the C=C taken out of the fat molecules. However, the matter is far from settled and the debate continues.

Another important **addition reaction** is the one used in the manufacture of ethanol. Ethanol has important uses as a solvent and a fuel (p. 190). It is formed when water (as steam) is added across the double bond in ethene. For this reaction to take place, the reactants have to be passed over a catalyst of phosphoric(V) acid (absorbed on silica pellets) at a temperature of 300 °C and pressure of 60 atmospheres (1 atmosphere = 1×10^5 pascals).

$$\text{ethene} + \text{steam} \xrightleftharpoons[\substack{\text{phosphoric(v)} \\ \text{acid catalyst}}]{\substack{300°C,\ 60\ atm}} \text{ethanol}$$

$$C_2H_4(g) + H_2O(g) \rightleftharpoons C_2H_5OH(g)$$

This reaction is reversible as is shown by the equilibrium (⇌) sign. The conditions have been chosen to ensure the highest possible yield of ethanol. In other words, the conditions have been chosen so that they favour the forward reaction.

For a further discussion of ethanol and alcohols generally see p. 204.

A test for unsaturated compounds

The addition reaction between bromine dissolved in an organic solvent, or water, and alkenes is used as a chemical test for the presence of a double bond between two carbon atoms. When a few drops of this bromine solution are shaken with the hydrocarbon, if it is an alkene, such as ethene, a reaction takes place in which bromine joins to the alkene double bond. This results in the bromine solution losing its red/brown colour. If an alkane, such as hexane, is shaken with a bromine solution of this type, no colour change takes place (Figure 12.14). This is because there are no double bonds between the carbon atoms of alkanes.

$$\text{ethane} + \text{bromine} \longrightarrow \text{dibromoethane}$$

$$C_2H_4(g) + Br_2(\text{in solution}) \longrightarrow C_2H_4Br_2(\text{in solution})$$

Questions

1 What is meant by the term 'addition reaction'?

2 Write a word and balanced chemical equation for the reaction between ethene and hydrogen chloride.

3 Write the structural formula for pentene.

Figure 12.14 The alkene decolorises bromine in 1,1,1-trichloroethane.

■ *Checklist*

After studying Chapter 12 you should know and understand the following terms.

■ **Alkanes** A family of saturated hydrocarbons with the general formula C_nH_{2n+2}. The term 'saturated', in this context, is used to describe molecules that have only single bonds. The alkanes can only undergo substitution reactions in which there is replacement of one atom in the molecule by another atom.

■ **Alkenes** A family of unsaturated hydrocarbons with the general formula C_nH_{2n}. The term 'unsaturated', in this context, is used to describe molecules which contain one or more double carbon–carbon bonds. Unsaturated compounds undergo addition reactions across the carbon–carbon double bonds and so produce saturated compounds. The addition of hydrogen across the carbon–carbon double bonds is used to reduce the amount of unsaturation during the production of margarines.

■ **Catalytic cracking** The decomposition of higher alkanes into alkenes and alkanes of lower relative molecular mass. The process involves passing the larger alkane molecules over a catalyst of aluminium and chromium oxides, heated to 500 °C.

■ **CFC** Abbreviation for chlorofluorocarbon, a type of organic compound in which some or all of the hydrogen atoms of an alkane have been replaced by fluorine and chlorine atoms. These substances are generally unreactive but they can diffuse into the stratosphere where they break down under the influence of ultraviolet light. The products of this photochemical process then react with ozone (in the ozone layer). Because of this, their use has been discouraged. They are now being replaced by hydrochlorofluorocarbons (HCFCs).

■ **Chain reaction** A reaction which is self-sustaining owing to the products of one step of the reaction assisting in promoting further reaction.

■ **Free radicals** Atoms or groups of atoms with unpaired electrons and which are therefore highly reactive. They can be produced by high-energy radiation such as ultraviolet light in photochemical reactions.

■ **Halogenoalkanes** Organic compounds in which one or more hydrogen atoms of an alkane have been substituted by halogen atoms such as chlorine.

■ **Hydrocarbon** A substance which contains atoms of carbon and hydrogen only.

■ **Isomers** Compounds which have the same molecular formula but different structural arrangements of the atoms.

■ **Oil refining** The general process of converting the mixture that is collected as crude oil into separate fractions. These fractions, known as petroleum products, are used as fuels, lubricants, bitumens and waxes. The fractions are separated from the crude oil mixture by fractional distillation.

■ **Organic chemistry** The branch of chemistry concerned with compounds of carbon found in living organisms.

■ **Test for unsaturation** A few drops of bromine dissolved in an organic solvent are shaken with the hydrocarbon. If it is decolorised, the hydrocarbon is unsaturated.

The petroleum industry
Additional questions

1 Explain the following.
  a Ethene is called an unsaturated hydrocarbon.
  b The cracking of larger alkanes into simple alkanes and alkenes is important to the petrochemical industry.
  c The conversion of ethene to ethanol is an example of an addition reaction.

2 The following question is about some of the reactions of ethene.

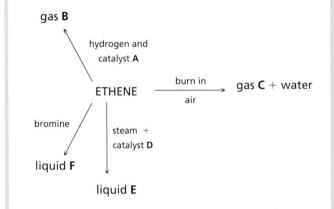

  a Give the names and formulae for substances **A** to **F**.
  b (i) Write a word and balanced chemical equation to represent the reaction in which liquid **E** is formed.
   (ii) What reaction conditions are required for the process to take place?
   (iii) Hydrogen is used in the production of margarine to remove unsaturation. Explain what you understand by this statement.
  c Name the homologous series that gas **B** belongs to.
  d Describe a chemical test which would allow you to identify gas **C**.

3 a Crude oil is a mixture of *hydrocarbons* which belong to the *homologous series* called the *alkanes*. This mixture can be separated into fractions by the process of *fractional distillation*. Some of the fractions obtained are used as *fuels*. Some of the other fractions are subjected to *catalytic cracking* in order to make *alkenes*.
   Explain the meaning of the terms in italics.
  b Alkanes can be converted into substances which are used as solvents. To do this the alkane is reacted with a halogen, such as chlorine, in the presence of ultraviolet light.
   (i) Write a word and balanced chemical equation for the reaction between methane and chlorine.
   (ii) Name the type of reaction taking place.
   (iii) Highly reactive chlorine atoms are produced in the presence of ultraviolet light. When atoms are produced in this way, what are they called?

   (iv) Write a balanced chemical equation for the reaction which takes place between CHF_3 and Cl_2 to produce a chlorofluorocarbon (CFC).
   (v) Why are CFCs such a problem?

4 Crude oil is a mixture of hydrocarbons. The refining of crude oil produces fractions which are more useful to us than crude oil itself. Each fraction is composed of hydrocarbons which have boiling points within a specific range of temperature. The separation is carried out in a fractionating column, as shown below.

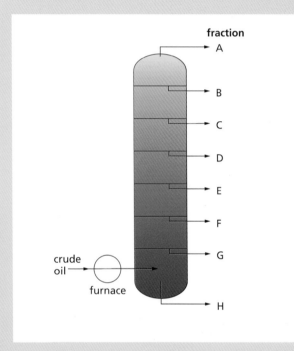

  a Which separation technique is used to separate the fractions?
  b Name each of the fractions **A** to **H** and give a use for each.
  c Why do the fractions come from the fractionating column in this order?
  d What is the connection between your answer to **c** and the size of the molecules in each fraction?
  e Which of the fractions will be the most flammable?

5 Alkanes and alkenes are hydrocarbons. They are composed of molecules which contain covalent bonds. For each of the molecules below, use a dot and cross diagram to show the bonding it contains.
  a Methane, CH_4.
  b Propene, C_3H_6.
  c Propane, C_3H_8.
  d Ethene, C_2H_4.

continued

6 Crude oil is an important source of organic chemical fuels. It is refined by fractional distillation. Use the information in the table below to answer the questions which follow.

Fraction	Boiling point/°C
A	40
B	80
C	200
D	350
E	above 350

a For each of the questions that follow, give the letter of the fraction which is most appropriate as an answer. You should also give a reason for your answer in each case.
(i) Which fraction would contain the most volatile substances?
(ii) Which of the fractions would collect at the bottom of the fractionating column?
(iii) Which fraction could be used as a fuel for cars?
(iv) Which fraction would contain the largest molecules?

b Some of the fractions undergo a further process called cracking to produce further substances.
(i) Explain what you understand by the term 'cracking'. What conditions are employed when cracking occurs?
(ii) Write a word and balanced chemical equation to show how octane can be produced by the cracking of $C_{15}H_{32}$.

7 a A hydrocarbon contains 92.3% by mass of carbon. Work out the empirical formula of this hydrocarbon.

b The relative molecular mass of this hydrocarbon was found by mass spectrometry to be 78. Work out its molecular formula. (A_r: H = 1, C = 12)

8 a Which of the following formulae represent alkanes, which represent alkenes and which represent neither?

$$CH_3, C_6H_{12}, C_5H_{12}, C_6H_6, C_9H_{20},$$
$$C_{12}H_{24}, C_{20}H_{42}, C_2H_4, C_8H_{18}, C_3H_7$$

b Draw all the possible isomers which have the molecular formula C_6H_{14}.

13 Energy sources

Fossil fuels

What is a fuel?
How are fossil fuels used?

Alternative sources of energy
Nuclear power
Hydroelectric power
Geothermal energy
Wave power
Tidal power
Wind power
Solar energy
Biomass and biogas
Hydrogen – the fuel of the future

Chemical energy
Combustion
Enthalpy of neutralisation (molar heat of neutralisation)

Change of state

Cells and batteries

Fuel cells

Checklist

Additional questions

■ *Fossil fuels*

Coal, oil and natural gas are all examples of **fossil fuels**. The term, fossil fuels, is derived from the fact that they are formed from dead plants and animals which were fossilised over 200 million years ago during the carboniferous era.

Coal was produced by the action of pressure and heat on dead wood from ancient forests which once grew in the swampland in many parts of the world under the prevailing weather conditions of that time. When dead trees fell into the swamps they were buried by mud. This prevented aerobic decay (which takes place in the presence of oxygen). Over millions of years, due to movement of the Earth's crust as well as to changes in climate, the land sank and the decaying wood became covered by even more layers of mud and sand. Anaerobic decay (which takes place in the absence of oxygen) occurred, and as time passed the gradually forming coal became more and more compressed as other material was laid down above it (Figure 13.1). Over millions of years, as the layers of forming coal were pushed deeper and the pressure and temperature increased, the final conversion to coal took place (Figure 13.2).

Different types of coal were formed as a result of different pressures being applied during its formation. For example, anthracite is a hard coal with a high carbon content, typical of coal produced at greater depths. Table 13.1 shows some of the different types of coal along with their carbon contents.

Oil and gas were formed during the same period as coal. It is believed that oil and gas were formed from the remains of plants, animals and bacteria that once lived in seas and lakes. This material sank to the bottom of these seas and lakes and became covered in mud, sand and silt which thickened with time.

Figure 13.1 Piece of coal showing a fossilised leaf.

Figure 13.2 Cutting coal today is extremely mechanised.

Table 13.1 The different coal types.

Type of coal	Carbon content/%
Anthracite	90
Bituminous coal	60
Lignite	40
Peat	20

Anaerobic decay took place, and, as the mud layers built up, high temperatures and pressures were created which converted the material slowly into oil and gas. As rock formed, earth movements caused it to buckle and split, and the oil and gas were trapped in folds beneath layers of non-porous rock or cap-rock (Figures 13.3 and 13.4).

Figure 13.3 Oil production in the North Sea.

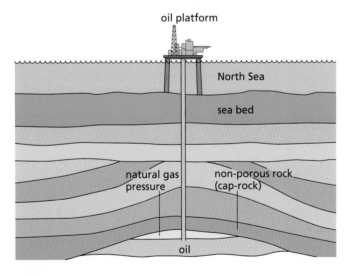

Figure 13.4 Natural gas and oil are trapped under non-porous rock.

Questions

1 Coal, oil and natural gas are all termed 'fossil fuels'. Why is the word 'fossil' used in this context?

2 a Name the process by which plants convert carbon dioxide and water into glucose.
  b What conditions are necessary for this process to occur?

3 Draw a flow diagram to represent the formation of coal, oil or gas.

What is a fuel?

A fuel is a substance which can be conveniently used as a source of energy. Fossil fuels produce energy when they undergo **combustion**.

fossil fuel + oxygen → carbon dioxide + water + energy

For example, natural gas burns readily in air (Chapter 12, p. 177).

methane + oxygen → carbon dioxide + water + energy
$$CH_4(g) + 2O_2(g) \rightarrow CO_2(g) + 2H_2O(l)$$

It should be noted that natural gas, like crude oil, is a mixture of hydrocarbons such as methane, ethane and propane, and may also contain some sulfur. The sulfur content varies from source to source (Chapter 16, p. 227). Natural gas obtained from the North Sea is quite low in sulfur.

The perfect fuel would be:

* cheap
* available in large quantities
* safe to store and transport
* easy to ignite and burn, causing no pollution
* capable of releasing large amounts of energy.

Solid fuels are safer than volatile liquid fuels like petrol and gaseous fuels like natural gas.

How are fossil fuels used?

A major use of fossil fuels is in the production of electricity. Coal, oil and natural gas are burned in power stations (Figure 13.5) to heat water to produce steam, which is then used to drive large turbines (Figure 13.6). A high proportion of the electricity generated worldwide is derived from fossil fuels. However, it should be noted that the relative importance of the three major fossil fuels is changing. Coal and oil are becoming less important while natural gas is increasingly important.

Figure 13.5 A power station.

In a power station, the turbine drives a generator to produce electricity which is then fed into the national grid (Figure 13.6). The national grid is a system for distributing electricity throughout a country.

Other major uses of the fossil fuels are:

- as a major feedstock (raw material) for the chemicals and pharmaceuticals industries
- for domestic and industrial heating and cooking
- as fuels for various forms of vehicle transport.

Figure 13.6 The way in which fuels are used to produce electricity.

Questions

1 'We have not yet found the perfect fuel.' Discuss this statement.

2 'Fossil fuels are a major feedstock for the chemical and pharmaceutical industries.' With reference to Chapter 14, give examples which support this statement.

■ *Alternative sources of energy*

Fossil fuels are an example of **non-renewable** resources, so called because they are not being replaced at the same rate as they are being used up. For example, we have approximately 55 years' supply of crude oil remaining from known reserves if we continue to use it at the current rate as a source of energy and chemicals (Table 13.2). It is important to use non-renewable fuels very carefully and to consider alternative, **renewable** sources of energy for use now and in the future.

Table 13.2 Estimates of how long our fossil fuels will last.

Fossil fuel	Estimated date it is expected to run out
Gas	2055
Oil	2065
Coal	2500

Nuclear power

Calder Hall power station in Cumbria, UK, on the site of the present-day nuclear power complex at Sellafield (Figure 13.7), opened in 1956 and was the first nuclear reactor in the world to produce electricity on an industrial scale.

Figure 13.7 The nuclear power complex at Sellafield, Cumbria, UK was the first to produce electricity in the world.

Nuclear reactors harness the energy from the fission of uranium-235. **Nuclear fission** occurs when the unstable nucleus of a radioactive isotope splits up, forming smaller atoms and producing a large amount of energy as a result. Scientists believe that the energy comes from the conversion of some of the mass of the isotope.

This fission process begins when a neutron hits an atom of uranium-235, causing it to split and produce three further neutrons. These three neutrons split three more atoms of uranium-235, which produces nine neutrons and so on. This initiates a **chain reaction** (Figure 13.8).

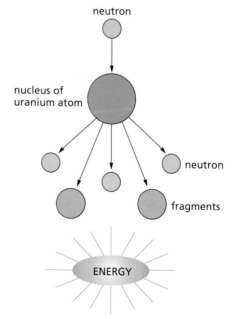

Figure 13.8 Chain reaction in uranium-235 fission.

In a reactor the fission process cannot be allowed to get out of control as it does in an atomic bomb. To prevent this, boron control rods can be pushed into different positions in the reactor to absorb some of the neutrons which are produced and so slow down the chain reaction. If this is done, the energy released from the reaction is obtained in a more controlled way. The energy is used to produce steam, which in turn is used to generate electricity (Figure 13.6, on p. 187).

However, there are problems. The main problem associated with a nuclear power station is that the reactor produces highly radioactive waste materials. These waste materials are difficult to store and cannot be disposed of very easily. Also, leaks of radioactive material have occurred at various sites throughout the world. Accidents at a small number of nuclear power stations, such as Chernobyl in the Ukraine (Figure 13.9) and Three Mile Island in the US, have led to a great deal of concern about their safety.

Figure 13.9 A nuclear accident happened at the Chernobyl power station in 1986.

Hydroelectric power

Hydroelectric power (HEP) is electricity generated from the energy of falling water (Figure 13.10). It is an excellent energy source, and electricity has been generated in this way for some time. For example, the Hoover Dam in the US provides HEP for the surrounding area. It is a very cheap source of electricity. Once you have built the power station, the energy is absolutely free. In some mountainous areas of the world, such as the Alps, HEP is the main source of electricity. One of the main advantages of this system is that it can be quickly used to supplement the national grid at times of high demand. A disadvantage of HEP schemes is that they often require valleys to be flooded and communities to be moved.

Figure 13.10 A hydroelectric power station.

Geothermal energy

Water is pumped into hot rocks in the Earth's crust far below ground level (Figure 13.11). The internal heat of the rocks converts the water to steam, which is used to drive turbines and hence generate electricity. This is a major source of electrical energy in Iceland.

Geothermal energy is a natural, non-polluting source of energy.

a Geothermal plant.

b Geothermal energy is used extensively for generating electricity in Iceland.

Figure 13.11

Wave power

In this method the energy of moving waves is used to generate electricity. Figure 13.12 shows the Salters' 'Duck' wave machine in operation. The vertical motion of the waves is converted to rotary motion, which is used to drive a generator producing electricity. A disadvantage of this method for generating large amounts of electricity is that vast numbers of strings of these ducks would be required.

Wave power is a non-polluting source of energy.

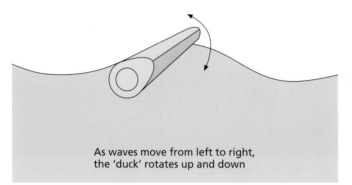

As waves move from left to right, the 'duck' rotates up and down

Figure 13.12 The Salters' 'Duck' wave machine.

Tidal power

The ebb and flow of the tides drives turbines built into a dam or barrage across an estuary where the height difference between high and low tides is large (Figure 13.13). There is a successful tidal power station across the River Rance near St Malo in northern France. The most likely place in Britain to put a tidal barrage is across the Severn Estuary near Bristol, but there would be environmental disadvantages with such a scheme. For example, there would be a threat to the wildlife around the estuary, since the mud banks are one of the few remaining sites in Britain for wildfowl and wading birds to nest.

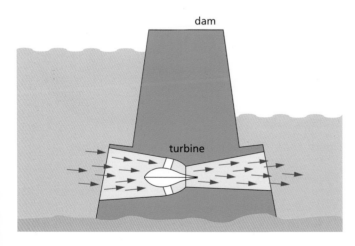

Figure 13.13 Tidal power – the ebb and flow of the tide drives the turbine set into the barrage or dam.

Wind power

This method uses the force of the wind to turn generators to produce electricity. Wind machines 24 m high are capable of generating 200 kW of electricity per machine. Large 'wind farms' have been developed in many parts of the world, for example in the US and on the Pennines in England (Figure 13.14).

The disadvantages of wind power are that the wind farms are somewhat unsightly, they require vast amounts of land, as the machines have to be carefully arranged so that their operation is not impaired, and they produce a lot of noise.

Figure 13.14 Wind farm at Ovenden Moor, Yorkshire, UK.

Solar energy

Two possible methods exist to use the energy of the Sun. In the first, the Sun's energy can be absorbed on to black-painted collector plates and used to heat water and homes (Figure 13.15). This method is relatively cheap. The second method involves the use of photovoltaic cells or photocells to generate electricity (Figure 13.16, overleaf). Disadvantages include the initial cost of cells as well as the fact that the Sun does not shine all the time. To solve this problem, solar cells are commonly linked to storage batteries.

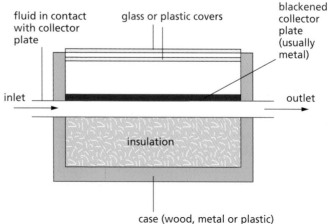

Figure 13.15 A single solar energy panel.

189

Figure 13.16 A bank of photocells.

Biomass and biogas

When any biological material, whether plant or animal, is converted into energy, this energy is called **biomass** energy. It can be taken from animal or plant materials in different ways:

- by burning it, for example wood (Figure 13.17)
- by pressing out oils that can be burned
- by fermenting it to produce fuels such as ethanol or methane.

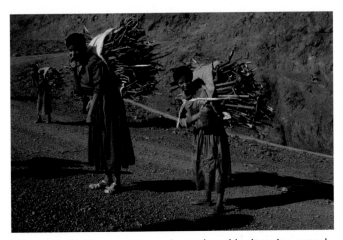

Figure 13.17 Biomass energy is produced by burning wood.

At least 50% of the world's population rely on wood as their main energy source.

In India there are millions of methane generators. Methane generated by the digestion of animal waste is called **biogas**. The biogas produced is used for cooking, heating and lighting. The by-product of this process is an excellent fertiliser.

Some countries have already experimented with ethanol as a fuel for cars. Up to 20% of ethanol can be added to petrol without the need to adjust the carburettor. Brazil, which has few oil reserves, produces ethanol by fermentation (breakdown by enzymes) of sugar cane and grain, and uses it as a petrol additive (Figure 13.18). The Brazilian government has cut down its petrol imports by up to 60% through using this alcohol/petrol mixture.

Figure 13.18 In Brazil cars use an ethanol/petrol mixture.

Hydrogen – the fuel of the future

It is widely believed that hydrogen will play an important part as a fuel in the future. It is already being trialled by major motor manufacturers as an alternative to fossil fuels such as petrol (Figure 13.19).

Hydrogen burns cleanly to form water:

$$\text{hydrogen} + \text{oxygen} \rightarrow \text{water}$$
$$2H_2(g) + O_2(g) \rightarrow 2H_2O(l)$$

Hence virtually no exhaust pollution is created. Also, hydrogen can be obtained in an almost unlimited supply from water by electrolysis. However, at present it is a costly alternative.

Figure 13.19 This car is powered by hydrogen gas.

Questions

1 Draw up a table showing the alternative sources of energy along with their advantages and disadvantages.

2 What is meant by the terms:
   a non-renewable energy sources?
   b renewable energy sources?

■ *Chemical energy*

We obtain our energy needs from the combustion of fuels, such as hydrocarbons, from the combustion of foods and from many other chemical reactions.

Combustion

When natural gas burns in a plentiful supply of air it produces a large amount of energy.

methane + oxygen → carbon + water + heat
                    dioxide   energy

$$CH_4(g) + 2O_2(g) \rightarrow CO_2(g) + 2H_2O(l) + \text{heat energy}$$

During this process, the **complete combustion** of methane, heat is given out. It is an **exothermic** reaction. If only a limited supply of air is available then the reaction is not as exothermic and the poisonous gas carbon monoxide is produced.

methane + oxygen → carbon + water + heat
                    monoxide   energy

$$2CH_4(g) + 3O_2(g) \rightarrow 2CO(g) + 4H_2O(l) + \text{heat energy}$$

This is the **incomplete combustion** of methane.

The energy changes that take place during a chemical reaction can be shown by an **energy level diagram**. Figure 13.20 shows the energy level diagram for the complete combustion of methane.

When any reaction occurs, the chemical bonds in the reactants have to be broken – this requires energy. This is an endothermic process. When the new bonds in the products are formed, energy is given out (Figure 13.21). This is an exothermic process. The **bond energy** is defined as the amount of energy in kilojoules (kJ) associated with the breaking or making of one mole of chemical bonds in a molecular element or compound.

Figure 13.20 Energy level diagram for the complete combustion of methane.

Using the data in Table 13.3, which tells us how much energy is needed to break a chemical bond and how much is given out when it forms, we can calculate how much energy is involved in each stage.

Bond breaking

Breaking 4 C—H bonds in methane requires

$4 \times 435 = 1740\,kJ$

Breaking 2 O=O bonds in oxygen requires

$2 \times 497 = 994\,kJ$

  Total $= 2734\,kJ$ of energy

Table 13.3 Bond energy data.

Bond	Bond energy (kJ mol⁻¹)
C—H	435
O=O	497
C=O	803
H—O	464
C—C	347
C—O	358

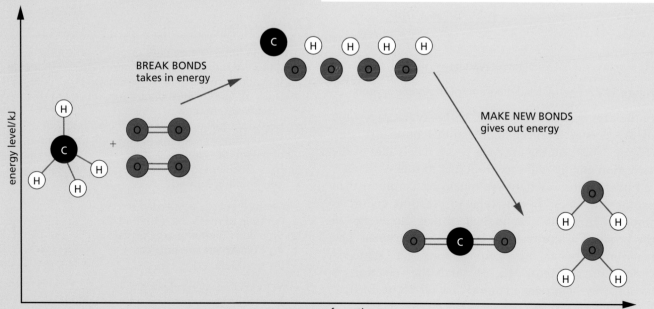

Figure 13.21 Breaking and forming bonds during the combustion of methane.

Making bonds

Making 2 C=O bonds in carbon dioxide gives out

$2 \times 803 = 1606\,kJ$

Making 4 O—H bonds in water gives out

$4 \times 464 = 1856\,kJ$

Total $= 3462\,kJ$ of energy

Energy difference

= energy required − energy given out when
   to break bonds   bonds are made

$= 2734 - 3462$

$= -728\,kJ$

The negative sign shows that the chemicals are losing energy to the surroundings, that is, it is an exothermic reaction. A positive sign would indicate that the chemicals are gaining energy from the surroundings. This type of reaction is called an **endothermic** reaction.

The energy stored in the bonds is called the **enthalpy** and is given the symbol H. The change in energy going from reactants to products is called the **change in enthalpy** and is shown as ΔH (pronounced 'delta H'). ΔH is called the **heat of reaction**.

For an exothermic reaction ΔH is negative and for an endothermic reaction ΔH is positive.

When fuels, such as methane, are burned they require energy to start the chemical reaction. This is known as the **activation energy**, E_A (Figure 13.22).

Figure 13.22 Energy level diagram for methane/oxygen.

In the case of methane reacting with oxygen, it is some of the energy involved in the initial bond breaking (Figure 13.21, on the previous page). The value of the activation energy will vary from fuel to fuel.

Endothermic reactions are much less common than exothermic ones. In this type of reaction energy is absorbed from the surroundings so that the energy of the products is greater than that of the reactants. The reaction between nitrogen and oxygen gases is endothermic (Figure 13.23), and the reaction is favoured by high temperatures (Chapter 15, p. 219).

nitrogen + oxygen → nitrogen(II) oxide

$N_2(g)\ +\ O_2(g)\ \rightarrow\ 2NO(g)\ +\Delta H$

Figure 13.23 Energy level diagram for nitrogen/oxygen.

Dissolving is often an endothermic process. For example, when ammonium nitrate dissolves in water the temperature of the water falls, indicating that energy is being taken from the surroundings. Photosynthesis and thermal decomposition are other examples of endothermic processes.

In equations it is usual to express the ΔH value in units of $kJ\,mol^{-1}$. For example:

$CH_4(g) + 2O_2(g) \rightarrow CO_2(g) + 2H_2O(l)\ \Delta H = -728\,kJ\,mol^{-1}$

This ΔH value tells us that when 1 mole of methane is burned in oxygen, 728 kJ of energy are released. This value is called the **enthalpy of combustion** of methane (or **molar heat of combustion** of methane).

Enthalpy of neutralisation (molar heat of neutralisation)

This is the enthalpy change that takes place when 1 mol of hydrogen ions ($H^+(aq)$) is neutralised.

$H^+(aq) + OH^-(aq) \rightarrow H_2O(l)\ \Delta H = -57\,kJ\,mol^{-1}$

This process occurs in the titration of an alkali by an acid to produce a neutral solution (Chapter 7, p. 108).

Questions

1 Using the bond energy data given in Table 13.3:
  a Calculate the enthalpy of combustion of ethanol, a fuel added to petrol in some countries.
  b Draw an energy level diagram to represent this combustion process.
  c How does this compare with the enthalpy of combustion of heptane (C_7H_{14}), a major component of petrol, of $-4853\,kJ\,mol^{-1}$?
  d How much energy is released per gram of ethanol and heptane burned?

2 How much energy is released if:
  a 0.5 mole of methane is burned?
  b 5 moles of methane are burned?
  c 4 g of methane are burned?
     (A_r: C = 12, H = 1)

3 How much energy is released if:
  a 2 moles of hydrogen ions are neutralised?
  b 0.25 mole of hydrogen ions is neutralised?
  c 1 mole of sulphuric acid is completely neutralised?

Change of state

In Chapter 1, p. 4, we discussed the melting and boiling of a substance. The heating curve for water is shown in Figure 1.11 on p. 5. For ice to melt to produce liquid water, it must absorb energy from its surroundings. This energy is used to break down the weak forces between the water molecules (intermolecular forces) in the ice. This energy is called the **enthalpy of fusion** and is given the symbol ΔH_{fusion}. Similarly, when liquid water changes into steam, the energy required for this process to occur is called the **enthalpy of vaporisation** and is given the symbol ΔH_{vap}. Figure 13.24 shows an energy level diagram representing both the fusion and the vaporisation processes.

Figure 13.24 Energy level diagram for the fusion and vaporisation of water.

Questions

1 Describe the energy changes which take place when the processes described in this section, with water, are reversed.
2 Using the knowledge you have obtained from Chapter 1, p. 5, give a full definition of the enthalpy of fusion and enthalpy of vaporisation for water.

Cells and batteries

A chemical cell is one which produces electrical energy from chemical energy. The electrical energy is produced as a result of the transfer of electrons which takes place during the chemical reaction via a redox process.

A simple type of chemical cell is that shown in Figure 13.25a. In this cell the more reactive metal zinc dissolves in the dilute sulfuric acid, producing zinc ions (Zn^{2+}(aq)) and releasing two electrons.

$$Zn(s) \rightarrow Zn^{2+}(aq) + 2e^-$$

These electrons produced at the zinc electrode cause it to become the negative terminal. They flow to the positive copper terminal through the external circuit via the bulb, which glows. Bubbles of hydrogen are seen when the electrons arrive at the copper electrode. The hydrogen gas is produced from the hydrogen ions in the acid, which collect the electrons appearing at the copper electrode.

$$2H^+(aq) + 2e^- \rightarrow H_2(g)$$

Slowly, the zinc electrode dissolves in the acid and the bulb will then go out. If the zinc is replaced by a more reactive metal, such as magnesium, then the bulb glows more brightly. Magnesium loses electrons more easily as it reacts faster with the dilute acid.

The difference in the reactivity between the two metals used in the cell creates a particular voltage reading on the voltmeter shown in Figure 13.25b. The more the two metals differ in reactivity, the larger is the voltage shown and delivered by the cell. This method can be used to confirm the order of reactivity of the metals (Chapter 9, p. 129). Other types of chemical cell in common use are dry cells used in radios, torches, and so on, and lead–acid accumulators used in motor vehicles. These are convenient and portable energy sources.

a A simple chemical cell.

b The voltage reflects the difference in reactivity of the metals.
Figure 13.25

193

Fuel cells

Scientists have found a much more efficient way of changing chemical energy into electrical energy, using a fuel cell (Figure 13.26). Fuel cells are like the chemical cells in the previous section, except that the reagents are supplied continuously to the electrodes. The reagents are usually hydrogen and oxygen. The fuel cell principle was first discovered by Sir William Grove in 1839.

When he was electrolysing water and he switched off the power supply, he noticed that a current still flowed but in the reverse direction. Subsequently, the process was explained in terms of the reactions at the electrodes' surfaces of the oxygen and hydrogen gases which had been produced during the electrolysis.

The hydrogen fuel cells used by NASA in the US space programme are about 70% efficient and, since the only product is water, they are pollution free. The aqueous NaOH electrolyte is kept within the cell by electrodes which are porous, allowing the transfer of O_2, H_2 and water through them (Figure 13.27). As O_2 gas is passed into the cathode region of the cell it is reduced:

$$O_2 + 2H_2O + 4e^- \rightarrow 4OH^-$$

The OH^- ions formed are removed from the fuel cell by reaction with H_2:

$$H_2 + 2OH^- \rightarrow 2H_2O + 2e^-$$

The electrons produced by this process pass around an external circuit to the cathode.

Figure 13.27 A diagrammatic view of a fuel cell.

Questions

1 Describe how simple chemical cells can be used to confirm the order of reactivity of the metals in the reactivity series.

2 The fuel cell was discovered during electrolysis experiments with water. It is the reverse process which produces the electricity. Write a balanced chemical equation to represent the overall reaction taking place in a fuel cell.

194 Figure 13.26 The space shuttle's computers use electricity produced by fuel cells.

■ *Checklist*

After studying Chapter 13 you should know and understand the following terms.

- **Aerobic decay** Decay which takes place in the presence of oxygen.

- **Anaerobic decay** Decay which takes place in the absence of oxygen.

- **Bond energy** An amount of energy associated with a particular bond in a molecular element or compound.

- **Chain reaction** A nuclear reaction which is self-sustaining as a result of one of the products causing further reactions.

- **Chemical cell** A system for converting chemical energy to electrical energy.

- **Combustion** A chemical reaction in which a substance reacts rapidly with oxygen with the production of heat and light.

- **Endothermic reaction** A chemical reaction which absorbs heat energy from its surroundings.

- **Enthalpy** Energy stored in chemical bonds, given the symbol H.

- **Enthalpy change** Given the symbol ΔH, it represents the difference between energies of reactants and products.

- **Enthalpy of combustion** The enthalpy change which takes place when one mole of a substance is completely burned in oxygen.

- **Enthalpy of fusion** The enthalpy change that takes place when one mole of a solid is changed to one mole of liquid at the same temperature.

- **Enthalpy of neutralisation** The enthalpy change which takes place when one mole of hydrogen ions is completely neutralised.

- **Enthalpy of vaporisation** The enthalpy change that takes place when one mole of liquid is changed to one mole of vapour at the same temperature.

- **Exothermic reaction** A chemical reaction that releases heat energy into its surroundings.

- **Fossil fuels** Fuels, such as coal, oil and natural gas, formed from the remains of plants and animals.

- **Non-renewable energy sources** Sources of energy, such as fossil fuels, which take millions of years to form and which we are using up at a rapid rate.

- **Nuclear fission** The disintegration of a radioactive nucleus into two or more lighter fragments. The energy released in the process is called nuclear energy.

- **Renewable energy** Sources of energy which cannot be used up or which can be made at a rate faster than the rate of use.

1 a State which of the following processes is endothermic and which is exothermic.
   (i) The breaking of a chemical bond.
   (ii) The forming of a chemical bond.
 b The table below shows the bond energy data for a series of covalent bonds.
   (i) Use the information given in the table to calculate the overall enthalpy change for the combustion of ethanol producing carbon dioxide and water.
   (ii) Is the process in (i) endothermic or exothermic?

Bond	Bond energy/kJ mol^{-1}
C—H	435
O=O	497
C=O	803
H—O	464
C—C	347
C—O	358

2 Explain the following.
 a Hydroelectric power is a relatively cheap source of electricity.
 b Geothermal energy is a non-polluting form of energy.
 c A disadvantage of wind power is that it causes noise pollution.
 d The by-product from the process by which methane is generated by the digestion of animal waste is an excellent fertiliser.
 e The fission of uranium-235 in a nuclear reactor is an example of a chain reaction.
 f Tidal- and wave-generated electricity has a major environmental disadvantage.

3 One of the first practical chemical cells was the Daniell cell invented by John Daniell in 1836. A diagram of this type of cell is shown below.

It is capable of generating about 1.1 volts and was used to operate small electrical items such as doorbells.
 a The electrode reaction taking place at a copper anode is:

$$Cu^{2+}(aq) + 2e^- \rightarrow Cu(s)$$

   Write an electrode equation for the process taking place at the cathode.
 b Which way would the electrons flow in the wire connected to the voltmeter – from 'copper to zinc' or 'zinc to copper'?
 c Why should copper(II) sulfate crystallise at the bottom of the outer container?
 d What is the function of the porous pot?
 e There are problems associated with the Daniell cell which have led to it being replaced by other types of cell. Give two reasons why Daniell cells are no longer in use today.

4 This question is about endothermic and exothermic reactions.
 a Explain the meaning of the terms endothermic and exothermic.
 b (i) Draw an energy level diagram for the reaction:

$$NaOH(aq) + HCl(aq) \rightarrow NaCl(aq) + H_2O(l) \ \Delta H = -57\,kJ\,mol^{-1}$$

   (ii) Is this reaction endothermic or exothermic?
   (iii) Calculate the energy change associated with this reaction if 2 moles of sodium hydroxide were neutralised by excess hydrochloric acid.
 c (i) Draw an energy level diagram for the reaction:

$$2H_2O(l) \rightarrow 2H_2(g) + O_2(g) \ \Delta H = +575\,kJ\,mol^{-1}$$

   (ii) Is this reaction endothermic or exothermic?
   (iii) Calculate the energy change for this reaction if only 9 g of water were converted into hydrogen and oxygen.

5 The following results were obtained from an experiment carried out to measure the enthalpy of combustion (heat of combustion) of ethanol. The experiment involved heating a known volume of water with the flame from an ethanol burner. The burner was weighed initially and after the desired temperature rise had been obtained.

Volume of water in glass beaker = 200 cm³
Mass of ethanol burner at start = 85.3 g
Mass of ethanol burner at end = 84.8 g
Temperature rise of water = 12 °C
(Density of water = 1 g cm⁻³)

Heat energy = mass of × 4.2 × temperature
given to   water/g J g⁻¹°C⁻¹ rise/°C
water

a Calculate the mass of ethanol burned.
b Calculate the amount of heat produced, in joules, in this experiment by the ethanol burning.
c Convert your answer to **b** into kilojoules.
d Calculate the amount of heat produced by 1 g of ethanol burning.
e What is the mass of 1 mole of ethanol (C_2H_5OH)? (A_r: H = 1; C = 12; O = 16)
f How much heat would be produced if 1 mole of ethanol had been burned? (This is the heat of combustion of ethanol.)
g Compare your value with the actual value of 1371 kJ mol⁻¹ and suggest two reasons for the difference in values.
h Write a balanced chemical equation to represent the combustion of ethanol.

6 The following results were obtained from a neutralisation reaction between 1 mol dm⁻³ hydrochloric acid and 1 mol dm⁻³ sodium hydroxide. This experiment was carried out to measure the heat of neutralisation of hydrochloric acid. The temperature rise which occurred during the reaction was recorded.

Volume of sodium hydroxide used = 50 cm³
Volume of acid used = 50 cm³
Temperature rise = 5 °C
(Density of water = 1 g cm⁻³)

Heat energy = mass of × 4.2 × temperature
given out   water/g J g⁻¹°C⁻¹ rise/°C
during
reaction

a Write a balanced chemical equation for the reaction.
b What mass of solution was warmed during the reaction?
c How much heat energy was produced during the reaction?
d How many moles of hydrochloric acid were involved in the reaction?
e How much heat would be produced if 1 mole of hydrochloric acid had reacted? (This is the heat of neutralisation of hydrochloric acid.)
f The heat of neutralisation of hydrochloric acid is −57 kJ mol⁻¹. Suggest two reasons why there is a difference between this and your calculated value.

7 Write down which factors are most important when deciding on a particular fuel for the purpose given:
a fuel for a cigarette lighter
b fuel for a camping stove
c fuel for an aeroplane
d fuel for an underground transport system
e fuel for a space shuttle
f fuel for domestic heating.

8 'Propagas' is used in some central heating systems where natural gas is not available. It burns according to the following equation:

$C_3H_8(g) + 5O_2(g) \rightarrow 3CO_2(g) + 4H_2O(l)$ $\Delta H = -2220$ kJ mol⁻¹

a What are the chemical names for 'propagas' and natural gas?
b Would you expect the heat generated per mole of 'propagas' burned to be greater than that for natural gas? Explain your answer.
c What is 'propagas' obtained from?
d Calculate:
(i) the mass of 'propagas' required to produce 5550 kJ of energy
(ii) the heat energy produced by burning 0.5 mole of 'propagas'
(iii) the heat energy produced by burning 11 g of 'propagas'
(iv) the heat energy produced by burning 2000 dm³ of 'propagas'.
(A_r: H = 1; C = 12; O = 16. One mole of any gas occupies 24 dm³ at room temperature and pressure.)

14 The (wider) organic manufacturing industry

Plastics and polymers
Other addition polymers
Condensation polymers
Thermosoftening and thermosetting plastics
Disposal of plastics

Some biopolymers
Hydrolysis of starch

Ethanol (C_2H_5OH) – an alcohol

Cholesterol – a steroid

Biotechnology
Baking
New applications

Ethanoic acid (CH_3COOH) – a carboxylic acid

Other carboxylic acids
Aspirin
Ascorbic acid (vitamin C)
Citric acid

Soaps and detergents
Soapless detergents

Amino acids
Protein analysis

DNA

Pharmaceuticals
Drug abuse

Checklist

Additional questions

Plastics and polymers

In 1933, Reginald Gibson and Eric Fawcett carried out a reaction involving ethene and another organic chemical called benzaldehyde (C_6H_5CHO) using a pressure of about 2000 atmospheres. They were hoping to make these two chemicals react to produce another organic substance from a homologous series called ketones. The reaction vessel leaked and some air (oxygen) got in and much more ethene had to be added. Upon opening the reaction vessel they found a white waxy solid instead of what they expected. They subsequently discovered that this solid was in fact a new type of substance. It was found to consist of very large molecules, each one made up of many thousands of ethene molecules (Figure 14.1). 'Many' in Greek is *poly*, and so the new substance was called '**poly(ethene)**' (or polythene). The modern plastics industry was born with this accidental discovery.

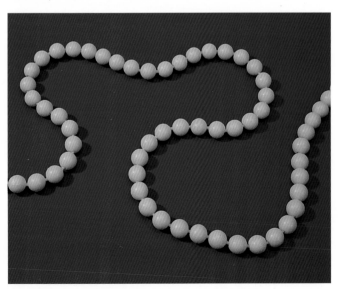

Figure 14.1 Molecules of ethene in this poly(ethene) chain are represented by individual poppet beads.

Poly(ethene) has many useful properties.

- It is easily moulded.
- It is an excellent electrical insulator.
- It does not corrode.
- It is tough.
- It is not affected by the weather.
- It is durable.

It was first used to insulate telephone cables and its unique electrical properties were essential during the development of radar. Its properties continued to be exploited, and today it can be found as a substitute for natural materials in plastic bags, sandwich boxes, washing-up bowls, wrapping film, milk-bottle crates and squeezy bottles.

We now manufacture poly(ethene) by heating ethene to a relatively high temperature under a high pressure in the presence of a catalyst.

$$n \begin{pmatrix} H & & H \\ & C=C & \\ H & & H \end{pmatrix} \longrightarrow \begin{pmatrix} H & H \\ -C-C- \\ H & H \end{pmatrix}_n$$

where *n* is a very large number. In poly(ethene) the ethene molecules have joined together to form a very long hydrocarbon chain (Figure 14.2). The ethene molecules are able to form chains like this because they possess carbon–carbon double bonds.

Figure 14.2 This model shows part of the poly(ethene) polymer chain.

Poly(ethene) is produced in three main forms:

- low density poly(ethene) (LDPE)
- linear low density poly(ethene) (LLDPE)
- high density poly(ethene) (HDPE).

The world production of all types of poly(ethene) is about 52 million tonnes per year.

Other alkene molecules can also produce substances like poly(ethene); for example, propene produces poly-(propene), which is used to make ropes and packaging. When small molecules like ethene join together to form long chains of atoms, called **polymers**, the process is called **polymerisation**. The small molecules, like ethene, which join together in this way are called **monomers**. A polymer chain, a very large molecule or a macromolecule, often consists of many thousands of monomer units and in any piece of plastic there will be many millions of polymer chains. Since in this polymerisation process the monomer units add together to form only one product, the polymer, the process is called **addition polymerisation**.

a Model of chloroethene, the PVC monomer.

Other addition polymers

Many other addition polymers have been produced. Often the plastics are produced with particular properties in mind, for example PVC (polyvinyl chloride or poly(chloroethene)) and PTFE (poly(tetrafluoroethene)). Both of these plastics have monomer units similar to ethene.

b Model of part of a PVC polymer chain.

Figure 14.3

$$
\begin{array}{cc}
H \quad\quad H & F \quad\quad F \\
\diagdown \quad \diagup & \diagdown \quad \diagup \\
C{=}C & C{=}C \\
\diagup \quad \diagdown & \diagup \quad \diagdown \\
H \quad\quad Cl & F \quad\quad F
\end{array}
$$

PVC monomer
(vinyl chloride or
chloroethene)

PTFE monomer
(tetrafluoroethene)

If we use chloroethene (Figure 14.3a), the polymer we make is slightly stronger and harder than poly(ethene) and is particularly good for making pipes for plumbing.

$$
n \left(\begin{array}{cc} H & H \\ & \\ C{=}C \\ & \\ H & Cl \end{array} \right) \longrightarrow \left(\begin{array}{cc} H & H \\ {-}C{-}C{-} \\ H & Cl \end{array} \right)_n
$$

monomer   polymer chain

PVC is the most versatile plastic and is the second most widely used, after poly(ethene). Worldwide 27 million tonnes are produced annually.

If we start from tetrafluoroethene (Figure 14.4a, overleaf) the polymer we make, PTFE, has some slightly unusual properties:

- it will withstand very high temperatures, of up to 260°C
- it forms a very slippery surface
- it is hydrophobic (water repellent)
- it is highly resistant to chemical attack.

These properties make PTFE an ideal 'non-stick' coating for frying pans and saucepans.

Every year 50000 tonnes of PTFE are made.

$$
n \left(\begin{array}{cc} F & F \\ & \\ C{=}C \\ & \\ F & F \end{array} \right) \longrightarrow \left(\begin{array}{cc} F & F \\ {-}C{-}C{-} \\ F & F \end{array} \right)_n
$$

monomer   polymer chain

The properties of some addition polymers along with their uses are given in Table 14.1 (overleaf).

a Model of tetrafluoroethene, the PTFE monomer.

b Model of part of the PTFE polymer chain.

Figure 14.4

Table 14.1 Some addition polymers.

Plastic	Monomer	Properties	Uses
Poly(ethene)	$CH_2{=}CH_2$	Tough, durable	Carrier bags, bowls buckets, packaging
Poly(propene)	$CH_3CH{=}CH_2$	Tough, durable	Ropes, packaging
PVC	$CH_2{=}CHCl$	Strong, hard (less flexible than poly(ethene))	Pipes, electrical insulation, guttering
PTFE	$CF_2{=}CF_2$	Non-stick surface, withstands high temperatures	Non-stick frying pans, soles of irons
Polystyrene	$CH_2{=}CHC_6H_5$	Light, poor conductor of heat	Insulation, packaging (especially as foam)
Perspex	$CH_2{=}C(CO_2CH_3)CH_3$	Transparent	Used as a glass substitute

Condensation polymers

Wallace Carothers discovered a different sort of plastic when he developed nylon in 1935. Nylon is made by reacting two different chemicals together, unlike poly(ethene) which is made only from monomer units of ethene. Poly(ethene), formed by addition polymerisation, can be represented by:

$$-A-A-A-A-A-A-A-A-A-A-$$

where A = monomer.

The starting molecules for nylon are more complicated than those for poly(ethene) and are called 1,6-diaminohexane and hexanedioic acid.

1,6–diaminohexane   +   hexanedioic acid
$H_2N(CH_2)_6NH_2$ $HOOC(CH_2)_4COOH$

$$\downarrow$$

$H_2N(CH_2)_6\textbf{NHOC}(CH_2)_4COOH$ + H_2O

amide link

The polymer chain is made up from the two starting molecules arranged alternately (Figure 14.5) as these molecules react and therefore link up. Each time a reaction takes place a molecule of water is lost, and because of this it is called **condensation polymerisation**. This differs from addition polymerisation, where there is only one product. Because an amide link is formed during the polymerisation, nylon is known as a **polyamide**. This is the same amide link as found in proteins. It is often called the peptide link. This type of polymerisation, in which two kinds of monomer unit react, results in a chain of the type:

$$-A-B-A-B-A-B-A-B-A-B-$$

Generally, polyamides have the structure

Figure 14.5 A nylon polymer chain is made up from the two molecules arranged alternately just like the two different coloured poppet beads in the photo.

When nylon is made in industry, it forms as a solid which is melted and forced through small holes (Figure 4.21, p. 60). The long filaments cool and solid nylon fibres are produced which are stretched to align the polymer molecules and then dried. The resulting yarn can be woven into fabric to make shirts, ties, sheets and parachutes or turned into ropes or racket strings for tennis and badminton rackets. The worldwide production of nylon is 5.4 million tonnes per year.

We can obtain different polymers with different properties if we carry out condensation polymerisation reactions between other monomer molecules. For example, if we react ethane-1,2-diol with benzene-1,4-dicarboxylic acid, then we produce a polymer called terylene.

$$\begin{array}{ccc}
\text{ethane-1,2-diol} & & \text{benzene-1,4-dicarboxylic acid} \\
HO(CH_2)_2OH & + & HOOC(C_6H_4)COOH
\end{array}$$

$$\downarrow$$

$$HO(CH_2)_2\mathbf{OCO}(C_6H_4)COOH \ + \ H_2O$$

ester link

This ester link is the same linkage as in fats. Generally, polyesters have the structure

Like nylon, terylene can be turned into yarn, which can then be woven. Terylene clothing is generally softer than that made from nylon but both are hard wearing. Because an ester link is formed during the polymerisation, terylene is known as a **polyester**.

Fats possess the same sort of linkage as terylene but have different units. For a further discussion see p. 207.

Thermosoftening and thermosetting plastics

Plastics can be put into one of two categories. If they melt or soften when heated (like poly(ethene), PVC and polystyrene) then they are called **thermoplastics** or **thermosoftening plastics**. If they do not soften on heating but only char and decompose on further heating, they are known as **thermosetting plastics**.

Thermoplastics are easily moulded or **formed** into useful articles. Once they are molten they can be injected or blown into moulds, and a variety of different-shaped items can be produced (Figure 14.6). Thermosetting plastics can be heated and moulded only once, usually by compression moulding (Figure 14.7).

Figure 14.8 shows the different molecular structures for thermosetting and thermosoftening plastics. Thermosetting plastics have polymer chains which are linked or bonded to each other to give a **cross-linked** structure, and so the chains are held firmly in place and no softening takes place on heating. Thermosoftening plastics do not have polymer chains joined in this way, so when they are subjected to heat their polymer chains flow over one another and the plastic softens.

Figure 14.6 Blow moulding in progress in the recycling of poly(ethene).

Figure 14.7 These objects are made from compression-moulded thermosetting plastic.

a In thermosetting plastic the chains are cross-linked.

b In thermosoftening plastic there is no cross-linking.

Figure 14.8

Disposal of plastics

In the last 30 years plastics have taken over as replacement materials for metals, glass, paper and wood as well as for natural fibres such as cotton and wool. This is not surprising since plastics are light, cheap, relatively unreactive, can be easily moulded and can be dyed bright colours. However, plastics have contributed significantly to the household waste problem (Figure 14.9).

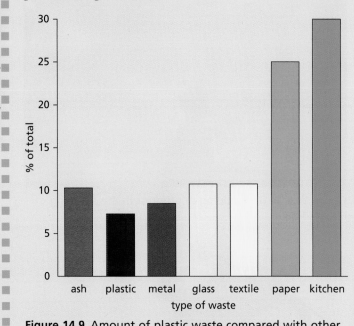

Figure 14.9 Amount of plastic waste compared with other household waste in EU countries.

In the recent past much of our plastic waste has been used to landfill disused quarries. However, all over the world these sites are getting harder to find and it is becoming more and more expensive. The alternatives to dumping plastic waste are certainly more economical and more satisfactory.

- Incineration – schemes have been developed to use the heat generated for heating purposes. However, problems with the combustion process (which can result in the production of toxic gases) mean that especially high temperatures have to be employed during the incineration process.
- Recycling – large quantities of black plastic bags and sheeting are produced for resale.
- Biodegradable plastics, as well as those polymers that degrade in sunlight (**photodegradable**, Figure 14.10a), have been developed. Other common categories of degradable plastics include synthetic biodegradable plastics which are broken down by bacteria, as well as plastics which dissolve in water (Figure 14.10b). The property that allows plastic to dissolve in water has been used in relatively new products, including soluble capsules containing liquid detergent. However, the vast majority of polymers are still non-biodegradable.

a This plastic bag is photodegradable.

b This plastic dissolves in water.

Figure 14.10

Questions

1 Write the general equation to represent the formation of polystyrene from its monomer.

2 Suggest other uses for the polymers shown in Table 14.1 on p. 200.

3 Draw the structure of the repeating units found in:
   a nylon
   b terylene.

4 Explain the differences between an addition polymer and a condensation polymer.

5 Give two advantages and two disadvantages of plastic waste (rubbish).

6 Draw the structure of the repeating unit of the addition polymer formed from $CH_3—CH=CH_2$.

7 Draw the structure of the monomer from which the addition polymer below has been produced.

$$\begin{array}{c} C_6H_5\ \ H \quad\quad C_6H_5\ \ H \quad\quad C_6H_5\ \ H \\ \ \ \ |\ \ \ \ \ |\quad\quad\ \ \ |\ \ \ \ \ |\quad\quad\ \ \ |\ \ \ \ \ | \\ -C—C—C—C—C—C- \\ \ \ \ |\ \ \ \ \ |\quad\quad\ \ \ |\ \ \ \ \ |\quad\quad\ \ \ |\ \ \ \ \ | \\ \ \ \ H\ \ \ H \quad\quad\ \ \ H\ \ \ H \quad\quad\ \ \ H\ \ \ H \end{array}$$

Some biopolymers

Starch is a **biopolymer** or **natural polymer**. It is a condensation polymer of glucose, a type of sugar. It is often produced as a way of storing energy and is formed as a result of photosynthesis in green plants.

$$\text{carbon dioxide} + \text{water} \xrightarrow[\substack{\text{sunlight/}\\\text{chlorophyll}\\\text{(in green plants)}}]{\text{photosynthesis}} \text{glucose} + \text{oxygen}$$

$$6CO_2(g) + 6H_2O(l) \longrightarrow C_6H_{12}O_6(aq) + 6O_2(g)$$

$$\text{glucose} \rightarrow \text{starch} + \text{water}$$
$$nC_6H_{12}O_6(aq) \rightarrow (C_6H_{10}O_5)_n(s) + nH_2O(l)$$

Starch can be represented by

—O—▭—O—▭—O—▭—O—

Both starch and glucose are **carbohydrates**, a class of naturally occurring organic compounds which can be represented by the general formula $(CH_2O)_x$. Along with fats and proteins, they are among the main constituents of food. Starch occurs in potatoes, rice and wheat. Glucose, from which starch is polymerised, belongs to a group of simple carbohydrates known as **monosaccharides**. They are sweet to taste and soluble in water. Starch belongs to the more complicated group of carbohydrates known as **polysaccharides**. Starch does not form a true solution and it does not have a sweet taste. With iodine it gives an intense blue colour (nearly black), which is used as a test for starch or iodine itself (Figure 14.11).

Figure 14.11 A dark blue–black colour is produced when dilute iodine solution is applied to starch, for example in a potato.

Hydrolysis of starch

Starch can be broken down in two ways, both of which take place in the presence of water. Hence the reactions are known as **hydrolysis** reactions. Hydrolysis of starch is the key reaction that enables us to use this energy source. If starch is boiled for about one hour with dilute hydrochloric acid, it is broken down into its monomers, glucose molecules.

$$\text{starch} + \text{water} \xrightarrow[\text{heat}]{\text{dilute acid}} \text{glucose}$$

$$(C_6H_{10}O_5)_n(s) + nH_2O(l) \longrightarrow nC_6H_{12}O_6(aq)$$

If starch is mixed with saliva and left to stand for a few minutes, it will break down to maltose, a **disaccharide** (that is two joined monosaccharides). The enzyme present in the saliva, called amylase, catalyses this hydrolysis reaction.

$$\text{starch} + \text{water in saliva} \xrightarrow{\text{amylase}} \text{maltose}$$

$$2(C_6H_{10}O_5)_n(s) + nH_2O(l) \longrightarrow nC_{12}H_{22}O_{11}(aq)$$

Enzymes are very efficient natural catalysts present in plants and animals. They do not require high temperatures to break down the starch to maltose. In humans, a salivary amylase breaks down the starch in our food. If you chew on a piece of bread for several minutes, you will notice a sweet taste in your mouth. The above hydrolysis reactions are summarised in Figure 14.12.

Figure 14.12 Starch produces glucose or maltose depending on the type of hydrolysis used. Using chromatography it is possible to separate and identify the products of the different styles of hydrolysis.

Questions

1 In the hydrolysis of starch, how, using a chemical test, could you tell whether all the starch had been hydrolysed?

2 Describe a method you could possibly use to identify the products of the different types of hydrolysis.

■ *Ethanol (C₂H₅OH) – an alcohol*

The alcohols form another homologous series with the general formula $C_nH_{2n+1}OH$. All the alcohols possess an –OH as the **functional group**. The functional group is the group of atoms responsible for the characteristic reactions of a compound. Table 14.2 shows the first four members along with their melting and boiling points.

Table 14.2 Some members of the alcohol family.

Alcohol	Formula	Melting point/°C	Boiling point/°C
Methanol	CH_3OH	−94	64
Ethanol	CH_3CH_2OH	−117	78
Propanol	$CH_3CH_2CH_2OH$	−126	97
Butanol	$CH_3CH_2CH_2CH_2OH$	−89	117

Ethanol is by far the most important of the alcohols and is often just called 'alcohol'. Ethanol can be produced by fermentation (p. 205) as well as by the hydration of ethene (Chapter 12, p. 180). It is a neutral, colourless, volatile liquid which does not conduct electricity. Ethanol burns quite readily with a clean, hot flame.

$$ethanol + oxygen \rightarrow carbon\ dioxide + water + energy$$

$$CH_3CH_2OH(l) + 3O_2(g) \rightarrow 2CO_2(g) + 3H_2O(g) + energy$$

As methylated spirit, it is used in spirit (camping) stoves. Methylated spirit is ethanol with small amounts of poisonous substances added to stop people drinking it. Some countries, like Brazil, already use ethanol mixed with petrol as a fuel for cars (Chapter 13, p. 190) and this use is increasing worldwide.

Many other materials, such as food flavourings, are made from ethanol. Ethanol is a very good solvent and evaporates easily. It is therefore used extensively as a solvent for paints, glues, aftershave and many other everyday products.

Figure 14.13 shows the arrangement of the atoms in the first four alcohols.

Figure 14.13 The molecules look like their corresponding models in the photographs.

Ethanol can be oxidised to ethanoic acid (an organic acid also called acetic acid) by powerful oxidising agents, such as warm acidified potassium dichromate(VI), or potassium manganate(VII). During the reaction the orange colour of potassium dichromate(VI) changes to a dark green (Figure 14.14) as the ethanol is oxidised to ethanoic acid.

$$ethanol + oxygen \xrightarrow{heat} ethanoic\ acid + water$$
(from potassium dichromate(VI))

$$CH_3CH_2OH(l) + 2[O] \longrightarrow CH_3COOH(aq) + H_2O(l)$$

Figure 14.14 Orange potassium dichromate(VI) slowly turns green as it oxidises ethanol to ethanoic acid. Potassium manganate(VII) turns from purple to colourless.

Question

1 Write a word and balanced chemical equation for:
 a the combustion of butanol
 b the oxidation of butanol.

Cholesterol – a steroid

Cholesterol is a naturally occurring and essential chemical. It belongs to a family of chemicals called steroids and also contains an alcohol group (Figure 14.15). Cholesterol is found in almost all of the tissues in the body, including nerve cells. Levels of cholesterol above normal (above 6.5 mmol/l) are associated with an increased risk of heart disease. Cholesterol hardens and blocks off arteries by building up layers of solid material (atheroma) inside the arteries (Figure 14.16). This is particularly serious if the arteries that supply the heart or brain are blocked. Simple tests are now available to monitor cholesterol levels and people with high levels can be treated and can follow special low-fat and low-cholesterol diets.

Figure 14.15 The structure of cholesterol.

a This artery is being blocked by atheroma, which may be related to high levels of cholesterol in the blood.

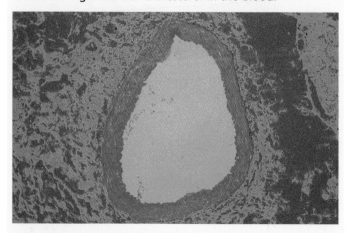

b This is a healthy artery.
Figure 14.16

Biotechnology

Biotechnology involves making use of micro-organisms or their components, such as enzymes, for the benefit of humans to produce, for example, foods such as yoghurt and bread. One of the oldest biotechnologies is that of **fermentation**. It involves a series of biochemical reactions brought about by micro-organisms or enzymes.

Fermentation in the laboratory can be carried out using sugar solution. A micro-organism called yeast is added to the solution. The yeast uses the sugar for energy during **anaerobic respiration** (respiration without oxygen), and so the sugar is broken down to give carbon dioxide and ethanol. The best temperature for this process to be carried out is at 37 °C.

$$\text{glucose} \xrightarrow{\text{yeast}} \text{ethanol} + \text{carbon dioxide}$$
$$C_6H_{12}O_{6(aq)} \longrightarrow 2C_2H_5OH_{(l)} + 2CO_{2(g)}$$

Figure 14.17 shows a simple apparatus for obtaining ethanol from glucose in the laboratory.

Figure 14.17 Fermenting glucose and yeast to produce ethanol. The bag is inflated during the experiment by CO_2.

Baking

To make bread, fresh yeast is mixed with warm sugar solution and the mixture added to the flour. This dough mixture is then put into a warm place to rise. The dough rises due to the production of carbon dioxide from **aerobic respiration** (respiration with oxygen) by the yeast. The products of this style of respiration are different to those of anaerobic respiration.

$$\text{sugar} + \text{oxygen} \xrightarrow{\text{yeast}} \text{carbon dioxide} + \text{water} + \text{energy}$$
$$C_6H_{12}O_{6(aq)} + 6O_{2(g)} \longrightarrow 6CO_{2(g)} + 6H_2O_{(l)} + \text{energy}$$

After the dough has 'risen', it is baked and the heat kills the yeast and the bread stops rising.

205

New applications

A large number of firms throughout the world are investing large sums of money in the newer biotechnology applications now in use.

- Enzymes can be isolated from micro-organisms and used to catalyse reactions in other processes. For example, proteases are used in biological detergents to digest protein stains such as blood and food. Also, catalase is used in the rubber industry to help convert latex into foam rubber.
- Our ability to manipulate an organism's genes to make it useful to us is called **genetic engineering**. This is being used, for example, to develop novel plants for agriculture as well as making important human proteins such as the hormones insulin and growth hormone.

However, a word of caution is necessary. The new biotechnologies may not be without dangers. For example, new pathogens (organisms that cause disease) might be created accidentally. Also, new pathogens may be created deliberately for use in warfare. As you can imagine, there are very strict guidelines covering these new biotechnologies, especially in the area of research into genetic engineering.

Question

1 What do you understand by the term 'biotechnology'? In your answer make reference to the making of bread.

Ethanoic acid (CH₃COOH) – a carboxylic acid

The carboxylic acids form another homologous series, this time with the general formula $C_nH_{2n+1}COOH$. All the carboxylic acids possess –COOH as their functional group. Table 14.5 shows the first four members of this homologous series along with their melting and boiling points. Figure 14.18 shows the actual arrangement of the atoms in these members of this family.

Table 14.5 Some members of the carboxylic acid series.

Carboxylic acid	Formula	Melting point/°C	Boiling point/°C
Methanoic acid	HCOOH	9	101
Ethanoic acid	CH₃COOH	17	118
Propanoic acid	CH₃CH₂COOH	−21	141
Butanoic acid	CH₃CH₂CH₂COOH	−6	164

Figure 14.18 The molecules look like the 3D models in the photographs.

Methanoic acid is present in stinging nettles and ant stings. Ethanoic acid, however, is the most well known as it is the main constituent of vinegar. Like other acids, ethanoic acid affects indicators and will react with metals such as magnesium. However, whereas the mineral acids such as hydrochloric acid are called strong acids, ethanoic acid is a weak acid (Chapter 7, p. 99). Even though it is a weak acid, it will still react with bases to form salts. For example, the salt sodium ethanoate is formed when ethanoic acid reacts with dilute sodium hydroxide.

ethanoic + sodium → sodium + water
acid   hydroxide   ethanoate

$$CH_3COOH(aq) + NaOH(aq) \rightarrow CH_3COONa(aq) + H_2O(l)$$

Ethanoic acid also undergoes other typical reactions of acids, in that it reacts with indicators, metals and carbonates in the usual way.

Ethanoic acid will react with ethanol, in the presence of a few drops of concentrated sulphuric acid, to produce ethyl ethanoate – an ester.

$$\text{ethanoic acid} + \text{ethanol} \underset{\text{H}_2\text{SO}_4}{\overset{\text{conc}}{\rightleftharpoons}} \text{ethyl ethanoate} + \text{water}$$

$$CH_3COOH(l) + CH_3CH_2OH(l)$$
$$\rightleftharpoons CH_3COOCH_2CH_3(aq) + H_2O(l)$$

This reaction is called **esterification**.

Members of the 'ester' family have strong and pleasant smells. Many of them occur naturally and are responsible for the flavours in fruits and the smells of flowers. They are used, therefore, in some food flavourings and in perfumes (Figure 14.19).

Figure 14.19 Perfumes contain esters.

Fats and oils are naturally occurring esters which are used as energy storage compounds by plants and animals (Figure 14.20). They possess the same linkage as terylene but have different units (see p. 201).

Figure 14.20 Olive oil is obtained from a plant whereas lard is made from animal fat.

Questions

1 Write the structural formula for propanoic acid.

2 Write word and balanced chemical equations for the esterification of propanoic acid with ethanol.

Other carboxylic acids

Aspirin

Aspirin (Figure 14.21) is one of the most frequently used painkillers in the world. It is also able to reduce inflammation and fever and a low dose taken on a daily basis over the age of 50 may prevent heart attacks. It is derived from another acid, salicylic acid, which can be obtained from willow bark. Salicylic acid has the same medicinal properties as aspirin and has been known since 1829. Salicylic acid, however, caused stomach bleeding. The conversion of salicylic acid to aspirin reduced these problems, but aspirin still has some adverse effects on the stomach if taken in excess.

Figure 14.21 The structure of aspirin.

Ascorbic acid (vitamin C)

Vitamin C, also known as ascorbic acid, is an essential vitamin (Figure 14.22). Vitamin C is required by the body in very small amounts and it is obtained from foods.

Figure 14.22 The structure of ascorbic acid (vitamin C)

Vitamin C is important to all animals, including humans, and is vital in the production of collagen. Collagen is important in the formation of connective tissues that give our body shape and help to support vital organs. Vitamin C prevents the disease scurvy. It is found in citrus fruits and brightly coloured vegetables, such as peppers and broccoli. Many people take vitamin C supplements, which are readily available from supermarkets and pharmacies (Figure 14.23, overleaf). Although vitamin C is destroyed by exposure to air and heat, the average person usually reaches the recommended daily allowance of 60 mg through food.

Figure 14.23 These fruits and supplements all contain vitamin C.

Citric acid

Citric acid is an example of a tricarboxylic acid, one which contains three —COOH groups (Figure 14.24). It is an important acid and is found in all citrus fruits, for example lemons and oranges.

citric acid

Figure 14.24 The structure of citric acid.

Soaps and detergents

Millions of tonnes of soaps and soapless detergents are manufactured worldwide every year. Soap is manufactured by heating natural fats and oils of either plants or animals with a strong alkali. These fats and oils, called triglycerides, are complicated ester molecules.

Fat is boiled with aqueous sodium hydroxide to form soap. The esters are broken down in the presence of water – hydrolysed. This type of reaction is called **saponification**. The equation given below is that for the saponification of glyceryl stearate (a fat).

glyceryl + sodium → sodium stearate + glycerol
stearate   hydroxide   (soap)

The cleaning properties of the soap depend on its structure and bonding. Sodium stearate consists of a long hydrocarbon chain which is hydrophobic (water hating) attached to an ionic 'head' which is hydrophilic (water loving) (Figure 14.25).

Figure 14.25 Simplified diagram of a soap molecule.

Covalent compounds are generally insoluble in water but they are more soluble in organic solvents. Ionic compounds are generally water soluble but tend to be insoluble in organic solvents. When a soap is put into water which has a greasy dish (or a greasy cloth) in it, the hydrophobic hydrocarbon chain on each soap molecule becomes attracted to the grease and becomes embedded in it (Figure 14.26).

Figure 14.26 Soap dissolves grease like this.

On the other hand, the hydrophilic ionic head group is not attracted to the grease but is strongly attracted to the water molecules. When the water is stirred, the grease is slowly released and is completely surrounded by the soap molecules. The grease is, therefore, 'solubilised' and removed from the dish. The soap is able to remove the grease because of the combination of the covalent and ionic bonds present.

Soapless detergents

In Chapter 8, p. 120, we discussed the way in which, in hard water areas, an insoluble scum forms when soap is used. This problem has been overcome by the development of synthetic **soapless detergents**. These new substances do not form scum with hard water since they do not react with Ca^{2+} and Mg^{2+} present in such water. Furthermore, these new soapless detergent molecules have been designed so that they are biodegradable. Bacteria readily break down these new molecules so that they do not persist in the environment.

Sodium alkyl benzene sulfonates were developed in the early 1970s. The structure of sodium 3-dodecylbenzene sulfate, $C_{18}H_{29}SO_3Na$, is given below.

The calcium and magnesium salts of this detergent molecule are water soluble, so the problem of scum is solved. Very many of our washing powders (and liquids) contain this type of substance. The manufacture of soapless detergents is explained in Chapter 16, p. 232.

Questions

1 What class of organic compound do substances like glyceryl stearate belong to?

2 What do you understand by the terms:
   a hydrophobic?
   b hydrophilic?
   c saponification?

3 What is the main advantage of detergents over soaps?

■ *Amino acids*

There are 20 different amino acids and they each possess two functional groups. One is the carboxylic acid group, –COOH. The other is the amine group, –NH₂. The two amino acids shown at the top of the next column are glycine and alanine.

Amino acids are the building blocks of **proteins**. Similar to nylon (see p. 200) proteins are polyamides, as they contain the –CONH– group, which is called the amide or, in the case of proteins, the peptide link. Proteins are formed by condensation polymerisation.

a **dipeptide**
(composed of two amino acids joined together)

Protein chains formed by the reaction of many amino acid molecules have the general structure shown below.

Further reaction with many more amino acids takes place at each end of each molecule to produce the final protein (Figure 14.27, overleaf). For a molecule to be a protein, there must be at least 100 amino acids involved. Below this number, they are called polypeptides. Proteins make up some 15% of our body weight.

Proteins fall broadly into two groups: they can be fibrous or globular.

- Fibrous proteins – these have linear molecules, are insoluble in water and resistant to alkalis and acids. Collagen (in tendons and muscles), keratin (in nails, hair, horn and feathers) and elastin (in arteries) are all fibrous proteins.
- Globular proteins – these have complicated three-dimensional structures and are soluble in water. They are easily affected by acids, alkalis and temperature increase, when they are said to be denatured. Casein (in milk), albumen (in egg white) and enzymes are examples of globular proteins.

A chain

H₂N — Gly Ile Val Glu Gln Cys Cys Thr Ser Ile Cys Ser Leu Tyr Gin Leu Glu Asn Tyr Cys Asn — COOH

1   5   10   15

B chain

H₂N — Phe Val Asn Gin His Leu Cys Gly Ser His Leu Val Glu Ala Leu Tyr Leu Val Cys Gly Glu

1   5   10   15   20

Arg

Gly

Phe

HOOC — Thr Lys Pro Thr Trp Phe

30   25

Figure 14.27 The structure of a protein – human insulin (the different coloured circles represent different amino acids in this protein).

Protein analysis

How can you determine which amino acids are present in a particular protein? This involves hydrolysis of the peptide (amide) bonds in the protein so that the individual amino acids are released. This can only be done by heating the protein with dilute hydrochloric acid. The mixture of amino acids is then separated by thin layer chromatography (TLC) (Figure 14.28) or electrophoresis. In both cases, a locating agent (Chapter 2, p. 24), such as ninhydrin, is used. This ensures that the spots of amino acid are visible.

a Testing for a protein.

b Adding dilute sodium hydroxide.

Figure 14.28 Amino acids can be separated and identified by TLC.

If you are trying to show only the presence of a protein, a quick test to carry out is known as the **Biuret test**. A mixture of dilute sodium hydroxide and 1% copper(II) sulfate solution is shaken with a sample of the material under test. If a protein is present, a purple colour appears after about three minutes (Figure 14.29).

c Adding 1% copper(II) sulfate.

d The purple colour shows the presence of a protein.

Figure 14.29

DNA

Deoxyribonucleic acid (DNA) belongs to a group of chemicals called the nucleic acids (Figure 14.30). They are also biopolymers. DNA controls the protein synthesis within your cells. When you eat a food containing proteins, such as meat or cheese, your digestive enzymes break down the proteins present into individual amino acids. The DNA in your cells controls the order in which the amino acids are repolymerised to make the proteins you need!

Figure 14.30 Crick and Watson based this model for DNA on X-ray studies and chemical analysis.

Genes are the units of heredity that control the characteristics of organisms. A gene is made of DNA. No two individuals have the same DNA sequence. DNA 'fingerprinting' has become a very powerful forensic science tool in the investigation of crime (Figure 14.31).

Figure 14.31 Identical patterns shown by DNA fingerprinting can identify the criminal.

Questions

1 Which two functional groups do amino acids possess?

2 How many amino acids have to be involved before the biopolymer is called a protein?

3 Name the process by which the individual amino acids in a protein are released by reaction with a dilute acid.

4 Explain how DNA fingerprinting may be used in paternity suits.

Pharmaceuticals

Pharmaceuticals are **drugs** that are prepared and sold with the intention of treating illness. A drug is any substance, natural or synthetic, which alters the way in which the body works. There are many categories of drugs. The following are some examples.

- Anaesthetics – these induce loss of feeling and/or consciousness, for example fluothane.
- Analgesics – these relieve pain, for example aspirin.
- Antibiotics – these are substances, for example penicillin, originally produced by micro-organisms, which are used to kill bacteria. However, most antibiotics are now made in chemical laboratories, for example carbenicillin.
- Sedatives – these induce sleep, for example barbiturates.
- Tranquillisers – these will give relief from anxiety, for example Valium.

There are, of course, many other types of drug available which have very specific uses. For example, methyldopa was developed to relieve hypertension (high blood pressure), and antihistamines were developed to help control travel sickness, hayfever and allergic reactions.

The pharmaceutical industry is one of the most important parts of the chemical industry and is a major consumer of the products of the petrochemical industry. It is a high-profit industry but with very high research and development costs. For example, it costs in excess of £100 million to discover, test and get a single drug on to the market.

Today, the pharmaceutical industry could be called the 'medicines by design' industry. Companies such as GlaxoSmithKline have teams of chemists and biochemists working almost around the clock to discover, test, check for safety and produce drugs that can deal with almost every known illness.

Table 14.6 (overleaf) shows the structures of a selection of some of the more common drugs available at the present time, along with their uses. The common names for these drugs are used, since their systematic, theoretical names are extremely complex.

Table 14.6 Commonly available drugs.

Name and structure	For the treatment of
Aspirin	Headaches, mild pain, heart disease
Paracetamol	Headaches, mild pain
Ibuprofen	Arthritis, fever, mild to moderate pain
Methyldopa	High blood pressure (hypertension)
Penicillin	A variety of bacterial infections
Valium	Feelings of anxiety or depression

Drug abuse

Some of the very useful drugs developed by chemists can be habit forming. For example, barbiturates (in sleeping tablets) and amphetamines (stimulants) fall into this category. Another drug that has created problems in the past is Valium, which is not itself addictive but when used in the long term makes people dependent on it. Severe psychological and physiological problems can arise.

It should be noted, however, that it is the opiates which cause **addiction** (Figure 14.32). Cocaine and heroin are just two examples of such substances. Consequences of the addiction include personal neglect, both of nutritional needs and of hygiene. For a short-term feeling of well-being ('fix'), the addict is prepared to do almost anything. Addicts often turn to a life of crime to fulfil their cravings for the opiates. Addicts, especially those injecting drugs, are at a high risk of HIV (human immunodeficiency virus) infection as they often share needles with other drug addicts, who may be HIV positive. Public awareness campaigns aim to educate everyone on the dangers in society, including drug abuse and its related risks.

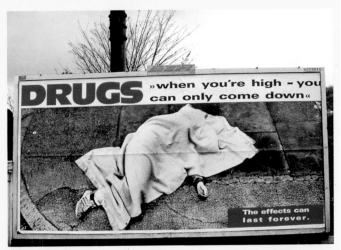

Figure 14.32 Increasing public awareness of the risks associated with drug abuse is very important.

Questions

1 Using the data given in Table 14.6, suggest which of the pharmaceuticals contains:
   a sulfur
   b an –OH group
   c an –NH_2 group
   d an ester group.

2 Drug abuse is a rapidly growing problem worldwide. Using the information given in this section as well as your research skills, make a list of the addictive drugs. Also explain the problems that drug abuse can cause.

■ *Checklist*

After studying Chapter 14 you should know and understand the following terms.

- **Addition polymer** A polymer formed by an addition reaction. For example, poly(ethene) is formed from ethene.

- **Alcohols** Organic compounds containing the –OH group. They have the general formula $C_nH_{2n+1}OH$. Ethanol is by far the most important of the alcohols and is often just called 'alcohol'.

- **Amino acids** These naturally occurring organic compounds possess both an –NH_2 group and a –COOH group on adjacent carbon atoms. There are 20 naturally occurring amino acids, of which glycine is the simplest.

- **Biodegradable plastics** Plastics designed to degrade (decompose) under the influence of bacteria.

- **Biopolymers** Natural polymers such as starch and proteins.

- **Biotechnology** Making use of micro-organisms in industrial and commercial processes. For example, the process of fermentation is brought about by the enzymes in yeast.

- **Biuret test** The test for proteins. A mixture of dilute sodium hydroxide and 1% copper(II) sulphate solution is shaken with the material under test. A purple colour appears after about three minutes if a protein is present.

- **Carbohydrates** A group of naturally occurring organic compounds which can be represented by the general formula $(CH_2O)_x$.

- **Carboxylic acids** A family of organic compounds containing the functional group –COOH. They have the general formula $C_nH_{2n+1}COOH$. The most important and well known of these acids is ethanoic acid, which is the main constituent in vinegar. Ethanoic acid is produced by the oxidation of ethanol.

- **Condensation polymer** A polymer formed by a condensation reaction (one in which water is given out). For example, nylon is produced by the condensation reaction between 1,6-diaminohexane and hexanedioic acid.

- **Cross-linking** The formation of side covalent bonds linking different polymer chains and therefore increasing the rigidity of, say, a plastic. Thermosetting plastics are usually heavily cross-linked.

- **DNA** Abbreviation for deoxyribonucleic acid. It belongs to a group of biopolymers called the nucleic acids. It is involved in the polymerisation of amino acids in a specific order to form the particular protein required by a cell.

- **Drug** Any substance, natural or synthetic, that alters the way in which the body works.

- **Drug abuse** This term usually applies to the misuse of addictive drugs, which include barbiturates and amphetamines, as well as the opiates, cocaine and heroin. These drugs create severe psychological and physiological problems through self-indulgence. This leads to a variety of personal problems for the user.

- **Esters** A family of organic compounds formed by the reaction of an alcohol with a carboxylic acid in the presence of concentrated H_2SO_4. This type of reaction is known as esterification. Esters are characterised by a strong and pleasant smell (many occur in nature and account for the smell of flowers).

- **Functional group** The atom or group of atoms responsible for the characteristic reactions of a compound.

- **HIV** Short for human immunodeficiency virus, from which AIDS (acquired immunodeficiency syndrome) can develop.

- **Hydrolysis** A chemical reaction involving the reaction of a compound with water. Acid hydrolysis usually involves dilute hydrochloric acid, and enzyme hydrolysis involves enzymes such as amylase.

- **Monomer** A simple molecule, such as ethene, which can be polymerised.

- **Monosaccharides** A group of simple carbohydrates. They are sweet to taste and are water soluble (for example, glucose).

continued

■ **Pharmaceuticals**   These are chemicals called drugs that are prepared and sold with the intention of treating disease (for example, methyldopa).

■ **Photodegradable plastics**   Plastics designed to degrade under the influence of sunlight.

■ **Polymer**   A substance possessing very large molecules consisting of repeated units or monomers. Polymers therefore have a very large relative molecular mass.

■ **Polymerisation**   The chemical reaction in which molecules (monomers) join together to form a polymer.

■ **Polysaccharides**   A group of more complicated carbohydrates. They generally do not form true solutions and do not have a sweet taste (for example starch).

■ **Proteins**   Polymers of amino acids formed by condensation reactions. They fall broadly into two categories: fibrous proteins (for example, keratin and collagen) and globular proteins (for example, casein and albumen).

■ **Soapless detergents**   These are soap-like molecules which do not form a scum with hard water. These substances have been developed from petrochemicals. Their calcium and magnesium salts are water soluble and they are biodegradable.

■ **Soaps**   Substances formed by saponification. In this reaction, the oil or fat (glyceryl ester) is hydrolysed by aqueous sodium hydroxide to produce the sodium salt of the fatty acid, particularly sodium stearate (from stearic acid). Soap will dissolve grease because of the dual nature of the soap molecule. It has a hydrophobic part (the hydrocarbon chain) and a hydrophilic part (the ionic head) and so will involve itself with both grease and water molecules. However, it forms a scum with hard water by reacting with the Ca^{2+} (or Mg^{2+}) present.

■ **Thermoplastics**   Plastics which soften when heated (for example poly(ethene), PVC).

■ **Thermosetting plastics**   Plastics which do not soften on heating but only char and decompose (for example, melamine).

The (wider) organic manufacturing industry

Additional questions

1 Explain the following.
 a The problem of plastic waste has been overcome.
 b The majority of detergents produced today are biodegradable.
 c In bread making, yeast is added to the mix and the dough left to stand for a period of time.
 d Poly(ethene) is a thermoplastic.

2 a A detergent molecule may be represented by the following simplified diagram.

ionic head

covalent hydrocarbon tail

Use this representation of a detergent molecule in a series of labelled diagrams to show how detergents can remove grease from a piece of greasy cloth.
 b Explain why detergents do not form a scum with hard water, whereas soaps do.
 c The modern detergents are biodegradable.
 (i) Explain what this statement means.
 (ii) Why is it necessary for detergents to be biodegradable?

3 A piece of cheese contains protein. Proteins are natural polymers made up of amino acids. There are 20 naturally occurring amino acids. The structures of two amino acids are shown below.

glycine

alanine

$$H_2N-\overset{\overset{\displaystyle H}{|}}{\underset{\underset{\displaystyle H}{|}}{C}}-COOH$$

$$H_2N-\overset{\overset{\displaystyle H}{|}}{\underset{\underset{\displaystyle CH_3}{|}}{C}}-COOH$$

 a Name the type of polymerisation involved in protein formation.
 b Draw a structural formula to represent the part of the protein chain formed by the reaction between the amino acids shown above.
 c What is the name given to the common linkage present in protein molecules?
 d Why is there such a huge variety of proteins?
 e Name and describe the features of the two broad groups of proteins.

4

starch → monosaccharide **A** → liquid **B** + carbon dioxide

dilute hydrochloric acid   yeast   37°C

heat + potassium dichromate(VI)

liquid **C** + water

liquid **B**, concentrated sulfuric acid

liquid **D** + water

 a Name, give the formula and one use of each of the substances **A** to **D**.
 b Write word and balanced chemical equations for the reactions involved in the formation of liquids **B**, **C** and **D**.
 c Starch is classified as a natural polymer or 'biopolymer'. Explain the meaning of this statement.
 d Name the processes by which:
 (i) starch is broken down
 (ii) liquid **B** is formed
 (iii) liquid **C** is formed
 (iv) liquid **D** is formed.

5 a Copy the following table and complete it by writing the structural formulae for methanol and methanoic acid.

Methane	Methanol	Methanoic acid
$H-\overset{\overset{\displaystyle H}{\textstyle|}}{\underset{\underset{\displaystyle H}{\textstyle|}}{C}}-H$		

 b Describe a simple chemical test that could be used to distinguish methanol from methanoic acid.
 c (i) Name the class of compound produced when methanol reacts with methanoic acid.
 (ii) Name the type of reaction taking place.
 (iii) Write a word and balanced chemical equation for this reaction.
 (iv) Give two uses related to the class of compound formed in this reaction.

continued

d The following reaction takes place when methanol is burned:

$$2CH_3OH(l) + 3O_2(g) \rightarrow 2CO_2(g) + 4H_2O(g) \quad \Delta H = -1452 \, kJ \, mol^{-1}$$

    (i) How much heat energy would be liberated by burning:
0.5 mol of methanol?
4.0 mol of methanol?
4 g of methanol?

    (ii) Calculate the volume of carbon dioxide produced at room temperature and pressure (rtp) when 16 g of methanol are burned.

    (A_r: H = 1; C = 12; O = 16. One mole of any gas at rtp occupies 24 dm^3.)

6 This question requires you to use the information in Table 14.6 (p. 212).

  a Which grouping of atoms is common to all three painkillers?

  b Which of the pharmaceuticals contains:
    (i) sulfur?
    (ii) the greatest number of nitrogen atoms?
    (iii) the carboxylic acid functional group?
    (iv) chlorine?
    (v) the greatest number of –CH$_3$ groups?

  c Which of the pharmaceuticals is used to combat:
    (i) bacterial infections?
    (ii) depression?
    (iii) heart disease?

  d The pharmaceuticals industry has often been termed the 'medicines by design' industry. Explain the meaning of this statement.

7 Why is it safe for us to use vinegar, which contains ethanoic acid, on food while it would be extremely dangerous for us to use dilute nitric acid for the same purpose?

15 Nitrogen

Nitrogen – an essential element
The Haber process
Obtaining nitrogen
Obtaining hydrogen
Making ammonia
Ammonia gas
Making ammonia in a laboratory
Physical properties of ammonia
Chemical properties of ammonia

Nitric acid
Manufacture of nitric acid
Properties of nitric acid
Artificial fertilisers
Manufacture of ammonium nitrate
Problems with fertilisers
Checklist
Additional questions

Nitrogen – an essential element

Nitrogen gas is all around us. It makes up 78% of the air. Each day each individual breathes thousands of litres of nitrogen into his/her lungs and then exhales it without any chemical change occurring.

Nitrogen gas has no smell, is colourless and is a very unreactive gas.

The nitrogen gas molecule comprises two nitrogen atoms covalently bonded together by a triple bond (Figure 15.1). This triple bond is very strong, requiring a large amount of energy to break it so that nitrogen can take part in chemical reactions. Only at very high temperatures or if an electrical spark is passed through nitrogen gas will it react with oxygen. This occurs in an internal combustion engine, where the spark from the spark plug is sufficient to convert some of the nitrogen to toxic nitrogen oxides (NO_x) which are then emitted as car exhaust fumes (Chapter 11, p. 167). Catalytic converters prevent the emission of nitrogen oxides by converting them back to nitrogen gas before they leave the exhaust pipe.

Nitrogen gas can be obtained industrially by the fractional distillation of liquid air. It is produced on a very large scale, since the gas has many very important industrial uses worldwide. The worldwide production of nitrogen each year is in excess of 36 million tonnes.

Nitrogen is an essential element necessary for the well-being of animals and plants. It is present in proteins, which are found in all living things. Proteins are essential for healthy growth. Animals obtain the nitrogen they need for protein production by feeding on plants and other animals. Most plants obtain the nitrogen they require from the soil.

Nitrogen is found in nitrogen-containing compounds called **nitrates**, which are produced as a result of the effect of lightning and from dead plants and animals as they decay. Nitrates are soluble compounds and can be absorbed by the plants through their roots. Nitrogen in the air is also converted into nitrates by some forms of bacteria. Some of these bacteria live in the soil. A group of plants called **leguminous** plants have these bacteria in nodules on their roots. These bacteria are able to take nitrogen from the atmosphere and convert it into a form in which it can be used to make proteins. This process is called **fixing nitrogen** and is carried out by plants such as beans and clover.

The vital importance of nitrogen to both plants and animals can be summarised by the **nitrogen cycle** (Figure 15.2, overleaf). If farm crops are harvested from the land rather than left to decay the soil becomes deficient in this important element. The nitrogen is removed in the harvested crops rather than remaining as the plants decay. In addition, nitrates can be washed from the soil by the action of rain (leaching). For the soil to remain fertile for the next crop, the nitrates need to be replaced. The natural process is by decay or by the action of lightning on atmospheric nitrogen. Without

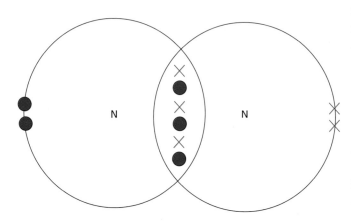

Figure 15.1 The nitrogen molecule.

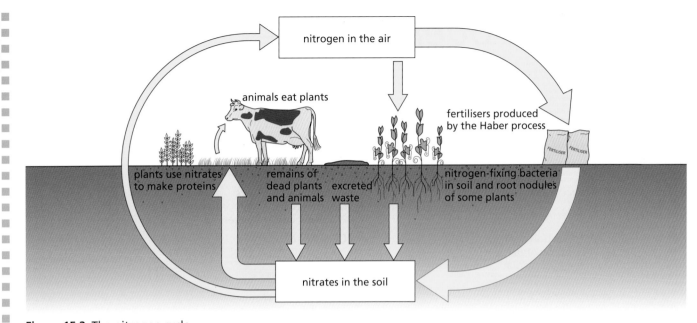

Figure 15.2 The nitrogen cycle.

the decay, however, the latter process is not efficient enough to produce nitrates on the scale required. Farmers often need to add substances containing these nitrates. Such substances include:

- farmyard manure – this is very rich in nitrogen-containing compounds which can be converted by bacteria in the soil to nitrates
- artificial fertilisers – these are manufactured compounds of nitrogen which are used on an extremely large scale to enable farmers to produce ever-bigger harvests for the increasing population of the world. One of the most commonly used artificial fertilisers is **ammonium nitrate**, which is made from ammonia gas and nitric acid, both nitrogen-containing compounds.

Questions

1 Why can most plants not use nitrogen gas directly from the atmosphere?

2 Nitrogen is essential in both plants and animals to form which type of molecule? Explain why these molecules are essential to life.

3 Explain why natural sources of the element nitrogen are not sufficient to produce the world's annual crop requirement.

The Haber process

The Haber process forms the basis of the artificial fertiliser industry, as it is used to produce ammonia gas. The process was developed by the German scientist Fritz Haber in 1913. He was awarded a Nobel Prize in 1918 for his work. The process involves reacting nitrogen and hydrogen. It was first developed to satisfy the need for explosives during World War I, as explosives can be made from ammonia.

Obtaining nitrogen

The nitrogen needed in the Haber process is obtained from the atmosphere.

Obtaining hydrogen

The hydrogen needed in the Haber process is obtained from the reaction between methane and steam.

$$\text{methane} + \text{steam} \rightleftharpoons \text{hydrogen} + \text{carbon monoxide}$$
$$CH_4(g) + H_2O(g) \rightleftharpoons 3H_2(g) + CO(g)$$

This process is known as **steam re-forming**. This reaction is a **reversible** reaction and special conditions are employed to ensure that the reaction proceeds to the right (the forward reaction), producing hydrogen and carbon monoxide. The process is carried out at a temperature of 750 °C, at a pressure of 30 atmospheres with a catalyst of nickel. These conditions enable the maximum amount of hydrogen to be produced at an economic cost.

The carbon monoxide produced is then allowed to reduce some of the unreacted steam to produce more hydrogen gas.

$$\text{carbon monoxide} + \text{steam} \rightleftharpoons \text{hydrogen} + \text{carbon dioxide}$$
$$CO(g) + H_2O(g) \rightleftharpoons H_2(g) + CO_2(g)$$

Making ammonia

In the Haber process, nitrogen and hydrogen in the correct proportions (1:3) are pressurised to approximately 200 atmospheres and passed over a catalyst of freshly produced, finely divided iron at a temperature of between 350 °C and 500 °C. The reaction in the Haber process is:

nitrogen + hydrogen $\rightleftharpoons$ ammonia
$$N_2(g) + 3H_2(g) \rightleftharpoons 2NH_3(g) \quad \Delta H = -92\,kJ\,mol^{-1}$$

The reaction is exothermic.

The industrial conditions employed ensure that sufficient ammonia is produced at a fast enough rate. The principles which underpin equilibrium processes such as this are discussed below and in Chapter 16 (the Contact process for manufacturing sulfuric acid, p. 229).

Under these conditions the gas mixture leaving the reaction vessel contains about 15% ammonia, which is removed by cooling and condensing it as a liquid. The unreacted nitrogen and hydrogen are recirculated into the reaction vessel to react together once more to produce further quantities of ammonia.

The 15% of ammonia produced does not seem a great deal. The reason for this is the reversible nature of the reaction. Once the ammonia is made from nitrogen and hydrogen, it decomposes to produce nitrogen and hydrogen. There comes a point when the rate at which the nitrogen and hydrogen react to produce ammonia is equal to the rate at which the ammonia decomposes.

This situation is called a **chemical equilibrium**. Because the processes continue to happen, the equilibrium is said to be **dynamic**. The conditions used ensure that the ammonia is made economically. Figure 15.3 shows how the percentage of ammonia produced varies with the use of different temperatures and pressures.

You will notice that the higher the pressure and the lower the temperature used, the more ammonia is produced. Relationships such as this were initially observed by Henri Le Chatelier, a French scientist, in 1888. He noticed that if the pressure was increased in reactions involving gases, the reaction which produced the fewest molecules of gas was favoured. If you look at the reaction for the Haber process you will see that, going from left to right, the number of molecules of gas goes from four to two. This is why the Haber process is carried out at high pressures. He also noticed that reactions which were exothermic produced more products if the temperature was low. Indeed, if the Haber process is carried out at room temperature you get a higher percentage of ammonia. However, in practice the rate of the reaction is lowered too much and the ammonia is not produced quickly enough for the process to be economical. An **optimum temperature** is used to produce enough ammonia at an acceptable rate. It should be noted, however, that the increased pressure used is very expensive in capital terms and so alternative, less expensive routes involving biotechnology (Chapter 14, p. 205) are being sought at the present time.

Worldwide, 140 million tonnes of ammonia are produced by the Haber process each year.

Figure 15.3 Yields from the Haber process.

Questions

1 Use the information given above and any other sources you may have to produce a flow diagram of the Haber process. Indicate the flow of the gases, and write equation(s) to show what happens at each stage. You may have to look up the boiling points of the gases involved for the stage in which the ammonia is separated from the reaction vessel mixture.

2 What problems do the builders of a chemical plant to produce ammonia have to consider when they start to build such a plant?

3 What problems are associated with building a plant which uses such high pressures as those required in the Haber process?

■ *Ammonia gas*

Making ammonia in a laboratory

Small quantities of ammonia gas can be produced by heating any ammonium salt, such as ammonium chloride, with an alkali, such as calcium hydroxide.

calcium   + ammonium → calcium + water + ammonia
hydroxide   chloride   chloride

$$Ca(OH)_2(s) + 2NH_4Cl(s) \rightarrow CaCl_2(s) + 2H_2O(g) + 2NH_3(g)$$

Water vapour is removed from the ammonia gas by passing the gas formed through a drying tower containing calcium oxide (Figure 15.4).

dry ammonia

mixture of dry solid
ammonium chloride
and calcium hydroxide

moist red
litmus paper

drying tower
packed with
calcium oxide

heat

Figure 15.4 Laboratory production of ammonia gas.

This reaction forms the basis of a chemical test to show that a compound contains the ammonium ion (NH_4^+). If any compound containing the ammonium ion is heated with sodium hydroxide, ammonia gas is given off which turns damp red litmus paper blue.

Physical properties of ammonia

Ammonia (Figure 15.5):

- is a colourless gas
- is less dense than air
- has a sharp or pungent smell
- is very soluble in water with about $680\,cm^3$ of ammonia in each $1\,cm^3$ of water (at $20\,°C$).

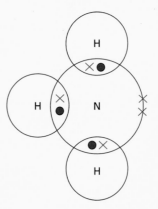

Figure 15.5 The ammonia molecule.

Chemical properties of ammonia

The reason ammonia is so soluble in water is that some of it reacts with the water. The high solubility can be shown by the 'fountain flask experiment' (Figure 15.6). As the first drop of water reaches the top of the tube all the ammonia gas in the flask dissolves, creating a much reduced pressure. Water then rushes up the tube to fill the space once occupied by the dissolved gas. This creates the fountain.

If the water initially contained some universal indicator, then you would also see a change from green to blue when it comes into contact with the dissolved ammonia. This shows that ammonia solution is a weak alkali, although dry ammonia gas is not. This is because a little of the ammonia gas has reacted with the water, producing ammonium ions and hydroxide ions. The hydroxide ions produced make the solution of ammonia alkaline.

ammonia + water ⇌ ammonium + hydroxide
                      ions   ions

$$NH_3(g) + H_2O(l) \rightleftharpoons NH_4^+(aq) + OH^-(aq)$$

The solution is only weakly alkaline because of the reversible nature of this reaction, which results in a relatively low concentration of hydroxide ions. Ammonia gas dissolved in water is usually known as aqueous ammonia.

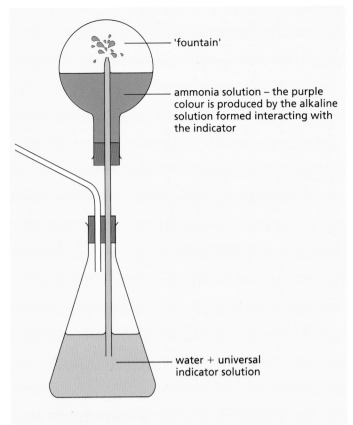

- 'fountain'
- ammonia solution – the purple colour is produced by the alkaline solution formed interacting with the indicator
- water + universal indicator solution

Aqueous ammonia can be used to identify salts of Cu^{2+}, Fe^{2+}, Fe^{3+}, Al^{3+}, Zn^{2+} and Ca^{2+} ions. The colour of the precipitate or solution formed identifies the metal present (Table 15.1).

Table 15.1 Identifying metals ions using aqueous ammonia.

Metal ion	With a few drops of ammonia solution	With excess ammonia solution
Cu^{2+}(aq)	Gelatinous blue precipitate	Precipitate dissolves to give a deep blue solution
Fe^{2+}(aq)	Dirty green precipitate	Dirty green precipitate remains
Fe^{3+}(aq)	Rust brown precipitate	Rust brown precipitate remains
Al^{3+}(aq)	White precipitate	White precipitate remains
Zn^{2+}(aq)	White precipitate	White precipitate dissolves to give a colourless solution
Ca^{2+}(aq)	Faint white precipitate	Faint white precipitate remains

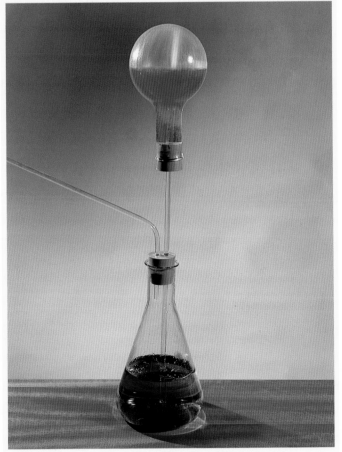

Figure 15.6 The fountain flask experiment.

Questions

1 Calcium oxide is used to dry ammonia gas in its laboratory preparation. Write a word and balanced chemical equation to show how calcium oxide can react with the water vapour to remove it from damp ammonia gas.

2 Explain why ammonia gas only acts as a weak alkali in the presence of water.

3 Describe how you could use aqueous ammonia to distinguish between three unlabelled bottles containing solutions of Al^{3+}(aq), Ca^{2+}(aq) and Cu^{2+}(aq).

Nitric acid

Manufacture of nitric acid

One of the largest uses of ammonia is in the production of nitric acid. The process, which was invented by Wilhelm Ostwald, has three stages.

1 A mixture of air and ammonia is heated to about 230 °C and is passed through a metal gauze made of platinum (90%) and rhodium (10%) (Figure 15.7). The reaction produces a lot of heat energy. This energy is used to keep the reaction vessel temperature at around 800 °C. The reaction produces nitrogen monoxide (NO) and water.

$$\text{ammonia} + \text{oxygen} \rightarrow \text{nitrogen} + \text{water}$$
$$\text{monoxide}$$
$$4NH_3(g) + 5O_2(g) \rightarrow 4NO(g) + 6H_2O(g)$$
$$\Delta H = -904\,kJ\,mol^{-1}$$

2 The colourless nitrogen monoxide gas produced from the first stage is then reacted with oxygen from the air to form brown nitrogen dioxide gas (NO_2).

$$\text{nitrogen monoxide} + \text{oxygen} \rightarrow \text{nitrogen dioxide}$$
$$2NO(g) + O_2(g) \rightarrow 2NO_2(g)$$
$$\Delta H = -114\,kJ\,mol^{-1}$$

3 The nitrogen dioxide is then dissolved in water to produce nitric acid.

$$\text{nitrogen} + \text{water} \rightarrow \text{nitric} + \text{nitrogen}$$
$$\text{dioxide} \qquad \text{acid} \qquad \text{monoxide}$$
$$3NO_2(g) + H_2O(l) \rightarrow 2HNO_3(aq) + NO(g)$$

Figure 15.7 Platinum–rhodium gauze is used as a catalyst in the production of nitric acid.

A small amount of nitrogen dioxide cannot be turned into nitric acid and so it is expelled into the atmosphere through a very tall chimney. This can cause pollution because it is an acidic gas.

The nitric acid produced is used in the manufacture of the following:

• artificial fertilisers, such as ammonium nitrate
• explosives, such as 2,4,6-trinitrotoluene (TNT)
• dyes
• artificial fibres, such as nylon.

It is also used in the treatment of metals.

Sixty million tonnes of nitric acid are produced each year. Of this figure 20 million tonnes are produced annually in Europe.

Properties of nitric acid

Dilute nitric acid

Nitric acid is a strong acid. As a dilute acid it shows most of the properties of an acid.

• It forms an acidic solution in water and will turn pH paper red.
• It will react with bases such as sodium hydroxide and zinc oxide to give salts, called **nitrates**, and water.

$$\text{sodium} + \text{nitric} \rightarrow \text{sodium} + \text{water}$$
$$\text{hydroxide} \quad \text{acid} \quad \text{nitrate}$$
$$NaOH(aq) + HNO_3(aq) \rightarrow NaNO_3(aq) + H_2O(l)$$

$$\text{zinc} + \text{nitric} \rightarrow \text{zinc} + \text{water}$$
$$\text{oxide} \quad \text{acid} \quad \text{nitrate}$$
$$ZnO(s) + 2HNO_3(aq) \rightarrow Zn(NO_3)_2(aq) + H_2O(l)$$

• It reacts with carbonates to give a salt, water and carbon dioxide gas.

$$\text{sodium} + \text{nitric} \rightarrow \text{sodium} + \text{water} + \text{carbon}$$
$$\text{carbonate} \quad \text{acid} \quad \text{nitrate} \qquad \text{dioxide}$$
$$Na_2CO_3(s) + 2HNO_3(aq) \rightarrow 2NaNO_3(aq) + H_2O(l) + CO_2(g)$$

• Most acids will react with metals to give a salt and hydrogen gas. However, since nitric acid is an oxidising agent it behaves differently and hydrogen is rarely formed. The exceptions to this are the reactions of cold dilute nitric acid with magnesium and calcium.

$$\text{magnesium} + \text{nitric} \rightarrow \text{magnesium} + \text{hydrogen}$$
$$\text{acid} \quad \text{nitrate}$$
$$Mg(s) + 2HNO_3(aq) \rightarrow Mg(NO_3)_2(aq) + H_2(g)$$

Dilute nitric acid reacts with copper, for example, to produce nitrogen monoxide instead of hydrogen.

$$\text{copper} + \text{nitric} \rightarrow \text{copper(II)} + \text{water} + \text{nitrogen}$$
$$\text{acid} \quad \text{nitrate} \qquad \text{monoxide}$$
$$3Cu(s) + 8HNO_3(aq) \rightarrow 3Cu(NO_3)_2(aq) + 4H_2O(l) + 2NO(g)$$

The nitrogen monoxide produced quickly reacts with oxygen from the air to give brown fumes of nitrogen dioxide gas.

Concentrated nitric acid

Concentrated nitric acid is a powerful **oxidising agent**. It will oxidise both metals and non-metals.

- It will oxidise carbon to carbon dioxide:

carbon + conc → carbon + nitrogen(IV) + water
nitric acid   dioxide   oxide

$$C(s) + 4HNO_3(l) \rightarrow CO_2(g) + 4NO_2(g) + 2H_2O(l)$$

- It will oxidise copper to copper(II) nitrate (Figure 15.8).

copper + conc → copper(II) + nitrogen + water
nitric acid   nitrate   dioxide

$$Cu(s) + 4HNO_3(l) \rightarrow Cu(NO_3)_2(aq) + 2NO_2(g) + 2H_2O(l)$$

Figure 15.8 Copper reacting with concentrated nitric acid.

Questions

1 What do you think happens to the nitrogen monoxide produced in the third step of the industrial preparation of nitric acid?

2 What is the maximum amount of nitric acid that could be produced from 500 tonnes of ammonia gas?

3 In reality the amount of nitric acid produced would be less than the amount calculated above. Explain why this would be so.

4 Write a word and balanced chemical equation to show the neutralisation reaction between nitric acid and ammonium hydroxide.

Artificial fertilisers

The two processes so far described, the production of ammonia and nitric acid, are extremely important in the production of many artificial fertilisers. As was explained earlier, the use of artificial fertilisers is essential if farmers are to produce sufficient crops to feed the ever-increasing population. Crops remove nutrients from the soil as they grow; these include nitrogen, phosphorus and potassium. Artificial fertilisers are added to the soil to replace these nutrients and others, such as calcium, magnesium, sodium, sulfur, copper and iron. Examples of nitrogenous fertilisers (those which contain nitrogen) are shown in Table 15.2.

Table 15.2 Some nitrogenous fertilisers.

Fertiliser	Formula
Ammonium nitrate	NH_4NO_3
Ammonium phosphate	$(NH_4)_3PO_4$
Ammonium sulfate	$(NH_4)_2SO_4$
Urea	$CO(NH_2)_2$

Artificial fertilisers can also make fertile land which was once unable to support crop growth. The fertilisers which add the three main nutrients (N, P and K) are called NPK fertilisers. They contain ammonium nitrate (NH_4NO_3), ammonium phosphate ($(NH_4)_3PO_4$) and potassium chloride (KCl) in varying proportions.

Manufacture of ammonium nitrate

Ammonium nitrate (Nitram) is probably the most widely used nitrogenous fertiliser. It is manufactured by reacting ammonia gas and nitric acid.

ammonia + nitric acid → ammonium nitrate

$$NH_3(g) + HNO_3(aq) \rightarrow NH_4NO_3(aq)$$

Problems with fertilisers

If artificial fertilisers are not used correctly, problems can arise. If too much fertiliser is applied to the land, rain washes the fertiliser off the land and into rivers and streams. This leaching encourages the growth of algae and marine plants. As the algae die and decay oxygen is removed from the water, leaving insufficient amounts for fish and other organisms to survive.

Questions

1 Calculate the percentage of nitrogen in each of the four fertilisers in Table 15.2.
(A_r: H = 1; N = 14; O = 16; P = 31; S = 32)

2 Write down a method that you could carry out in a school laboratory to prepare a sample of ammonium sulfate fertiliser.

■ *Checklist*

After studying Chapter 15 you should know and understand the following terms.

■ **Artificial fertiliser**   A substance added to soil to increase the amount of elements such as nitrogen, potassium and phosphorus. This enables crops grown in the soil to grow more healthily and to produce higher yields.

■ **Chemical equilibrium**   A dynamic state. The concentrations of the reactants and products remain constant because the rate at which the forward reaction occurs is the same as that of the back reaction.

■ **Nitrogen cycle**   The system by which nitrogen and its compounds, both in the air and in the soil, are interchanged.

■ **Nitrogen fixation**   The direct use of atmospheric nitrogen in the formation of important compounds of nitrogen. Bacteria present in root nodules of certain plants are able to take nitrogen directly from the atmosphere to form essential protein molecules.

■ **Optimum temperature**   A compromise temperature used in industry to ensure that the yield of product and the rate at which it is produced make the process as economical as possible.

■ **Reversible reaction**   A chemical reaction which can go both ways. This means that once some of the products have been formed they will undergo a chemical change once more to re-form the reactants. The reaction from left to right, as the equation for the reaction is written, is known as the forward reaction and the reaction from right to left is known as the back reaction.

Nitrogen
Additional questions

1 Study the following reaction scheme.

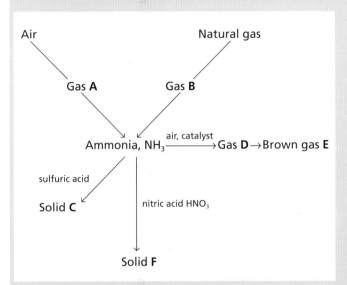

a Identify the substances **A** to **F** by giving their names and chemical formulae.
b How is gas **A** obtained from the air?
c Write a word and balanced chemical equation for the formation of ammonia gas from gases **A** and **B**.
d Write an equation for the formation of solid **C** from ammonia.
e Give a use for:
(i) solid **C**
(ii) nitric acid
(iii) solid **F**.

2 a Write word and balanced chemical equations for the reaction of dilute nitric acid with:
(i) sodium hydroxide
(ii) copper(II) oxide
(iii) magnesium carbonate.
b Describe how you would obtain dry crystals of potassium nitrate from dilute nitric acid and potassium hydroxide solution.

3 An international chemical firm has decided that it needs to build a new plant to produce the fertiliser ammonium sulfate, starting from the raw materials.
a What raw materials would be needed for this process?
b What factors would a chemical engineer look for before she decided where to build the new plant?
c At this type of plant there is always the danger of the leakage of some ammonia gas. What effect would a leak of ammonia gas have on:
(i) people living in the surrounding area?
(ii) the pH of the soil in the surrounding area?
(iii) the growth of crops in the surrounding area?

4

Fertiliser	Formula	% nitrogen
Ammonia solution	NH_3	82.4
Calcium nitrate	$Ca(NO_3)_2$	17.1
Nitram	NH_4NO_3	35.0
Sodium nitrate	$NaNO_3$	
Potassium nitrate	KNO_3	

a Copy and complete the above table by calculating the percentage of nitrogen in the fertilisers sodium nitrate and potassium nitrate.
(A_r: H = 1; N = 14; O = 16; Na = 23; K = 39; Ca = 40)
b Including the data you have just calculated, which of the fertilisers contains:
(i) the largest percentage of nitrogen?
(ii) the smallest percentage of nitrogen?
c Give the chemical name for the fertiliser that goes by the name Nitram.
d Ammonia can be used directly as a fertiliser but not very commonly. Think of two reasons why ammonia is not often used directly as a fertiliser.
e Nitram fertiliser is manufactured by the reaction of nitric acid with ammonia solution according to the equation:

$$NH_3(aq) + HNO_3(aq) \rightarrow NH_4NO_3(aq)$$

A bag of Nitram may contain 50 kg of ammonium nitrate. What mass of nitric acid would be required to make it?

5 Ammonia gas is made industrially by the Haber process, which involves the reaction between the gases nitrogen and hydrogen. The amount of ammonia gas produced from this reaction is affected by both the temperature and the pressure at which the process is run. The graph below shows how the amount of ammonia produced from the reaction changes with both temperature and pressure. The percentage yield of ammonia indicates the percentage of the nitrogen and hydrogen gases that are actually changed into ammonia gas.

continued

a Write a word and balanced chemical equation for the reversible reaction between nitrogen and hydrogen to produce ammonia using the Haber process.

b What is meant by the term a 'reversible reaction'?

c Use the graphs to say whether more ammonia is produced at:
(i) higher or lower temperatures
(ii) higher or lower pressures.

d What is the percentage yield of ammonia if the conditions used to run the process are:
(i) a temperature of 350 °C and a pressure of 100 atmospheres?
(ii) a temperature of 550 °C and a pressure of 350 atmospheres?

e The conditions in industry for the production of ammonia are commonly of the order of 200 atmospheres and 450 °C. What is the percentage yield of ammonia using these conditions?

f Why does industry use the conditions stated in part e if it is possible to obtain a higher yield of ammonia using different conditions?

6 The following results were obtained from a neutralisation reaction between potassium hydroxide and nitric acid. The dilute nitric acid had been found in a cupboard in the school laboratory. The student carrying out the experiment was interested in finding out the concentration of the nitric acid.

The student used 25 cm^3 of 0.15 mol dm^{-3} potassium hydroxide solution in a conical flask to which was added the indicator phenolphthalein. The dilute nitric acid was added from a burette until the indicator just changed colour. The student repeated the experiment four times. Her results are shown below.

	Rough	1	2	3
Final burette reading/cm^3	10.25	13.20	13.90	12.65
Initial burette reading/cm^3	0.00	3.10	3.80	2.50
Volume used/cm^3				

a Copy and complete the table by calculating the volume of dilute nitric acid used in each titration.

b From the three most accurate results, calculate the average volume of dilute nitric acid required to neutralise the 25 cm^3 of potassium hydroxide solution.

c (i) Write a word and balanced chemical equation for the reaction which has taken place.
(ii) Write down the number of moles of nitric acid and potassium hydroxide shown reacting in the equation.

d (i) Calculate the number of moles of potassium hydroxide present in 25 cm^3 of solution.
(ii) Calculate the number of moles of nitric acid neutralised.

e Calculate the molarity of the dilute nitric acid.

7 Explain the following.

a Dry litmus paper does not change colour when added to dry ammonia gas.

b Ammonia gas cannot be collected over water but can be collected by downward displacement of air.

c Soil often needs artificial fertilisers added to it if it is to continuously support the growth of healthy crops.

d When colourless nitrogen(II) oxide gas is formed, it almost instantaneously turns brown.

8 Nitrogen oxides are emitted from car exhaust pipes. To animals and plants these gases are very harmful.

a Normally, nitrogen gas and oxygen gas do not react. Why, in a car engine, do they react to form these dangerous gases?

b Where does the nitrogen which forms these nitrogen oxides come from?

c How is the production of these nitrogen oxides being minimised in newer cars?

16 Sulfur

Sulfur – the element
The Frasch process
Uses of sulfur
Properties of sulfur
Sulfur dioxide

Acid rain

Industrial manufacture of sulfuric acid – the Contact process

Properties of sulfuric acid
Dilute sulfuric acid
Sulfates
Concentrated sulfuric acid

Manufacture of a soapless detergent

Checklist

Additional questions

Sulfur – the element

Sulfur is a non-metallic element which has a very important role in the chemical industry. It is a yellow solid which is found in large quantities but in various forms throughout the world (Figure 16.1).

It is found in metal ores such as copper pyrites ($CuFeS_2$) and zinc blende (ZnS) and in volcanic regions of the world. Natural gas and oil contain sulfur and its compounds, but the majority of this sulfur is removed as it would cause environmental problems. Sulfur obtained from these sources is known as 'recovered sulfur' and it is an important source of the element. It is also found as elemental sulfur in sulfur beds in Poland, Russia and the US (Louisiana). These sulfur beds are typically 200 m below the ground. Sulfur from these beds is extracted using the **Frasch process**, named after its inventor Hermann Frasch.

The Frasch process

Superheated water at 170 °C and hot compressed air are forced underground through pipes, forcing water and molten sulfur to the surface (Figure 16.2). Sulfur is insoluble in water and so the two substances emerging from the pipes are easily separated. The sulfur is kept molten and sold in this form. The sulfur obtained from this process is about 99.5% pure and can be used directly.

Figure 16.1 Sulfur – rhombic (top) and monoclinic.

Figure 16.2 The Frasch process.

SULFUR

Uses of sulfur

The vast majority of sulfur is used to produce perhaps the most important industrial chemical, sulfuric acid. Sulfur is also used to **vulcanise** rubber, a process which makes the rubber harder and increases its elasticity. Relatively small amounts are used in the manufacture of matches, fireworks and fungicides, as a sterilising agent and in medicines.

Properties of sulfur

Sulfur:

- is a yellow, brittle solid at room temperature
- does not conduct electricity
- is insoluble in water.

Sulfur will react with both metals and non–metals.

- It reacts with magnesium metal to form magnesium sulfide.

$$magnesium + sulfur \rightarrow magnesium\ sulfide$$
$$Mg(s) + S(s) \rightarrow MgS(s)$$

- It reacts with oxygen to produce sulfur dioxide gas.

$$sulfur + oxygen \rightarrow sulfur\ dioxide$$
$$S(s) + O_2(g) \rightarrow SO_2(g)$$

Sulfur dioxide

Sulfur dioxide is a colourless gas produced when sulfur or substances containing sulfur, for example crude oil or natural gas, are burned in oxygen gas. It has a choking smell and is extremely posionous. The gas dissolves in water to produce an acidic solution of sulfurous acid.

$$sulfur\ dioxide + water \rightleftharpoons sulfurous\ acid$$
$$SO_2(g) + H_2O(l) \rightleftharpoons H_2SO_3(aq)$$

It is one of the major pollutant gases and is the gas principally responsible for **acid rain**. However, it does have some uses: as a bleaching agent for paper manufacture, in fumigants and in the preservation of food by killing bacteria.

Question

1 'Sulfur is a non-metallic element.' Discuss this statement, giving physical and chemical reasons to support your answer.

228

■ *Acid rain*

Rainwater is naturally acidic since it dissolves carbon dioxide gas from the atmosphere as it falls. Natural rainwater has a pH of about 5.7. In recent years, especially in central Europe, the pH of rainwater has fallen to between pH 3 and pH 4.8. This increase in acidity has led to extensive damage to forests (Figure 16.3), lakes and marine life.

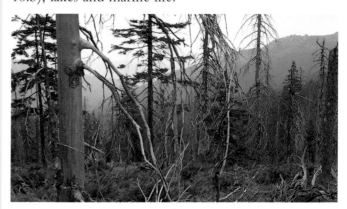

Figure 16.3 This forest has been devastated by acid rain.

In addition it has led to the increased corrosion of exposed metals and to damage of buildings and statues made from limestone or marble (Figure 16.4). The sulfuric acid in the rain reacts with the limestone, which is eaten away by the chemical process.

$$limestone + sulfuric\ acid \rightarrow calcium\ sulfate + water + carbon\ dioxide$$
$$CaCO_3(s) + H_2SO_4(aq) \rightarrow CaSO_4(s) + H_2O(l) + CO_2(g)$$

Figure 16.4 Acid rain is responsible for much of the damage to this temple on the Acropolis in Athens.

The amount of sulfur dioxide in the atmosphere has increased dramatically over recent years. There has always been some sulfur dioxide in the atmosphere, from natural processes such as volcanoes and rotting vegetation. Over Europe, however, around 80% of the sulfur dioxide in the atmosphere is formed from the combustion of fuels containing sulfur (Figure 16.5). After dissolving in rain to produce sulfurous acid, it further reacts with oxygen to form sulfuric acid.

Nitric acid (HNO_3) in the atmosphere, formed from NO_2 gas dissolving in the rainwater, also contributes to the production of acid rain.

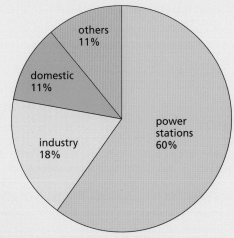

Figure 16.5 Sources of sulfur dioxide.

Figure 16.7 A Contact process plant used for making sulfuric acid.

Questions

1 How could the amount of sulfur dioxide being produced by the above sources be reduced?

2 Devise an experiment which you could carry out in the school laboratory to determine the amount of sulfur in two different types of coal.

Industrial manufacture of sulfuric acid – the Contact process

The major use of sulfur is in the production of sulfuric acid. This is probably the most important industrial chemical, and the quantity of it produced by a country has been linked with the economic stability of the country. Many millions of tonnes of sulfuric acid are produced worldwide each year. It is used mainly as the raw material for the production of many substances (Figure 16.6).

The process by which sulfuric acid is produced is known as the **Contact process** (Figure 16.7).

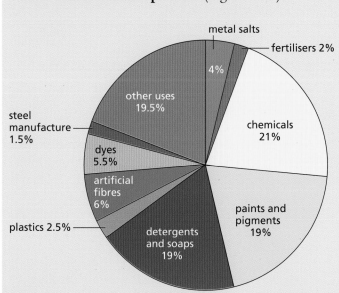

Figure 16.6 Products made from sulfuric acid include fertilisers, paints and pigments.

The process has the following stages.

- Sulfur dioxide is first produced, primarily by the reaction of sulfur with air.

sulfur + oxygen → sulfur dioxide

$$S(s) + O_2(g) \rightarrow SO_2(g)$$

- Any dust and impurities are removed from the sulfur dioxide produced, as well as any unreacted oxygen. These 'clean' gases are heated to a temperature of approximately 450°C and fed into a reaction vessel, where they are passed over a catalyst of vanadium(v) oxide (V_2O_5). This catalyses the reaction between sulfur dioxide and oxygen to produce sulfur trioxide (sulfur(VI) oxide, SO_3).

sulfur + oxygen ⇌ sulfur
dioxide   trioxide

$$2SO_2(g) + O_2(g) \rightleftharpoons 2SO_3(g)$$
$$\Delta H = -197 \, kJ \, mol^{-1}$$

This reaction is reversible and so the ideas of Le Chatelier (Chapter 15, p. 219) can be used to increase the proportion of sulfur trioxide in the equilibrium mixture. The forward reaction is exothermic and so would be favoured by low temperatures. The temperature of 450°C used is an optimum temperature which produces sufficient sulfur trioxide at an economical rate. Since the reaction from left to right is also accompanied by a decrease in the number of molecules of gas, it will be favoured by a high pressure. In reality, the process is run at atmospheric pressure. Under these conditions, about 96% of the sulfur dioxide and oxygen are converted into sulfur trioxide. The heat produced by this reaction is used to heat the incoming gases, thereby saving money.

- If this sulfur trioxide is added directly to water, sulfuric acid is produced. This reaction, however, is very violent and a thick mist is produced.

sulfur trioxide + water → sulfuric acid

$$SO_3(g) + H_2O(l) \rightarrow H_2SO_4(l)$$

This acid mist is very difficult to deal with and so a different route to sulfuric acid is employed. Instead, the sulfur trioxide is dissolved in concentrated sulfuric acid (98%) to give a substance called **oleum**.

sulfuric acid + sulfur trioxide → oleum

$$H_2SO_4(aq) + SO_3(g) \rightarrow H_2S_2O_7(l)$$

The oleum formed is then added to the correct amount of water to produce sulfuric acid of the required concentration.

oleum + water → sulfuric acid

$$H_2S_2O_7(l) + H_2O(l) \rightarrow 2H_2SO_4(l)$$

Questions

1 Produce a flow diagram to show the different processes which occur during the production of sulfuric acid by the Contact process. Write balanced chemical equations showing the processes which occur at the different stages, giving the essential raw materials and conditions used.

2 Both the following reactions are reversible:
   (i) $X_2(g) + O_2(g) \rightarrow 2XO(g)$
   (ii) $2XO(g) + O_2(g) \rightarrow 2XO_2(g)$
   Suggest a reason why an increase in pressure:
   a does not favour reaction (i)
   b increases the amount of XO_2 produced in reaction (ii)

Properties of sulfuric acid
Dilute sulfuric acid

Dilute sulfuric acid is a typical strong **dibasic** acid. A dibasic acid is one with two replaceable hydrogen atoms which may produce two series of salts – normal and acid salts (Chapter 7, pp. 100 and 104).

It will react with bases such as sodium hydroxide and copper(II) oxide to produce normal salts, called **sulfates**, and water.

- With sodium hydroxide:

sodium + sulfuric → sodium + water
hydroxide   acid   sulfate

$$2NaOH(aq) + H_2SO_4(aq) \rightarrow Na_2SO_4(aq) + H_2O(l)$$

- With copper(II) oxide:

copper(II) + sulfuric → copper(II) + water
oxide   acid   sulfate

$$CuO(s) + H_2SO_4(aq) \rightarrow CuSO_4(aq) + H_2O(l)$$

It also reacts with carbonates to give normal salts, carbon dioxide and water, and with reactive metals to give a normal salt and hydrogen gas. The reaction between zinc and sulfuric acid is often used to prepare hydrogen gas in the laboratory (Figure 16.8 opposite).

zinc + sulfuric acid → zinc sulfate + hydrogen

$$Zn(s) + H_2SO_4(aq) \rightarrow ZnSO_4(aq) + H_2(g)$$

Figure 16.8 The laboratory preparation of hydrogen gas.

The preparation of the acid salt with sodium hydroxide requires twice the volume of acid as that used in the preparation of the normal salt. Therefore, if $25\,cm^3$ of dilute sulfuric acid were required to form the normal salt from a given volume of alkali of a particular concentration then $50\,cm^3$ of the same acid solution would be required to produce the acid salt, sodium hydrogensulfate, from the same volume of alkali.

$$\text{sodium hydroxide} + \text{sulfuric acid} \longrightarrow \text{sodium hydrogensulfate} + \text{water}$$
$$NaOH(aq) + H_2SO_4(aq) \longrightarrow NaHSO_4(aq) + H_2O(l)$$

Sulfates

The salts of sulfuric acid, sulfates, can be identified by a simple test–tube reaction. To test for a sulfate, add a few drops of dilute hydrochloric acid to your unknown followed by a few drops of barium chloride. If a sulfate is present, a white precipitate of barium sulfate forms.

$$\text{barium ions} + \text{sulfate ions} \rightarrow \text{barium sulfate}$$
$$Ba^{2+}(aq) + SO_4^{2-}(aq) \rightarrow BaSO_4(s)$$

Many sulfates have very important uses, as can be seen from Table 16.1.

Table 16.1 Uses of some metal sulfates.

Salt	Formula	Use
Ammonium sulfate	$(NH_4)_2SO_4$	Fertiliser
Barium sulfate	$BaSO_4$	'Barium meal' used in diagnostic medical X-ray studies
Calcium sulfate	$CaSO_4.\frac{1}{2}H_2O$	'Plaster of Paris' used to set bones
Magnesium sulfate	$MgSO_4$	In medicine it is used as a laxative

Concentrated sulfuric acid

Concentrated sulfuric acid is a powerful **dehydrating agent** – it will take water from a variety of substances. One such substance is cane sugar, or sucrose (Figure 16.9).

$$\text{sucrose (sugar)} \xrightarrow{\text{conc } H_2SO_4} \text{carbon} + \text{water}$$
$$C_{12}H_{22}O_{11}(s) \longrightarrow 12C(s) + 11H_2O(l)$$

Concentrated sulfuric acid will also take water from hydrated copper(II) sulfate crystals, leaving only anhydrous copper(II) sulfate. If a few drops of concentrated sulfuric acid are added to some blue hydrated copper(II) sulfate crystals, they slowly turn white as the water of crystallisation is removed by the acid. Eventually, only a white powder – anhydrous copper(II) sulfate – remains.

$$\text{hydrated copper(II) sulfate} \xrightarrow{\text{conc } H_2SO_4} \text{anhydrous copper(II) sulfate} + \text{water}$$
$$CuSO_4.5H_2O(s) \longrightarrow CuSO_4(s) + 5H_2O(l)$$

Figure 16.9 The concentrated sulfuric acid has removed the elements of water from the sugar, leaving black carbon.

Concentrated sulfuric acid should be treated very carefully, because it will also remove water from flesh! It is a very corrosive substance and should always be handled with care.

Diluting concentrated sulfuric acid must be done with great care because of its affinity for water. The concentrated sulfuric acid should always be added to the water, not the other way around.

Questions

1 If you were given an unlabelled bottle which was thought to be dilute sulfuric acid, how would you show that the solution contained sulfate ions (SO_4^{2-}(aq)), how would you show that it was an acid and how would you determine the concentration of the acid?

2 Write balanced chemical equations for the reactions between dilute sulfuric acid and:
   a zinc oxide
   b potassium carbonate
   c aluminium.

Manufacture of a soapless detergent

A more recent use of sulfuric acid is in the production of soapless detergents. These are detergents that can be used more effectively than soap in hard water areas, are fairly cheap to make and are gradually replacing soaps (Chapter 14, p. 208).

The general process involves the reaction of a long, straight-chain alkene, such as dodecene, with benzene.

benzene + dodecene → dodecylbenzene
$C_6H_6(l) + CH_3(CH_2)_9CH=CH_2(l) \rightarrow C_6H_5(CH_2)_{11}CH_3(l)$

The molecular formula of dodecylbenzene is $C_{18}H_{30}$. This compound is then reacted with concentrated sulfuric acid to give a compound that is known as a sulfonic acid.

dodecylbenzene + sulfuric acid → dodecylbenzene sulfonic acid + water
$C_{18}H_{30}(l) + H_2SO_4(l) \rightarrow C_{18}H_{29}SO_3H(aq) + H_2O(l)$

Finally, this is reacted with the alkali sodium hydroxide, NaOH.

dodecylbenzene sulfonic acid + sodium hydroxide → sodium dodecylbenzene sulfonate + water
$C_{18}H_{29}SO_3H(aq) + NaOH(aq) \rightarrow C_{18}H_{29}SO_3^-Na^+(aq) + H_2O(l)$
(soapless detergent)

Soapless detergents such as this are to be found in most washing powders.

Checklist

After studying Chapter 16 you should know and understand the following terms.

■ **Contact process**   The industrial manufacture of sulfuric acid using the raw materials sulfur and air.

■ **Dibasic acid**   An acid which contains two replaceable hydrogen atoms per molecule of the acid, for example sulfuric acid, H_2SO_4.

■ **Frasch process**   The process of obtaining sulfur from sulfur beds below the Earth's surface. Superheated water is pumped down a shaft to liquefy the sulfur, which is then brought back to the surface.

■ **Soapless detergents**   Substances which are more effective than soap at producing lathers, especially in hard water areas. They are large organic molecules, produced using sulfuric acid.

■ **Sulfate**   A salt of sulfuric acid formed by the reaction of the acid with carbonates, bases and some metals. It is possible to test for the presence of a sulfate by the addition of dilute hydrochloric acid and some barium chloride solution. A white precipitate of barium sulfate is formed if a sulfate is present.

1 What do you understand by the following terms?
  a Dehydrating agent.
  b Oxidising agent.
  c Optimum temperature.
  d Acid rain.

2 Explain the following.
  a Chemical plants that produce sulfuric acid are often located on the coast.
  b Even though more sulfuric acid could be produced using high pressures, normal atmospheric pressure is used.
  c Natural rubber cannot be used to produce car tyres but vulcanised rubber can.
  d Sulfur dioxide gas is regarded as a pollutant.
  e Coal-fired and oil-fired power stations produce sulfur dioxide. Some of them are being fitted with flue gas desulfurisation (FGD) units.

3 A type of coal contains 0.5% of sulfur by mass.
  a Write an equation for the formation of sulfur dioxide gas when this coal is burned.
  b What mass of sulfur is contained in 1500 tonnes of coal?
  c What mass of sulfur dioxide gas would be formed if 1500 tonnes of coal were burned?
  d What volume would this mass of sulfur dioxide gas occupy, measured at room temperature and pressure (rtp)? (A_r: O = 16; S = 32. One mole of a gas occupies 24 dm^3 at rtp.)

4 Fossil fuels, such as oil, coal and natural gas, all contain some sulfur. When these fuels are burned they produce many different gases. Concern has grown in recent years about the effects of one of these gases, sulfur dioxide. When sulfur dioxide dissolves in rainwater it forms an acidic solution which has become known as acid rain.
  Money has been made available to solve the problem of acid rain. Attempts are being made to clean gases being released from power stations and to look into ways in which the effects of acid rain can be reversed.
  The table below and Figure 16.5 (p. 229) give some data about the emission of sulfur dioxide.

	Million tonnes per year
USA	26
Russia/Ukraine	18
Germany	7
UK ·	5
Canada	5
France	3
Poland	3
Italy	3
Other countries	30

  a Using the figures in the table, produce a bar chart to show the amount of sulfur dioxide produced by each of the countries listed.
  b What percentage of the world's sulfur dioxide is produced in:
    (i) France?
    (ii) North America?
  c Using the information above, explain why countries such as the US, Russia and Germany are at the top of the list of sulfur dioxide producers.
  d If the total amount of sulfur dioxide produced by Canada is 5 million tonnes per year, what amount is produced by:
    (i) power stations?
    (ii) domestic users?
    (iii) industry?

5 Study the following reaction scheme:

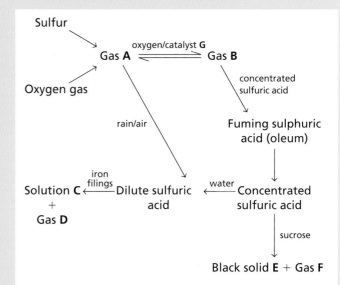

  a Identify the substances **A** to **F** by giving their names and formulae.
  b Write a balanced chemical equation for the formation of gas **B**.
  c (i) Describe a chemical test, and give the positive result of it, to identify gas **D**.
    (ii) Describe a chemical test, and give the positive result of it, to identify gas **F**.
  d How would you obtain solid **C** from the solution **C**?
  e Which pathway shows the formation of acid rain?
  f In which way is the concentrated sulfuric acid acting in its reaction with sucrose?
  g Where does the oxygen gas come from to form gas **A**?

6 In a neutralisation experiment, 25 cm³ of dilute sulfuric acid was required to react completely with 40 cm³ of a solution of 0.25 mol dm⁻³ potassium hydroxide.
   a Write a balanced chemical equation for the reaction between dilute sulfuric acid and potassium hydroxide.
   b Calculate the number of moles of potassium hydroxide solution used in the reaction.
   c How many moles of dilute sulfuric acid would this number of moles of potassium hydroxide react with?
   d Calculate the concentration of the dilute sulfuric acid.
   e Which indicator could have been used to determine when neutralisation had just occurred?

7 Describe how you would prepare some crystals of hydrated copper(II) sulfate from copper(II) oxide and dilute sulfuric acid. Draw a diagram of the apparatus you would use and write a balanced chemical equation for the reaction.

8 When sulfur is extracted from sulfur beds below the Earth's surface, superheated water is pumped down a shaft into the beds to melt the solid sulfur.
   a What is meant by superheated water?
   b Why does the water have to be superheated? Why would boiling water not work?
      Most sulfur obtained from these sulfur beds, in countries such as Poland and France, is exported.
   c In which form do you think the sulfur is loaded on to sulfur tankers? Explain your answer.
   d What hazards do you think are faced by people working in industries that use sulfur? Give reasons for your answer.

Experimental chemistry

Scientists find out about the nature of materials by carrying out experiments in a laboratory. Many of these experiments require apparatus that you have used in your study of chemistry to date. Certainly a knowledge and understanding of the use of this scientific apparatus is required for successful experimentation and investigations that you may carry out in your further study of chemistry. Much of the work involves **accurate** measurements with particular pieces of apparatus in particular experiments many of which are shown in the section below.

Apparatus used for measurement in chemistry

Measurement of time

Figure 1 This stopwatch can be used to measure the time passed in a chemical reaction

Experiments involving rates of reaction will require the use of an accurate stopwatch – one that measures to a hundredth of a second. The units of time are hours (h), minutes (min) and seconds (s).

Measurement of temperature

Figure 2 A thermometer can be used to measure temperature.

The most commonly used thermometers used in a laboratory are alcohol in glass. However, mercury in glass thermometers can be used but should be handled with great care. The mercury inside them is poisonous and should not be handled if a thermometer breaks. The units of temperature are those of the Celsius scale. This scale is based on the temperature at which water freezes and boils, that is:

the freezing point of water is 0 °C whilst
the boiling point of water is 100 °C.

For accuracy the thermometer should be capable of being read to one tenth of °C. The usual thermometer used is that shown in the photograph that measures accurately between 10 ° and 110 °C. When reading the thermometer always ensure that your eye is at the same level as the liquid meniscus in the thermometer to ensure there are no parallax effects. The meniscus is the way that the liquid curves at the edges of the capillary in which the liquid is held in the thermometer.

Measurement of mass

Figure 3 An electronic balance can be used to measure the mass of reagents.

There are many different electronic balances which can be used. The important detail with any of them is that they are accurate to one hundredth of a gram. The units for measuring mass are grams (g) and kilograms (kg).

1 kg = 1000 g

When using an electronic balance you should wait until the reading is steady before taking it.

Measurement of volume

Figure 4 The apparatus shown in the photograph is generally used in different experiments to measure volume accurately.

Different experiments involving liquids will require one or other or all the various measuring apparatus for volume. The volume of a liquid is a measure of the amount of space that it takes up. The units of volume are litres (l) and cubic centimetres (cm^3)

1 litre $= 1000$ cm^3

However, some of the manufacturers of apparatus used for measuring volume use millilitres (ml). This is not a problem, however, since 1 cm^3 = 1 ml.

When reading the volume using one of the pieces of apparatus it is important to ensure that the apparatus is vertical and that your eye is level with the top of the meniscus of the liquid being measured.

■ *Suggestions for practical work and assessment*

1 Objectives for experimental skills and investigations

The IGCSE chemistry syllabus suggests that students should be able to:

1 Use techniques, apparatus and materials (including the following of a sequence of instructions where appropriate).
2 Make and record observations, measurements and estimates.
3 Interpret and evaluate experimental observations and data.
4 Plan and carry out investigations, evaluate methods and suggest possible improvements (including the selection of techniques, apparatus and materials).

All candidates must take a practical paper based on the criteria above.

You must take one of the following papers:

Paper 4 (Coursework), or
Paper 5 (Practical Test), or
Paper 6 (Alternative to Practical)

Some examples of questions from Papers 5 and 6 are given at the end of this section. The practical assessment is 20% of the available marks.

The following is a list of practical exercises some of which could be used to both support the assessment objectives given above and enrich the study of chemistry; in addition it should suggest plenty of opportunities for the enhancement and assessment of practical skills. The list is neither exhaustive nor prescriptive: the actual selection of practical work, whether from this list or from elsewhere, has to be governed by local factors such as the facilities available, safety considerations and the ability of the class.

Safety

In the suggested practical exercises, materials are used which, although familiar in many cases, are of a potentially hazardous nature, and appropriate care and precautions should be taken. We believe that the experiments can be carried out safely in school laboratories. However, it is the responsibility of the teacher to make the final decision depending on the circumstances at the time. Goggles should be worn at all times. In certain cases disposable gloves and fume cupboards will be required.

Teachers must ensure that they follow the safety guidelines set down by their employers, and a risk assessment must be completed for *any* experiment that is carried out. Teachers should draw students' attention to the hazards involved in the particular exercise to be performed.

Redox

The concept of redox, either in its elementary form (that is, as the loss or gain of oxygen) or as electron transfer, is a constant theme throughout the syllabus and the associated practical work. The list suggests plenty of varied examples, designated by **R**, to reinforce the theory.

Chapter 1   All about matter

Solids, liquids and gases

- A simple demonstration to illustrate the three states of matter and their interconversions – ice, water and steam.
- Demonstrate the compressibility of gases and the incompressibility of liquids by using a syringe to show that the volume of a gas (e.g. air) decreases with pressure whereas that of a liquid (e.g. water) does not.

Changes of state

- Measure the melting points of both a pure sample and an impure sample of a solid; this will demonstrate that melting point is a means of assessing purity.
  Measure the temperature of melting ice, then repeat after adding salt to the ice.
  Measure the melting point of octadecanoic acid using a water bath (Figure 1.12, p. 5), then repeat using a sample that is contaminated with a trace of salol (phenyl 2-hydroxybenzoate).
  Use a liquid paraffin bath or electrical heating apparatus to measure the melting point of an unknown compound and decide whether it is a pure sample.
- Boiling point as a criterion of purity. Use a test-tube and thermometer to measure the boiling point of a suitable liquid. A simple example would involve heating calcium chloride solution in a test-tube with a few anti-bumping granules. Note that the bulb of the thermometer must be in the liquid itself to measure its boiling point (typically 102 °C in this case) and not in the vapour above the liquid (where it would register 100 °C).
- The fractional distillation of a mixture of water and ethanol (Figure 2.22, p. 22) illustrates that a pure liquid boils at one temperature but a mixture of liquids boils over a range of temperatures.
- A liquid will boil at a lower temperature when the external pressure is reduced, e.g. by using a water pump. The variation of boiling point with external pressure can be demonstrated with a volatile liquid, such as methanol; it is advisable to introduce some form of cold trap to reduce the amount of vapour entering the waste.
  An alternative is to allow boiling water to cool and so stop boiling, and then reduce the pressure: it will start to boil again at a temperature below 100 °C.
- Iodine, ammonium chloride or dry ice can be used to demonstrate sublimation. Heat the solid, ammonium chloride or iodine, in a test-tube to show that it disappears from the bottom of the tube and condenses higher up the tube, where it is cooler.

Heating and cooling curves

- Students will already be familiar with melting and boiling, so a more profitable approach might be to supply them with data to construct and interpret these curves.

Diffusion

- Brownian motion can be demonstrated using pollen or smoke. Smoke cells are commercially available.
- Illustrate gaseous diffusion using a gas jar of bromine vapour and a gas jar of air as described in the text (Figure 1.13, p. 6).
- An additional demonstration of gaseous diffusion is similar, but uses a gas jar of hydrogen *above* a gas jar of carbon dioxide. Both gas jars can be tested for the presence of carbon dioxide. It is preferable to set up this experiment twice, then, using the second set, hydrogen can be shown to be in both jars.
- Demonstrate the ammonia and hydrochloric acid experiment described in the text (Figure 1.14, p. 6). This also illustrates the dependence of the rate of diffusion on the relative molecular mass of a gas, M_r.
- An alternative is an arrangement of manometer and porous pot which effectively compares the rate of diffusion of a gas with that of air (Figure 5). This can be used to show that carbon dioxide diffuses slower than air and hydrogen diffuses faster.

a The pressure inside the porous pot rises because hydrogen diffuses in faster than air diffuses out.

b The pressure inside the porous pot falls because air diffuses out faster than carbon dioxide diffuses in.

Figure 5 Comparing the rate of diffusion of gases with that of air.

- Diffusion of a coloured solute in water. A large crystal of copper(II) sulfate is added to a measuring cylinder of water and is observed over several days.

Gas laws

- Most chemistry syllabuses do not include the quantitative verification of the gas laws but if this is required most physics departments can supply the necessary apparatus.
- Simple demonstrations with a syringe full of air or balloons filled with air will usually provide a sufficient insight into these laws.
  Use a sealed gas syringe of air to show that if the pressure is increased then the volume decreases.

Similarly, place a syringe partially filled with air into hot water and the volume will increase.

Behind a safety screen, heat a sealed container of air, such as a syrup tin or lightly corked *hard glass* test-tube, to demonstrate that the pressure will increase with temperature if the volume of the gas remains constant.

Alternatively, connect a flask of air to a simple U-tube manometer to demonstrate that the pressure inside the flask increases when it is heated.

Chapter 2 Elements, compounds and mixtures

Elements

- Display samples of as many elements as possible, either on information cards or on a copy of the periodic table. This is an effective observation exercise.
- Set up a circus of activities which includes both collecting data from computers or data books (e.g. melting and boiling points and density) and practical exercises on comparing electrical conductivities. These activities will illustrate the physical differences between metals and non-metals.

Compounds and mixtures

- Demonstrate the combination of elements to form compounds, for example magnesium and oxygen, iron and chlorine, hydrogen and oxygen, zinc and sulfur, carbon and oxygen, and aluminium and iodine.
- Demonstrate the iron and sulfur experiment described in the text (p. 17).

Separating mixtures

- Filtration, decanting and centrifuging to separate a solid from a liquid (p. 19–20).
- Evaporation to obtain a solute from a solution (p. 20).
- Crystallisation of impure benzoic acid. Benzoic acid is contaminated with a *trace* of a blue dye and then dissolved in the minimum amount of boiling water. Allow to cool and separate crystals by filtration. The blue colour will have disappeared or become fainter, showing that crystallisation has removed the impurity.
- Simple distillation to separate a solvent from a solution (p. 21).
- Separation of immiscible liquids using a separating funnel (p. 22).
- Fractional distillation of a mixture of ethanol and water will separate two miscible liquids and illustrates that an impure liquid does not boil at a specific temperature but over a range of temperatures (p. 22).
- Illustrate the use of chromatography as a test of purity and an analytical tool. Suitable examples include coloured inks, food colouring, pigments from flowers or grass, metallic cations and identification of sugars. At least one exercise should involve developing a chromatogram and at least one should involve the measurement of R_f values.

Gels, sols, foams and emulsions

- Set up a circus of gels, sols, emulsions and foams, e.g. 'jelly', milk, pumice stone, polyurethane foam, bread, emulsion paint, cola, hair cream, aerosol dispenser, salad cream, emulsion paint. A silica gel can be made from sodium silicate and hydrochloric acid. Classify the examples according to type of colloidal system.
- Make a colloidal sol of iron(III) hydroxide by adding aqueous iron(III) chloride to boiling water, or a colloidal sulfur sol by adding dilute aqueous sodium thiosulfate to hydrochloric acid. For both of these sols, it can be shown that the solid phase is not separated by filtration.

Chapter 3 Atomic structure and the periodic table

Group 1 – the alkali metals

- Teachers can demonstrate that lithium, sodium and potassium are soft metals and good conductors of electricity, and students can find the densities and melting points of these metals using data books.
- The metals are burnt in air or oxygen and then it is shown that their oxides are alkaline. Using a deflagrating spoon, the hot metal is placed into a gas jar of oxygen, the colour of the flame is noted and the pH of the white powder formed is measured.
- The reactions of lithium, sodium and potassium with cold water can be demonstrated to show that alkaline solutions are formed and to illustrate the difference in their reactivities.
- Sodium is burnt in chlorine to demonstrate the formation of sodium chloride.

Group 2 – the alkaline earth metals

- Similar exercises as for group 1 can be conducted with magnesium and calcium. In addition, it can be demonstrated that magnesium reacts slowly with cold water but will burn in steam (Figure 9.3, p. 130).

Group 7 – the halogens

- Investigate their physical properties by inspection and using data books.
- Demonstrate their displacement reactions. **R** Add chlorine water to potassium bromide solution.
  Add chlorine water to potassium iodide solution.
  Add bromine water to potassium iodide solution.
- Demonstrate the formation of halides. Use small quantities in all of these reactions, which must be performed in an efficient fume cupboard.
  Heat aluminium turnings in a current of chlorine or add hot aluminium powder to a gas jar of chlorine. Add hot aluminium powder to bromine vapour. Add one drop of water to a intimate mixture of aluminium powder and iodine.

Reaction of copper (use Dutch metal) and chlorine. Reaction of phosphorus and chlorine. Put a small piece of white phosphorus on a spoon into a gas jar of chlorine.

Hydrogen and chlorine. Make a mixture of these gases by electrolysis and expose it to a bright light from a photographic flash or burning magnesium.

Metals

- Most metals react with dilute acids. Add a small piece of a metal to dilute hydrochloric acid and test for hydrogen. Use magnesium, aluminium, zinc, iron and copper.
- Differences in the physical properties of the transition elements, such as melting points and densities, can be established from data books.
- Transition metals react with oxygen to form insoluble oxides that are not alkaline. The reactions of these metals with acids will illustrate their lower reactivity.
- An observational exercise using compounds of chromium, manganese, cobalt, nickel and copper will emphasise that these metals have coloured compounds.
- Flame colours (p. 43).

Chapter 4 Bonding and structure

Ionic bonding

- Look up the melting and boiling points of ionic compounds in data books.
- Show that most ionic compounds dissolve in water.
- Demonstrate that an ionic compound, potassium iodide or lithium chloride, conducts electricity in the molten state but not as a solid.

Covalent bonding

- Look up the melting and boiling points of simple covalent compounds in data books.
- Show that they are not good conductors of electricity in the liquid phase – demonstrate this with hexane, ethanol or phosphorus trichloride.
- Show that graphite is both soft and a good conductor of electricity.
- Use a glass cutter to illustrate the hardness of diamond.

Glasses and ceramics

- Heat a piece of glass tubing to demonstrate that it gradually softens, will bend and can be pulled into a finer tube. These changes show that glass is a supercooled liquid, not a solid.

Metallic bonding

- The properties of typical metals will have been covered by the practical work on the transition elements in Chapter 3.

Chapter 5 Chemical calculations

Moles and gases

- Measure the molar volume at rtp for hydrogen. Use a flask with a delivery tube and collect the gas over water in an inverted measuring cylinder or burette (Figure 6). Put an excess of dilute hydrochloric acid, say 100 cm^3 of 2 mol dm^{-3} acid, in the flask and add a piece of magnesium ribbon whose mass is known accurately and is about 0.1 g. (It is useful to measure the mass of a 1 m length of ribbon and then the mass of a smaller length can be calculated.) Measure the volume of hydrogen collected in the measuring cylinder or burette.

Figure 6 Measuring the molar volume of a gas at rtp.

- Measure the molar volume at rtp for carbon dioxide. Repeat the experiment above, using a known mass of a carbonate and collecting the carbon dioxide in a syringe.

Chapter 6 Electrolysis and its uses

- Introductory practical. Electrolytes only conduct in the liquid phase, either molten or in aqueous solution. Experiments can be carried out to illustrate the differences between strong, weak and non-electrolytes. Use a circuit similar to that in Figure 6.2 (p. 80), but with an ammeter in preference to the lamp, and measure the conductivity of a selection of liquids, for example sodium chloride solution, hydrochloric acid, sodium hydroxide solution, water, dilute ethanoic acid, aqueous ammonia, paraffin and ethanol.
- Electrolysis as decomposition, for example of copper chloride solution, zinc bromide solution, hydrochloric acid and acidified water. Include tests for chlorine, oxygen and hydrogen. **R**

240

- Electrolysis of aqueous solutions of the following using inert electrodes: sodium chloride, copper(II) sulfate, sodium sulfate and sodium hydroxide.
- Electrolysis of aqueous copper(II) sulfate using copper electrodes.
- Examples of electroplating, such as nickel plating. Use a nickel anode, a copper cathode and nickel sulfate solution as the electrolyte in a cell similar to that in Figure 6.23 (p. 91).
- Quantitative experiments with aqueous copper(II) sulfate and copper electrodes. Measure the masses of two clean copper electrodes before and after electrolysis. Include a variable resistance in the circuit and note the current and the time for which it flows. This experiment can either be used to study the effect of variables (current and time), or as an introduction to moles of electrons and hence the masses formed or used in electrolysis.

Chapter 7 Acids, bases and salts

Acids and alkalis

- With simple indicators, find the acid colour, the alkaline colour and the neutral colour. Hold the acid tube and the alkaline tube up to a light together and look through both to see the neutral colour.
- Dilution experiments using universal indicator and $0.1 \, \text{mol dm}^{-3}$ solutions of hydrochloric acid, sodium hydroxide, ethanoic acid and ammonia. These will establish the pH scale and promote an understanding of the distinction between strong and weak electrolytes.
- Demonstrate that a solution of hydrogen chloride in dry hexane does not exhibit acid properties. It does not conduct electricity, nor does it react with magnesium.

Formation of salts

- Reactions of a typical acid with metals. Use hydrochloric and/or sulfuric acid with magnesium, aluminium, zinc, iron and copper. **R**
- Reactions of typical acids with bases and carbonates. Use a selection of alkalis, insoluble bases and carbonates with hydrochloric, sulfuric and nitric acids.
- Small-scale preparation of a salt from a metal – magnesium sulfate-7-water (p. 101).
- Preparation of a salt by titration – sodium chloride or sodium sulfate-10-water (p. 102).
- Preparation of a salt from an insoluble carbonate and an acid – nickel sulfate-7-water (p. 102).
- Preparation of a salt from an acid and an insoluble base – copper(II) sulfate-5-water (p. 103).
- Tests for anions – nitrate, chloride, bromide, iodide, carbonate and sulfate (pp. 104–5).
- Preparation of a number of insoluble salts by precipitation on a test-tube scale – an exercise in practical skills, observation, recording and writing equations.

- One formal preparation of an insoluble salt, such as lead iodide, to include filtering, washing and drying (p. 103).

Crystal hydrates

- Thermal decomposition of a hydrated salt – copper(II) sulfate-5-water. Add water carefully to the anhydrous salt (Figure 7.21, p. 106).
- Find x in $MgSO_4.xH_2O$ or $BaCl_2.xH_2O$ by thermal decomposition. Weigh a clean, dry crucible and lid. Add between 2 g and 3 g of the hydrated compound and re-weigh, then heat to constant mass. From the results calculate x.

Solubility of salts in water

- Prepare a hot saturated solution of potassium nitrate and allow it to cool. Potassium nitrate crystals will form. This illustrates the idea of a saturated solution and that solubility changes with temperature.
- Measure the solubility of a salt in water. Decant some saturated solution into a pre-weighed dish, re-weigh and evaporate to dryness. Finally weigh the dish and solute.

Titration

- By titration find the concentration of a solution of an alkali using a standard acid, hydrochloric acid, as described in the text (p. 108).
- Using the same method (but a different calculation), find the concentration of a solution of an alkali using sulfuric acid.
- By titration find the mass of one mole of an acid, H_2Y, given a solution of H_2Y of known concentration in $g \, dm^{-3}$. A suitable acid would be ethanedioic acid, with phenolphthalein as the indicator.

Chapter 8 Inorganic carbon chemistry

Carbonates and hydrogencarbonates

- Heat a piece of calcium carbonate. Allow it to cool and carefully add water dropwise. Then add excess water and filter. Keep the filtrate, a solution of calcium hydroxide. Make carbon dioxide by adding an acid to calcium carbonate. Bubble this gas through the previously prepared calcium hydroxide solution until no further change is observed. Boil the resulting solution of calcium hydrogencarbonate.
- Action of heat, if any, on sodium, zinc and copper carbonates.
- Reaction of acids with a selection of carbonates.
- Formation of insoluble carbonates by precipitation. Add aqueous sodium carbonate to solutions of copper sulfate, calcium chloride and nickel sulfate.
- Establish the identity of the solid product formed by the thermal decomposition of sodium hydrogen-carbonate. First measure the pH of a dilute solution of the product and observe the reaction of the solid product with an acid.

Confirm the identity by quantitative means, and by conducting mole calculations establish which one of the following is the correct equation for the thermal decomposition:

$$NaHCO_3 \rightarrow NaOH + CO_2$$
$$2NaHCO_3 \rightarrow Na_2CO_3 + CO_2 + H_2O$$
$$2NaHCO_3 \rightarrow Na_2O + 2CO_2 + H_2O$$

Weigh a clean, dry crucible and lid. Add between 2 g and 3 g of sodium hydrogencarbonate and heat to constant mass. The mass of the sodium hydrogencarbonate and that of the solid product are now known. From the results determine which equation is correct by comparing the following ratios:

$$\frac{\text{mass of NaHCO}_3}{84} : \frac{\text{mass of solid product}}{40}$$

$$\frac{\text{mass of NaHCO}_3}{84} : \frac{\text{mass of solid product}}{106}$$

$$\frac{\text{mass of NaHCO}_3}{84} : \frac{\text{mass of solid product}}{62}$$

Hardness in water

- Standard experiments with permanently and temporarily hard water. Measure the number of drops of soap solution needed to form a permanent lather in 10 cm^3 samples of both types of hard water. Boil both types and allow to cool. Then measure the number of drops of soap solution needed to produce a permanent lather in 10 cm^3 samples of the treated samples of hard water. The study can be extended to include the addition of sodium carbonate and treatment with ion exchange resins.
- Compare soap and soapless detergents. Add soap solution dropwise with shaking to samples of hard and soft water. Repeat using a soapless detergent.

Carbon dioxide

- Laboratory preparation of carbon dioxide and its properties (Figure 8.22, p. 123).
- Demonstrate that carbon and carbon-containing compounds form carbon dioxide on combustion. Burn a variety of materials and test for carbon dioxide with limewater. Suitable materials are a candle, wood shavings, charcoal, paraffin etc.

Chapter 9 Metal extraction and chemical reactivity

Typical reactions of metals

- Reaction of metals with dilute acids – Chapter 7. **R**
- Reaction of metals with air and oxygen – see Chapter 3. **R**
- Reaction of metals with cold water or steam – see Chapter 3. **R**

Competition reactions

- Reduction of metal oxides by other metals (Figure 9.6, p. 131). Demonstrate the Thermit reaction and the reactions between the metals magnesium, zinc, iron and copper and their oxides. This will establish an order of reactivity for these metals. Some of these reactions are very violent so the use of small quantities and a rehearsal before the class demonstration are essential. **R**
- Reduction of metal oxides by methane/hydrogen. It is safer to use methane, which is passed over heated copper oxide. **R**
- Metal/metal displacement reactions in aqueous solution. Clean pieces of a metal are added to the aqueous nitrate of another metal. The students look for evidence of displacement and so establish an order of reactivity. Suitable metals are magnesium, zinc, lead, copper and silver. **R**

Identifying metal ions

- Reactions of cations in aqueous solution with sodium hydroxide – Al^{3+}, Fe^{2+}, Fe^{3+}, Cu^{2+}, Ca^{2+}, Zn^{2+}. An excess of sodium hydroxide solution is added slowly to a small volume of the solution containing the cation. This is an exercise in observation, organisational and recording skills and in the ability to write chemical equations – word, molecular and ionic.
- Reactions of the same cations with aqueous ammonia. This is essentially a repeat of the exercise above but with less emphasis on equation writing.
- Identification of unknown cations using sodium hydroxide solution and aqueous ammonia.

Amphoteric hydroxides and oxides

- Hydroxides are classified as basic or amphoteric. The qualitative exercises above will provide a practical basis for the underpinning theory. A useful extension exercise is to provide unknown cations in solution and for the students to ascertain if their hydroxides are basic or amphoteric.
- Another informative practical involves adding dilute acid dropwise to an aluminate or zincate and observing the changes in reverse.

$$Al^{3+} \underset{H^+}{\overset{OH^-}{\rightleftharpoons}} Al(OH)_3 \underset{H^+}{\overset{OH^-}{\rightleftharpoons}} Al(OH)_4^-$$

- Oxides are classified as basic, acidic, neutral or amphoteric. An exercise with unknown oxides is conducted using the following tests:

  Does the oxide dissolve in water? If so, measure the pH.

  Do the insoluble oxides react with 4 mol dm^{-3} nitric acid?

  Do the insoluble oxides react with 4 mol dm^{-3} sodium hydroxide?

Action of heat on metal compounds

- Action of heat on hydroxides. It is easier to use metal hydroxides that have a different colour from their oxides, such as the hydroxides of iron, lead and copper.
- Action of heat on nitrates. For safety reasons this is best demonstrated as there are some serious hazards — molten potassium and sodium nitrates/nitrite, decrepitation with lead nitrate and the emission of the poisonous gas, nitrogen dioxide.
- Action of heat on carbonates — see Chapter 8.

Rusting of iron

- Experiment on rusting as described in the text (Figure 9.21, p. 138). **R**
- Set up a cell with two iron electrodes connected through a voltmeter and bubble oxygen onto one electrode. From the polarity of the cell, students can deduce the direction of the electron flow and analyse the cell reactions in terms of electron transfer. **R**
- Set up a zinc/iron cell to demonstrate sacrificial protection. Analyse the cell reactions as above. **R**

Chapter 10 Atmosphere and oceans

- Find the percentage of oxygen in the air as described in the text (Figure 10.5, p. 150).
- Tests to show the presence of water using cobalt chloride paper and anhydrous copper(II) sulfate.
- Tests to show that water is pure: it melts at 0 °C or boils at 100 °C.

Chapter 11 Rates of reaction

- Reaction of calcium carbonate and acid (Figure 11.3, p. 163) — the effect of concentration and surface area.
- Reaction of sodium thiosulfate and acid (Figure 11.7, p. 164) — the effect of concentration.
- Decomposition of hydrogen peroxide, catalysed by manganese(IV) oxide — the effect of concentration and mass of catalyst on the rate and on the volume of product.
  First use a mixture of 2 cm^3 of 20 vol hydrogen peroxide and 48 cm^3 of water, with 0.2 g of manganese(IV) oxide. Then use a mixture of 4 cm^3 of 20 vol hydrogen peroxide and 46 cm^3 of water with 0.2 g of manganese dioxide. Finally use a mixture of 2 cm^3 of 20 vol hydrogen peroxide and 48 cm^3 of water with 0.4 g of manganese dioxide. Collect the oxygen over water in an inverted measuring cylinder or burette and obtain sets of readings of volume against time.
- Add a 2 cm length of magnesium ribbon to an excess of 2 mol dm^{-3} hydrochloric acid and measure the time taken for the metal to react. Repeat, with the same sized piece of magnesium but a different concentration of acid. (Another version of this exper-

iment is to make temperature the variable and use hydrochloric acid at different temperatures.)
- Add a 2 cm length of magnesium ribbon to an excess of 2 mol dm^{-3} ethanoic acid. Measure the time taken for the metal to react and compare this result with the one above.
- Reaction of metal and acid — the effect of concentration and surface area on the reaction rate and on the volume of product.
  Collect the hydrogen and obtain data on the volume evolved against time. The apparatus used could be a flask with delivery tube and a gas syringe or burette. A length of magnesium ribbon is suspended above excess acid and is allowed to fall into the acid as the timing starts.
  The experiment could be repeated using different lengths of ribbon and different concentrations of acid. Also, a mass of magnesium *powder* equal to that of the ribbon could be suspended above the acid in a small tube.
- Decomposition of sodium chlorate(I), catalysed by a cobalt(II) salt.

$$2NaClO_{(aq)} \rightarrow 2NaCl_{(aq)} + O_2$$

Use the same apparatus as in the previous exercise to obtain data on the volume of oxygen evolved against time. This can be analysed to investigate the effect of concentration on the reaction rate and on the volume of product.
- Identical mixtures of sodium thiosulfate solution and hydrochloric acid are used at different temperatures in the experiment described on pp. 164, 165. The results illustrate how rate depends on temperature when the concentrations are kept constant.
- Compare a silver halide that has been exposed to light with a sample kept in the dark. Prepare a suspension of silver bromide and divide it into two portions. Wrap one portion in aluminium foil and hold the other near a bright light.
- Divide a mixture of bromine and hexane into two portions. Expose one to bright light while the other is kept in the dark. Only the sample exposed to the light will evolve hydrogen bromide, and its brown colour will fade.
- Biology-type experiments on green leaves and light. It is probable that the students will already be familiar with these.
- The effect of enzymes on the decomposition of hydrogen peroxide. Add a piece of raw liver to 20 vol hydrogen peroxide.
- Add a few drops of washing-up liquid to 20 vol hydrogen peroxide. Divide this mixture into two portions and add manganese(IV) oxide to one portion. Compare the rate of decomposition of hydrogen peroxide in the two samples.

Chapter 12 The petroleum industry

- Simple distillation of a synthetically manufactured 'crude oil' obtained from chemical suppliers.
- Inspection of some of the fractions of the primary distillation of crude oil − colour, viscosity, boiling point and flammability.
- Demonstrate combustion of some alkanes. Use a Bunsen burner to show complete and incomplete combustion of methane. Burn a range of alkanes to show the variation in ease of ignition − pentane and hexane are highly flammable but liquid paraffin and paraffin wax need pre-heating and/or a wick.
- Reaction of alkanes with bromine – see Chapter 11.
- Cracking of alkanes as described in the text (Figure 12.12, p. 180).
- Test for unsaturation using bromine in a suitable solvent (Figure 12.15, p. 181).

Chapter 13 Energy sources

- Show examples of exothermic and endothermic reactions.
  Dissolve ammonium nitrate in water – an example of an endothermic change.
  Dissolve anhydrous copper(II) sulfate in water – an example of an exothermic change.
- Determination of the enthalpy of combustion of ethanol, as described in question 5 on p. 197.
- Show that the reaction between zinc and copper(II) ions is exothermic.
  Add zinc dust to copper(II) sulfate solution. Note the temperature of the solution before and after the addition of the zinc. Here chemical energy is transformed into heat energy.
  Set up a zinc/copper cell and note its polarity and voltage (Figure 13.25, p. 193). Here chemical energy is changed into electrical energy. Repeat using other cells, for example Cu/Ag, Fe/Cu and Mg/Cu. This will demonstrate that the reactivity series can be established using cells. **R**
- To simulate a fuel cell, electrolyse water for a few minutes until both electrodes are covered with bubbles of gas. Replace the power supply with a voltmeter. This will show that the cell produces a voltage when the hydrogen on one electrode reacts with the oxygen on the other by the transfer of electrons through the external circuit. **R**

Chapter 14 The (wider) organic manufacturing industry

Plastics and polymers

- Demonstration of thermosoftening and thermosetting plastics. Heat small samples of poly(ethene) and melamine.
- The 'nylon rope trick'. The diamine, 1,6-diaminohexane, is dissolved in water to which some sodium carbonate has been added. A solution of a diacid chloride, adipoyl chloride, in cyclohexane is added and a nylon thread can be pulled from the interface between the two phases.
- Making a solid foam. The ingredients for making a polyurethane foam can be purchased as a demonstration kit.

Starch and glucose

- Tests for starch and glucose. Test for starch using dilute aqueous iodine and for glucose with Fehling's solution or Benedict's reagent.
- Hydrolysis of starch (p. 203). Use the tests above to show that glucose is formed and starch is used up when aqueous starch is boiled with dilute hydrochloric acid.

Ethanol

- Combustion of ethanol in a small spirit lamp to demonstrate that it burns with a small blue flame.
- Add ethanol to water to show that the two liquids are miscible.
- Show examples of ethanol as a solvent.
- Oxidation of ethanol by the air. Use a water pump to bubble air through dilute aqueous ethanol.
- Add ethanol dropwise to a warm mixture of potassium dichromate(VI) and dilute sulfuric acid.
- Formation of esters using ethanol and higher alcohols. Mix $1\,cm^3$ of ethanol and $1\,cm^3$ of glacial ethanoic acid. Add 2 drops of concentrated sulfuric acid and warm the mixture for a few minutes. Then pour the mixture into a beaker of cold water. The distinctive smell of an ester can be detected. Repeat the procedure using higher alcohols – propanol, butanol and pentanol.
- Making ethanol by fermentation (Figure 14.17, p. 205).
- Fractional distillation of a mixture of water and ethanol mixture, using the apparatus shown in Figure 2.22 (p. 22). Note that this procedure may be illegal in some countries.

Ethanoic acid

- Show that ethanoic acid is an acid.
  Reaction with metals – magnesium and zinc powders.
  Reaction with carbonates – aqueous sodium carbonate and copper(II) carbonate.
  Reaction with bases – sodium hydroxide solution, copper(II) oxide and magnesium oxide.
  Reaction with an indicator – litmus.
- Show that ethanoic acid is a weak acid – a comparative study with hydrochloric acid.
  Measure the pH of $0.1\,mol\,dm^{-3}$ solutions of both acids.
  Measure the conductivity of $0.1\,mol\,dm^{-3}$ solutions of both acids.

Measure the time taken for a short length of magnesium ribbon to react with 1.0 mol dm^{-3} solutions of both acids.

Soaps and detergents
- Making soap from vegetable oil and alkali. Boil a vegetable oil or animal fat with aqueous sodium hydroxide. Cool and add salt to precipitate the soap.

Amino acids
- Identification of amino acids by chromatography. Use a $4 : 1 : 2$ mixture of butanol, ethanoic acid and water as a solvent. Dry the chromatogram and develop with a ninhydrin spray.
- Biuret test for proteins. Warm the sample in water, add concentrated sodium hydroxide solution and a drop of very dilute copper(II) sulfate solution.

Chapter 15 Nitrogen

Reversible reactions
- Acid/base indicators.
- Action of heat on copper(II) sulfate-5-water.
- Add sodium hydroxide solution to a solution of a zinc, an aluminium or a lead salt. The hydroxide will precipitate and then dissolve in excess alkali to form an aluminate, zincate or plumbate. On the addition of dilute nitric acid the hydroxide will reappear, only to dissolve with the addition of more nitric acid.

$$\text{metallic cation} \underset{\text{H}^-}{\overset{\text{OH}^-}{\rightleftharpoons}} \text{metal hydroxide} \underset{\text{H}^+}{\overset{\text{OH}^-}{\rightleftharpoons}} \text{metallic anion}$$

Equilibria
- Ice/water and other changes of state.
- Saturated solutions in equilibrium with excess solute.
- Bismuth chloride and water. Prepare a solution of bismuth chloride in the minimum of hydrochloric acid and add to a large excess of water.

$$\underset{\text{white}}{\text{BiCl}_3 + \text{H}_2\text{O} \rightleftharpoons \text{BiOCl(s)} + 2\text{HCl}}$$

On the addition of hydrochloric acid, the white precipitate will become fainter and then disappear. The addition of sodium hydroxide solution will cause the precipitate to reappear. This illustrates different positions of the same equilibrium.
- Iodine will react with chlorine to form iodine monochloride, and with an excess of chlorine the following equilibrium will be established.

$$\text{Cl}_2 + \underset{\text{brown}}{\text{ICl(l)}} \rightleftharpoons \underset{\text{yellow}}{\text{ICl}_3\text{(s)}}$$

The removal of chlorine by blowing air through the U-tube or by inverting it results in the formation of the brown liquid, while the addition of chlorine results in the formation of the yellow solid.
- $\text{Fe}^{3+}\text{(aq)} + \text{Ag(s)} \rightleftharpoons \text{Fe}^{2+}\text{(aq)} + \text{Ag}^+\text{(aq)}$

- $\text{FeCl}_3 + 3\text{KCNS} \rightleftharpoons \text{Fe(CNS)}_3 + 3\text{KCl}$
  Although this equation is not strictly correct, at this level it is preferable to the ionic equation involving FeCNS^{2+}.
- $\text{N}_2\text{O}_4 \rightleftharpoons 2\text{NO}_2$
  Altering the temperature using ice-water or hot water will change the position of equilibrium and illustrate its dependence on the exothermic or endothermic nature of the reaction. Similarly, changing the pressure by compressing a sample of the equilibrium mixture in a gas syringe will also move the position of equilibrium.
- $\text{NH}_4\text{Cl} \rightleftharpoons \text{NH}_3 + \text{HCl}$
  Heat ammonium chloride to show the influence of temperature on the position of equilibrium.

Ammonia gas
- Laboratory preparation of ammonia (Figure 15.4, p. 220).
- The fountain flask experiment (Figure 15.6, p. 221).
- Show that ammonia is a base. Neutralise acids by adding 2 mol dm^{-3} aqueous ammonia to dilute acids in the presence of universal indicator.
- Show that ammonia is a weak base. Measure the pH of 0.1 mol dm^{-3} aqueous ammonia and its conductivity. Compare the pH and conductivity with those of 0.1 mol dm^{-3} sodium hydroxide.
- Precipitation of metal hydroxides – see Chapter 9.
- Test for the ammonium ion (p. 220).

Nitric acid
- Reactions of both dilute and concentrated nitric acid with metals – copper and zinc. **R**
- Reaction with bases – NaOH, CuO and ZnO.
- Reaction with carbonates – Na_2CO_3 and CuCO_3.
- Reaction with carbon (p. 223).

Chapter 16 Sulfur

Sulphur and sulphur compounds
- Burn sulfur in air or oxygen. Show that an acidic gas is formed.
- Formation of metal sulfides. Demonstrate the reactions of sulfur with iron, zinc and magnesium – care is needed.
- Preparation of sulfur trioxide. Pass dry oxygen and sulfur dioxide over heated platinised mineral wool. Collect sulfur trioxide as a solid in a cooled receiver.
- Reaction of sulfur trioxide with water – care is needed. Test the solution to show that it is acidic (using both universal indicator and magnesium powder), and that it contains sulfate ions (by adding acidified barium chloride solution).

Dilute sulfuric acid
- Reaction with magnesium and zinc. **R**
- Reaction with bases – CuO and NaOH(aq).
- Reaction with carbonates – NiCO_3, CoCO_3 and $\text{Na}_2\text{CO}_3\text{(aq)}$.

- Precipitation of insoluble sulfates of barium, lead and calcium.
- Preparation of sodium sulfate-10-water by titration. Use the same method as on p. 100, using dilute sulfuric acid and dilute aqueous sodium hydroxide.
- Preparation of sodium hydrogensulfate by titration. Use the results from the previous exercise but add double the titre of acid to the same volume of alkali.
- Comparison of the salts sodium sulfate and sodium hydrogensulfate. Compare their crystal shapes.

Measure the pH of their aqueous solutions. Add magnesium powder to both salts in solution. Add excess dilute hydrochloric acid and barium chloride solution to both solutions.

Concentrated sulfuric acid
- Reactions with copper(II) sulfate-5-water and sugar.
- Oxidation of metals, such as copper.
- Preparation of a soapless detergent from a vegetable oil.

IGCSE exam questions (practical papers)

1 You are going to investigate a mixture of calcium hydroxide and water.
Read **all** the instructions below carefully **before** starting the experiment.

Instructions
Shake the mixture of calcium hydroxide and water provided in the bottle.
After one minute of shaking, filter the mixture into a beaker.
Start Experiment 1 when enough solution has been collected. Continue to filter the mixture.

Experiment 1
By using a measuring cylinder, measure 25 cm³ of the solution (filtrate) into the conical flask provided.
Carry out the titration as follows.
Fill the burette to the 0.0 cm³ mark with the solution **M** of hydrochloric acid.
Add 3 or 4 drops of phenolphthalein to the flask.
Add solution **M** slowly to the flask until the colour just disappears. Record the burette readings in the table.
Pour the solution away and rinse the conical flask.

Experiment 2
Empty the contents of the burette down the sink.
Rinse the burette with the solution **N** of hydrochloric acid.
Repeat Experiment 1 using the solution **N** of hydrochloric acid.
Record your results in the table.

Normally you would be required to carry out repeat titrations. However, owing to time considerations you are only required to carry out **one** titration for each experiment

Table of results

Burette readings/cm³	Experiment 1	Experiment 2
final reading		
initial reading		
difference		

(6)

a Describe the appearance of the mixture of calcium hydroxide and water. (1)
b How did the colour of the solution in the flask change?
from _____ to _____ (2)
c What type of chemical reaction occurs when hydrochloric acid reacts with calcium hydroxide? (1)
d (i) In which experiment was the greater volume of hydrochloric acid used? (1)
(ii) Compare the volumes of acid used in Experiments 1 and 2. (2)
(iii) Suggest an explanation for the difference in volumes. (2)

e Predict the volume of hydrochloric acid **M** which would be needed to react completely if Experiment 1 was repeated with 50 cm³ of calcium hydroxide solution and explain your answer. (3)
f Suggest **one** change you could make to the **apparatus** used in the experiments to obtain more accurate results (1)
g From the list below choose the correct word to complee the sentence.

        not   slightly   very

Calcium hydroxide is _____ soluble in water. (1)

       (UCLES, IGCSE, Paper 5, May/June 2005)

2 You are provided with liquid **A**.

Carry out the following tests on **A**, recording all of your observations in the table. Do **not** write any conclusions in the table.

Tests	Observations
a Describe the appearance and smell of **A**.	(2)
b Test the pH of the solution using indicator paper.	Colour pH (2)
c Divide the liquid into five test-tubes. (i) To the first portion, add the piece of magnesium ribbon provided. Note any observations and test the gas. (ii) To the second portion of liquid **A**, add slowly a spatula measure of anhydrous sodium carbonate. Test the gas with limewater. (iii) To the third portion of liquid **A**, add a spatula measure of solid **B**. Boil **gently** for 2 minutes. Note any observations By using a teat pipette transfer the solution to another test tube. To this solution add excess aqueous ammonia. (iv) To the fourth portion of the liquid add about 1 cm³ of ethanol. Ask your supervisor to add a few drops of concentrated sulfuric acid to the mixture. Boil the mixture gently. Pour the mixture into a beaker half full of water. Note your observations (v) To the fifth portion of liquid **A** add a few drops of dilute sulfuric acid and about 1 cm³ of potassium dichromate solution. Boil **gently** and note any observation.	(3) (2) (2) (2) (1)

d (i) Name the gas given off in test **c**(i). **(1)**
   (ii) Name the gas given off in test **c**(ii) **(1)**
e Use your observations in test **c**(iii) to say what ion is present in solid **B**. **(2)**
f What conclusions can you draw about liquid **A**? **(2)**

Notes for use in qualitative analysis

Test for anions

Anion	Test	Test result
carbonate (CO_3^{2-})	add dilute acid	effervescence, carbon dioxide produced
chloride (Cl^-) [in solution]	acidify with dilute nitric acid, then add aqueous silver nitrate	white ppt.
iodide (I^-) [in solution]	acidify with dilute nitric acid, then aqueous lead(II) nitrate	yellow ppt.
nitrate (NO_3^-) [in solution]	add aqueous sodium hydroxide then aluminium foil; warm carefully	ammonia produced
sulfate (SO_4^{2-}) [in solution]	acidify with dilute nitric acid, then aqueous barium nitrate	white ppt.

Test for aqueous cations

Cation	Effect of aqueous sodium hydroxide	Effect of aqueous ammonia
aluminium (Al^{3+})	white ppt., soluble in excess giving a colourless solution	white ppt., insoluble in excess
ammonium (NH_4^+)	ammonia produced on warming	-
calcium (Ca^{2+})	white., insoluble in excess	no ppt., or very slight white ppt.
copper (Cu^{2+})	light blue ppt., insoluble in excess	light blue ppt., soluble in excess giving a dark blue solution
iron(II) (Fe^{2+})	green ppt., insoluble in excess	green ppt., insoluble in excess
iron(III) (Fe^{3+})	red–brown ppt., insoluble in excess	red–brown ppt., insoluble in excess
zinc (Zn^{2+})	white ppt., soluble in excess giving a colourless solution	white ppt., soluble in excess giving a colourless solution

Test for gases

Gas	Test and test results
ammonia (NH_3)	turns damp red litmus paper blue
carbon dioxide (CO_2)	turns limewater milky
chlorine (Cl_2)	bleaches damp litmus paper
hydrogen (H_2)	'pops' with a lighted splint
oxygen (O_2)	relights a glowing splint

(UCLES, IGCSE, Paper 5, May/June 2005)

3 You are going to investigate the temperature changes that occur when two compounds, **A** and **B**, react with hydrochloric acid.

Read **all** the instructions below carefully before starting the experiments.

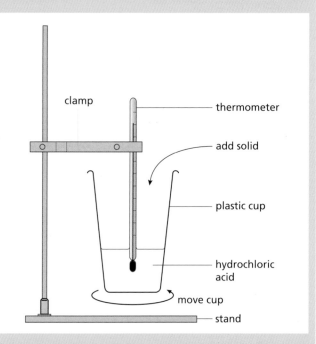

Instructions

Experiment 1

Using a measuring cylinder, add 30 cm³ of the hydrochloric acid provided to the plastic cup.

In the table record the initial temperature of the acid.

Start the timer, and using a spatula, add some of the solid **A** to the cup. Stir the mixture by moving the cup (not the thermometer) until the fizzing stops (see diagram above). Now add some more of **A**. Continue adding **A** in this way until all of solid **A** has been added.

During the addition of **A**, record the temperature of the mixture every 30 seconds/half minute. It should take less than five minutes to add all of the solid. Continue taking temperature readings for the full five minutes.

Experiment 3

Using the second plastic cup provided, a fresh supply of hydrochloric acid and solid **B** (instead of **A**), repeat experiment 1.

Table of results

Experiment 1

Time/minutes	0	0.5	1	1.5	2	2.5	3	3.5	4	4.5	5
Temperature/°C											

(3)

Experiment 2

Time/minutes	0	0.5	1	1.5	2	2.5	3	3.5	4	4.5	5
Temperature/°C											

(3)

a Plot your results for both experiments in the grid below. For each set of results draw a smooth line graph. Indicate clearly which line represents Experiment 1 and which line Experiment 2.

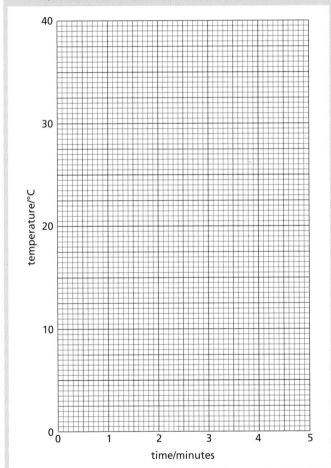

b **From your graphs,**
  (i) Find the temperature of the reaction mixture after the hydrochloric acid had reacted for 2 minutes 15 seconds with
  solid **A**,
  solid **B**. **(2)**
  (ii) What type of chemical reaction occurs when
  solid **A**,
  solid **B**. **(2)**
  reacts with hydrochloric acid?

c Give two observations in Experiment 1, apart from temperature change. **(2)**
d Suggest what type of compound solids **A** and **B** are. Explain your answer. **(2)**
e If the plastic cup and final reaction mixture were left for one hour, predict the temperature at this time for
  (i) solid **A** and hydrochloric acid,
  (ii) solid **B** and hydrochloric acid.
  Explain your answers. **(3)**

(UCLES, IGCSE, Paper 5, May/June 2004)

4 You are provided with a mixture of two solid compounds **C** and **D**, containing the same metal cation.

C is partially soluble in water and **D** is soluble.

Carry out the following tests on **C** and **D**, recording **all** of your observations in the table. Do **not** write any conclusions in the table.

Tests	Observations
a Note the appearance of the mixture of **C** and **D**. **(1)**	
b Add the mixture of **C** and **D** to 10 cm³ of distilled water in a boiling-tube. Stopper the tube and shake for 1 minute. Filter the mixture. Keep the filtrate.	
c Divide the filtrate from **b** into five approximately equal portions. (i) To the first portion add drops of aqueous sodium hydroxide, a little at a time, with shaking. **(2)** Add excess aqueous sodium hydroxide. **(1)** (ii) To the second portion add excess aqueous ammonia, a little at a time. **(1)** (iii) To the third portion add about 1 cm³ of dilute sodium hydroxide and aluminium powder. Carefully boil the mixture and test the gas with damp indicator paper. **(3)** (iv) Test the pH of the fourth portion with Indicator paper. **(1)** (v) Using a straw blow into the fifth portion. **(1)**	

d What conclusions can you draw about the identity of the metal cation in **C** and **D**? **(1)**
e Name the gas given off in **c**(iii). **(1)**
f Suggest an explanation for the observation in **c**(v) **(2)**
g What conclusions can you draw about the identity of the anions in solid **C** and **D**? **(2)**

(UCLES, IGCSE, Paper 5, May/June 2004)

5 You are going to investigate the speed of reaction between aqueous potassium bromate and potassium iodide solution.

Read **all** the **Instructions** below carefully before starting the experiments.

Instructions

Fill the burette up to the 0.0 cm³ mark with the aqueous potassium iodide.

Put 5 test-tubes in a rack. Using the burette, add 6 cm³ of aqueous potassium iodide to each test-tube to be used in the 5 following experiments

Experiment 1

You are going to measure 5 different solutions into a small beaker. Use the 10 cm³ measuring cylinder to measure all the solutions. The cylinder does **not** need rinsing between additions.

Using the measuring cylinder pour 12 cm³ of the aqueous potassium bromate into the small beaker. Use the 10 cm³ measuring cylinder to add 2 cm³ of hydrochloric acid and 4 cm³ of water to the beaker. Now add 5 cm³ of starch solution and 1 cm³ of sodium thiosulfate solution to the beaker.

Place the baker on the insert.

Att 6 cm³ of aqueous potassium iodide from a test-tube to the mixture in the beaker and start your timer. Stop the timer when you can no longer read the words on the insert when looking down through the beaker.

Record the time in the table.

Pour away the contents of the beaker and rinse the beaker with distilled water.

Table of results

Experiment	Volume		Time/s
	Potassium bromate/cm³	water/cm³	
1	12	4	
2	10	6	
3	8	8	
4	6	10	
5	4	12	

Experiment 2

Using a measuring cylinder pour into the beaker 10 cm³ of the potassium bromate solution. Follow the instructions exactly as given for Experiment 1, using the same volumes of the other reagents, but this time add 6 cm³ of water to the beaker.

Record your time in the table.

Experiments 3, 4 and 5

Repeat Experiment 1 using the volumes of aqueous potassium bromate and water specified in the table of results. Record the times in the table.

a Plot your results on the grid. Draw a smooth line graph.

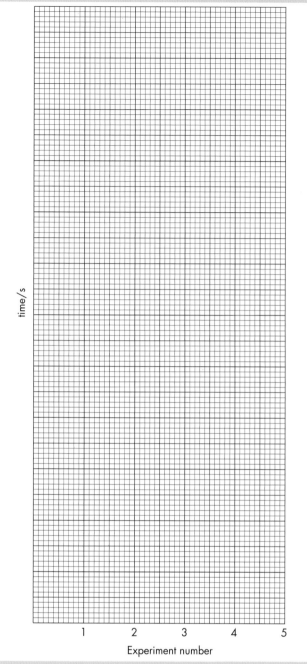

b Describe how the appearance of the mixture in the beaker changed as you timed the reaction. **(2)**
c From your graph estimate the time of the reaction if Experiment 1 was repeated using 7 cm³ of potassium bromate and 9 cm³ of water **(2)**
   Show clearly on your graph how you worked out your answers. **(1)**
d (i) Which Experiment is the quickest? **(1)**
   (ii) Explain why this Experiment is the quickest. **(2)**
e (i) State **two** sources of error in the Experiments. **(2)**
   (ii) Suggest **two** improvements to reduce the sources of error in the experiments **(2)**
   (UCLES, IGCSE, Paper 5, October/November 2003)

6 You are provided with an aqueous solution of substance **X**.

Carry out the following tests on **X**, record all of your observations in the table. Do **not** write any conclusions in the table.

Tests	Observations
a colour of solution **X**.	(1)
b (i) By using a teat pipette add drops of aqueous sodium hydroxide to about 2 cm³ of the solution in a test-tube. Now add excess aqueous sodium hydroxide to the test-tube. (ii) Carefully warm the mixture Test any gas given off with damp indicator paper.	(3) (2)
c Repeat **b**(i) using aqueous ammonia instead of aqueous sodium hydroxide.	(1)
d Acidify about 2 cm³ of solution **X** with dilute sulfuric acid. Add two pieces of zinc. Warm the mixture gently. Test the gas given off. Leave the mixture to react for 10 minutes. After 10 minutes decant the liquid and repeat test **b**(i).	(3) (2)
e Add a few drops of hydrochloric acid to about 2 cm³ of solution **X** in a test-tube. Add about 1 cm³ of barium chloride solution to the mixture.	(2)

f What conclusions can you draw about substance **X**? **(4)**

(UCLES, IGCSE, Paper 5, October/November 2003)

7 A small piece of limestone was heated strongly and left to cool. A few drops of cold water were added. The solid expanded and gave off steam.

- limestone
- water
- lime (calcium oxide)

a Complete the empty boxes to identify the pieces of apparatus labelled. **(3)**
b What type of chemical reaction takes place when water is added? **(1)**

(UCLES, IGCSE, Paper 6, May/June 2005)

8 The diagram shows the apparatus used to find out the effect of an electric current on a concentrated aqueous solution of sodium chloride.

a On the diagram label the electrodes **(1)**
b Give three observations when the circuit is switched on. **(3)**
c (i) Name the product at the positive electrode (anode). **(1)**
   (ii) State a test for this product and the result of the test. **(2)**

(UCLES, IGCSE, Paper 6, May/June 2005)

9 In a set of experiments zinc was reacted with sulfuric acid to form hydrogen. The apparatus below was used.

- gas syringe
- sulfuric acid
- excess zinc

The same mass of zinc was used each time. The volume of acid used was different each time. Use the syringe diagrams to record the volume of hydrogen produced each time in the table.

Table of results

Volume of sulfuric acid/cm³	Syringe diagram	Volume of hydrogen/cm³
0		
5		
15		
20		
25		
30		
35		
40		

251

a Plot the results on the grid below. Draw a smooth line graph. (4)

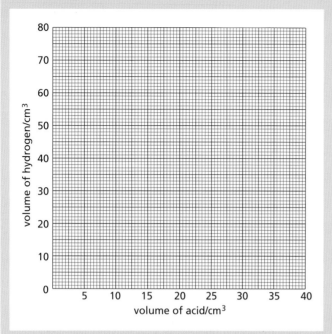

(4)

b Use the graph to find the volume of sulfuric acid that will produce 33 cm³ of gas. (1)
c What volume of gas is produced if 10 cm³ of sulfuric acid is used? (1)

(UCLES, IGCSE, Paper 6, May/June 2005)

10 A student investigated an aqueous solution of calcium hydroxide and water.

Two experiments were carried out.

Experiment 1

By using a measuring cylinder 25 cm³ of the aqueous solution of calcium hydroxide was placed in a flask. Phenolphthalein indicator was added to the flask. A burette was filled to the 0.0 cm³ mark with solution **M** of hydrochloric acid.

Solution **M** was added slowly to the flask until the colour just disappeared. Use the burette diagram to record the volume in the table and complete the column.

Experiment 2

Experiment 1 was repeated using a different solution, **N**, of hydrochloric acid.

Use the burette diagrams to record the volumes in the table and complete the table.

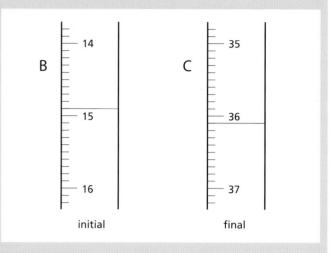

initial   final

Table of results

Burette readings/cm³	Experiment 1	Experiment 2
final reading		
initial reading	0.0	
difference		

(4)

a What type of chemical reaction occurs when hydrochloric acid reacts with calcium hydroxide? (1)
b (i) In which experiment was the greater volume of hydrochloric acid used? (1)
   (ii) Compare the volumes of acid used in Experiments 1 and 2. (2)
   (iii) Suggest an explanation for the difference in volumes. (2)
c Predict the volume of hydrochloric acid **M** that would be needed to react completely if Experiment 1 was repeated with 50 cm³ of calcium hydroxide Solution. Explain your answer. (3)
d Suggest **one** change you could make to the **apparatus** used in the experiments to obtain more accurate results. (1)

(UCLES, IGCSE, Paper 6, May/June 2005)

11 A sample of a solution of acid **A** was analysed.

The tests on **A** and some of the observations are in the following table.

Complete the observations in the table.

Tests	Observations
a The pH of the solution was tested using indicator paper	colour orange pH 4

b The solution was divided into three test-tubes
  (i) To the first portion was added a piece of magnesium ribbon. The gas was tested with a lighted splint.   **(2)**
  (ii) To the second portion of **A** was added sodium carbonate. The gas was tested with limewater.   **(2)**
  (iii) To the third portion of liquid **A** was added a spatula measure of solid **B**. By using a teat pipette the solution was transferred to another test tube. Excess aqueous ammonia was added.
	green solution formed
	
	dark blue solution formed

c What does test **a** tell you about the type of acid in solution **A**?   **(1)**

d (i) Name the gas given off in test **b**(i).   **(1)**

  (ii) Name the gas given off in test **b**(ii).   **(1)**

e Explain the observations in test **b**(iii).   **(2)**

(UCLES, IGCSE, Paper 6, May/June 2005)

12 The label below is from a bottle of concentrated lemon drink.

Concentrated lemon drink

Ingredients: Water, sugar, citric acid, preservatives, potassium sorbate (artificial sweetener). Yellow colourings E120 and E104.

a What is meant by the term *concentrated*?   **(1)**

b Predict the pH of the lemon drink.   **(1)**

c Describe an experiment to show that two different yellow colourings are present in the drink.   **(4)**

(UCLES, IGCSE, Paper 6, May/June 2005)

12 Samples of concrete were placed in solutions of different pH. The graph shows the percentage corrosion of the samples.

a Draw a smooth line graph on the grid   **(1)**

b Which point on the grid appears to be inaccurate? Explain your reason for identifying this point.   **(2)**

c What happens to the percentage corrosion as the pH changes from 1 to 7?   **(1)**

(UCLES, IGCSE, Paper 6, May/June 2005)

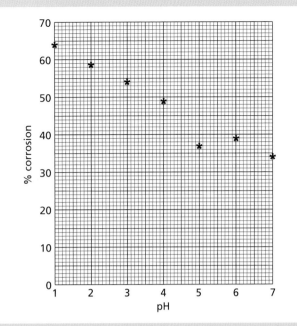

13 An aqueous solution of hydrogen peroxide decomposes very slowly to form oxygen. The speed of decomposition can ben increased by using a catalyst. Two possible catalysts are the solids copper(II) and chromium(III) oxide.

Plan an investigation to find out which of these two oxides is the better catalyst for this decomposition.   **(6)**

(UCLES, IGCSE, Paper 6, May/June 2005)

14 A student reacted sulfuric acid with copper (II) oxide. The diagram shows the procedure followed.

Copper oxide was added until all the sulfuric acid had reacted

50 cm³ of dilute sulfuric acid was measured into a beaker

dilute sulfuric acid

WARM

a Complete the boxes to identify the pieces of apparatus labelled.   **(3)**

b What is the colour of the solution formed?   **(1)**

c Describe how crystals could be quickly obtained from the solution.   **(2)**

(UCLES, IGCSE, Paper 6, October/November 2005)

15 The diagram shows the apparatus used to pass an electric current through concentrated hydrochloric acid.

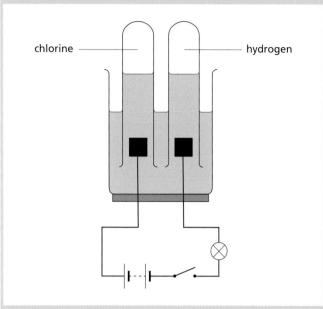

chlorine   hydrogen

a Label the electrodes. **(1)**
b Give two observations when the current is switched on. **(2)**
c Give a test for the product at the negative electride (cathode), and the result. **(2)**
(UCLES, IGCSE, Paper 6, October/November 2005)

16 The green pigment chlorophyll can be obtained from grass.

Step 1   The grass is crushed with sand.

Step 2   The grass is ground with ethanol until the solution is saturated.

Step 3   The solution is separated from the rest of the mixture.

Step 4   The colours in the solution are separated.
a What apparatus is used in Step 1? **(2)**
b Suggest why the grass is ground with ethanol rather than water in Step 2. **(1)**
c Name the separation method in Step 3. **(1)**
d Describe how Step 4 is carried **(4)**
(UCLES, IGCSE, Paper 6, October/November 2005)

17 A student investigated the speed of reaction between aqueous hydrochloric acid and marble chips (calcium carbonate).

The apparatus below was used.

hydrochloric acid

marble chip

5 test-tubes were put in a rack. To each test-tube was added 10 cm³ of different solutions of aqueous hydrochloric acid and a marble chip. The marble chips were the same size.

Experiment 1

By using a measuring cylinder 10 cm³ of the solution **P** of aqueous hydrochloric acid was placed in the first test-tube. A marble chip was added and the volume of gas collected after two minutes was measured. Use the gas syringe diagram to record the volume.

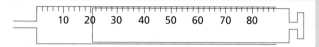

Experiment 2

Experiment 1 was repeated using the solution **Q** of aqueous hydrochloric acid. Use the diagram to record the volume of gas collected in the table.

Experiments 3, 4 and 5

Experiment 1 was repeated using the solutions **R**, **S** and **T** of aqueous hydrochloric acid in the third, fourth and fifth test-tubes.

Use the diagrams to record the volumes in the table.

Experiment 3

Experiment 4

Experiment 5

Table of results

Experiment	Solution of hydrochloric acid	Volume of gas collected/cm³
1	P	
2	Q	
3	R	
4	S	
5	T	

(3)

a Plot your results on the grid. Draw a straight line graph. **(4)**

b Which result appears inaccurate? Give a reason for your choice. **(2)**

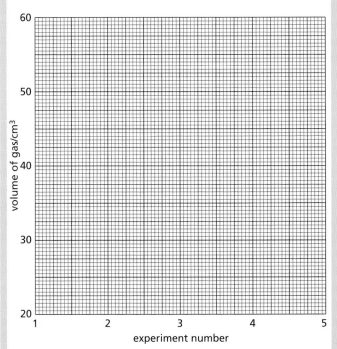

volume of gas/cm³ (y-axis) vs experiment number (x-axis)

c (i) Which Experiment had the fastest rate of reaction? **(1)**

(ii) Suggest why this Experiment was the fastest. **(2)**

d How would the student know which of the reactants in the Experiment was in *excess*? **(2)**

e (i) State **one** possible source of error in the Experiments. **(1)**

(ii) Suggest **one** improvement to reduce this source of error in the Experiments. **(1)**

(UCLES, IGCSE, Paper 6, October/November 2005)

18 A solid compound **X** was analysed. Solid **X** was an aluminium salt. The tests on **X** and some of the observations are in the following table.

Complete the observations in the table.

c What does test **a** tell you about the gas given off? **(1)**

d What conclusions can you draw about **X** from tests **b**(iii) and (iv)? **(2)**

e Identify the gas in **b**(v). **(1)**

f What conclusions can you draw about substance **X**? **(2)**

(UCLES, IGCSE, Paper 6, October/November 2005)

Tests	Observations
a One spatual measure of **X** was placed into a hard-gas test-tube. The solid was heated gently then strongly. The gas was tested with pH indicator paper.	condensation at top of tube paper went red
Distilled water was added to **X** and shaken to dissolve. The solution was divided into five portions in test-tubes. **b** (i) To the first portion, drops of aqueous sodium hydroxide were added. Excess aqueous sodium hydroxide was then added.	**(3)**
(ii) To the second portion drops of aqueous ammonia were added. Excess ammonia was then added.	**(3)**
(iii) To the third portion of solution, hydrochloric acid and barium chloride solution were added.	no visible change
(iv) To the fourth portion of solution, nitric acid and lead nitrate solution were added.	no visible change
(v) To the fifth portion, aqueous sodium hydroxide and a spatula measure of aluminium granules were added. The mixture was warmed and the gas tested with indicator paper.	pungent gas paper went blue, pH 10

19 The diagram shows how ethene can be made by cracking liquid alkanes.

liquid alkane on mineral fibre — aluminium oxide — ethene — water

a Indicate with arrows on the diagram where heat is applied. **(2)**

b Give a chemical test, and the results, to show the presence of ethene. **(2)**

c Why must the delivery tube be removed from the water before the heat is turned off? **(2)**

(UCLES, IGCSE, Paper 6, October/November 2005)

20 Some plants do not grow well in acidic soil. A farmer gives you a small sample of soil from a corner of one of his fields.

a Plan an investigation to find out the pH of the soil sample. You are provided with Universal indicator solution and common laboratory apparatus. **(5)**

b Why would further experiments be necessary to inform the farmer which plants should be grown in each of his fields? **(2)**

(UCLES, IGCSE, Paper 6, October/November 2005)

21 An experiment was carried out to find the pH of samples of soil from a farmer's field.

a Identify the pieces of apparatus labelled A, B and C. **(3)**

b Why was the soil crushed? **(2)**

c Why should soil samples be taken from different parts of the field? **(1)**

d Suggest why it is important to know the pH of soil. **(1)**

(UCLES, IGCSE, Paper 6, May/June 2004)

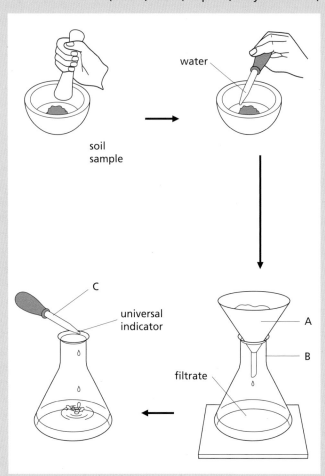

22 The four tubes show an investigation of rusting. Each one of these four tubes contains a nail the the reagents indicated.

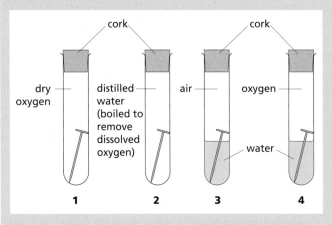

a Predict the order in which rust would appear, first and second. **(1)**

b Explain your prediction. **(2)**

(UCLES, IGCSE, Paper 6, May/June 2004)

23 Lead bromide was placed in a tube and connected to an electrical circuit as shown below.

The lead bromide was heated until molten. A brown gas was given off.

a State one other expected observation. **(1)**

b (i) Suggest a suitable material for the electrodes.
   (ii) Indicate on the diagram the negative electrode (cathode). **(2)**

c Name the brown gas. At what electrode will the gas be given off? **(2)**

d Why is this experiment carried out in a fume cupboard? **(1)**

(UCLES, IGCSE, Paper 6, May/June 2004)

24 A student investigated the temperature changes that occur when two compounds **A** and **B**, react with hydrochloric acid. The apparatus below as used.

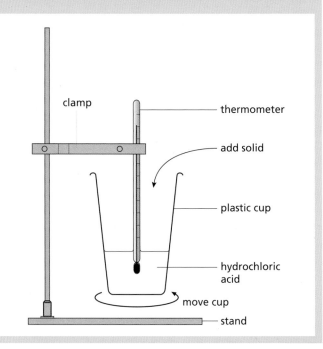

Experiment 1

By using a measuring cylinder, 30 cm³ of hydrochloric acid was added to the plastic cup.

Use the thermometer diagram to record the initial temperature of the acid in the table. The timer was started, and some of the solid **A** was added to the cup. Immediate effervescence occurred. The mixture was stirred by moving the cup until the fizzing stopped.

More of **A** was then added and the student continued adding **A** in this way until all of solid **A** had been added.

Use the thermometer diagrams to record the temperature of the mixture every half minute.

Experiment 2

Experiment 1 was repeated using solid **B**. Use the thermometer diagrams to record the temperatures in the table.

Table of results

Experiment 1

Time/min	0.0	0.5	1.0
Thermometer diagram	25 / 20 / 15	30 / 25 / 20	30 / 25 / 20
Temperature/°C			
	1.5	2.0	2.5
Thermometer diagram	30 / 25 / 20	35 / 30 / 25	35 / 30 / 25
Temperature/°C			
Time/min	3.0	3.5	4.0
Thermometer diagram	35 / 30 / 25	35 / 30 / 25	30 / 25 / 20
Temperature/°C			
Time/min	4.5	5.0	
Thermometer diagram	30 / 25 / 20	30 / 25 / 20	
Temperature/°C			

(2)

Experiment 2

Time/min	0.0	0.5	1.0
Thermometer diagram	35 / 30 / 25	25 / 20 / 15	20 / 15 / 10
Temperature/°C			
	1.5	2.0	2.5
Thermometer diagram	20 / 15 / 10	20 / 15 / 10	15 / 10 / 5
Temperature/°C			
Time/min	3.0	3.5	4.0
Thermometer diagram	15 / 10 / 5	20 / 15 / 10	20 / 15 / 10
Temperature/°C			
Time/min	4.5	5.0	
Thermometer diagram	20 / 15 / 10	20 / 15 / 10	
Temperature/°C			

(2)

a Plot the results from both experiments on the grid overleaf. For each set of results draw a smooth line graph. Indicate clearly which lines represents Experiment 1 and which line Experiment 2. **(6)**

b **From your graphs;**
(i) Find the temperature of the reaction mixture after the hydrochloric acid had reacted for 2 minutes 15 seconds with solid **A** and solid **B**. **(2)**
(ii) What type of chemical reaction occurs when solid **A** and solid **B** reacts with hydrochloric acid? **(2)**

c Suggest what type of compound solids **A** and **B** are. Explain your answer. **(2)**

d If the plastic cup and final reaction mixture are left for one hour, predict the temperature at this time for
(i) solid **A** and hydrochloric acid,
(ii) solid **B** and hydrochloric acid.
Explain your answers. **(3)**

(UCLES, IGCSE, Paper 6, May/June 2004)

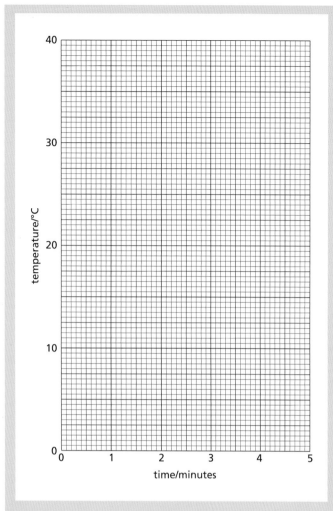

25 A mixture of two calcium compounds **C** and **D** was tested.

C is partially soluble in water and **D** is soluable in water.

Complete the observations in the table.

Tests	Observations
The mixture of **C** and **D** was added to distilled water in a boiling tube. The tube was shaken. The mixture was filtered.	

a The filtrate was divided into five equal portions.
   (i)   To the first portion was added drops of aqueous sodium hydroxide, a little at a time, with shaking. **(2)**
        Excess aqueous sodium hydroxide was added. **(1)**
  (ii)   To the second portion was added excess aqueous ammonia, a little at a time. **(1)**
 (iii) To the third portion was added dilute sodium hydroxide and aluminium powder. The mixture was boiled and the gas tested with damp litmus paper.
 (iv) The pH of the fourth portion was tested with Indicator paper.
 (iv) Carbon dioxide was bubbled through the fifth portion.
	
	red litmus went blue
	
	pH about 10
	solution turned milky/cloudy

b Name the gas given off in **a**(iii). **(1)**
c Suggest an explanation for the observation in **a**(v) **(1)**
d What conclusions can you draw about the identity of the anions in solid **C** and **D**? **(2)**

(UCLES, IGCSE, Paper 6, May/June 2004)

26 Copper oxide was reacted with hydrogen using the apparatus shown below.

dry hydrogen

excess hydrogen burning in air

heat

ice

colourless liquid

a Indicate on the diagram with an arrow where the copper oxide is placed. **(1)**
b The colour of the copper oxide would change from _____ to _____ **(2)**
c What is the purpose of the ice? **(2)**

(UCLES, IGCSE, Paper 6, May/June 2004)

27 Describe a chemical test to **distinguish** between each of the following pairs of substances. An example is given.

potassium chloride and potassium iodide

test: add aqueous lead (II) nitrate

result: potassium chloride gives a white precipitate,
         potassium iodide gives a yellow precipitate.

a water and ethanol
test
result with water
result with ethanol **(2)**

b sulfuric aicd and aqueous sodium sulfate
test
result with sulfuric acid
result with aqueous sodium sulfate **(2)**

c hydrochloric acid and nitric acid
test
result with hydrochloric acid
result with nitric acid **(2)**

(UCLES, IGCSE, Paper 6, May/June 2004)

28 Is manganese (IV) oxide a catalyst?

A catalyst is a substance that speeds up a chemical reaction and remains unchanged.

Hydrogen peroxice, H_2O_2 breaks down to form oxygen. This reaction is very slow without a catalyst. Describe an experiment to show that manganese (IV) oxide is a catalyst for this reaction.

You are provided with the following items.
Hydrogen peroxide solution
Manganese (IV) oxide
Measuring cylinder
Balance
Beaker
Filtration apparatus
Splints/Bunsen burner
Distilled water **(6)**

(UCLES, IGCSE, Paper 6, May/June 2004)

29 The apparatus below was used to make hydrogen. Dilute hydrochloric acid was added to zinc.

a Identify the pieces of apparatus labelled **A** and **B**. **(2)**
b Complete the boxes **(1)**
c Give a test, and the results, for hydrogen. **(2)**

(UCLES, IGCSE, Paper 6, October/November 2004)

30 The addition of calcium carbonate to exess dilute nitric acid produces carbon dioxide. The volume of carbon dioxide given off at 20 second intervals was recorded and plotted on the grid.

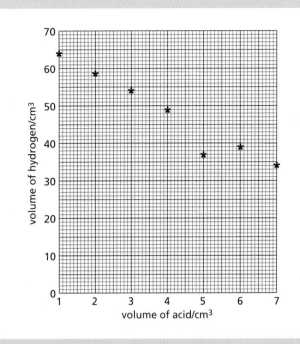

a Draw a smooth line graph on the grid. **(1)**
b Circle the result which appears to be incorrect? Why have you selected this result? **(2)**
c Why does the reaction slow down? **(1)**

(UCLES, IGCSE, Paper 6, October/November 2004)

31 A liquid alkane was passed over heated aluminium oxide to make ethene.

a What is the purpose of the mineral wool? **(1)**
b What is this type of chemical reaction called? **(1)**
c Give a test, and the results, for ethene. **(2)**
d What precaution should be taken in the experiment when the heat is removed? Explain. **(2)**

(UCLES, IGCSE, Paper 6, October/November 2004)

32 A student investigated what happened when sodium thiosulfate dissolved in water.

Experiment 1

By using a measuring cylinder, 20 cm³ of distilled water was poured into a polystyrene cup. Use the thermometer diagram to record the temperature of the water in the table.

1 g of powdered sodium thiosulfate was added to the cup and the mixture stirred with a thermometer. Use the thermometer diagram to record the temperature of the solution.

Experiment 2

Experiment 1 was repeated using 2 g of powdered sodium thiosulfate. Record the temperature in the table.

Experiments 3, 4 and 5

Experiment 1 was repeated using 3 g, 4 g and 5 g of powdered sodium thiosulfate respectively. Record the temperatures in the table.

Mass of sodium thiosulfate/g	Temperature/°C			
	Initial		**Final**	
0	(thermometer: 30, 25, 20)		(shaded)	(shaded)
1	(thermometer: 30, 25, 20)		(thermometer: 25, 20, 15)	
2	(thermometer: 30, 25, 20)		(thermometer: 20, 15, 10)	
3	(thermometer: 25, 20, 15)		(thermometer: 20, 15, 10)	
4	(thermometer: 25, 20, 15)		(thermometer: 15, 10, 5)	
5	(thermometer: 25, 20, 15)		(thermometer: 15, 10, 5)	

(5)

a Plot the results of the experiments on the grid below. Draw a straight line graph. Clearly label the graph. **(5)**

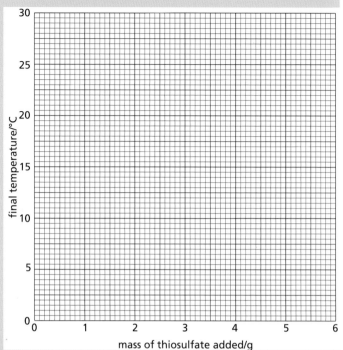

final temperature/°C vs mass of thiosulfate added/g

b (i) **Use your graph** to estimate the temperature of the reaction mixture if 3.5 g of powdered sodium thiosulfate were added to 20 cm³ of water.
Indicate clearly on the graph how you obtained your answer. **(2)**
(ii) From your graph work out the temperature of the reaction mixture if 6 g of powdered sodium thiosulfate were added to 20 cm³ of water.
Indicate clearly how you used your graph. **(2)**

c What type of chemical reaction occurs when sodium thiosulfate dissolves in water? **(1)**

d Explain how the temperature changes would differ in the experiments if 40 cm³ of water were used. **(2)**

e Explain why the sodium thiosulfate was powdered before being used. **(2)**

f Predict what the temperature of the reaction mixture in *Experiment 5* would be after 1 hour. Explain your answer. **(2)**

g Suggest **one** change you could make to the **apparatus** used in the experiments to obtain more accurate results. **(1)**

(UCLES, IGCSE, Paper 6, October/November 2004)

33 Salt **E**, which is ammonium chloride was tested.

Record all observations in the table.

Tests	Observations
a Describe the appearance of **E** (2)	
b Using a spatula salt **E** was placed in a hard glass test-tube. Inside the top of the tube was suspended a piece of damp blue litmus paper next to a piece of damp red litmus paper. **E** was heated gently until gas came out of the tube.	red litmus went blue then blue litmus went red
c **E** was dissolved in water to make an aqueous solution. The solution was divided into three test-tubes (i) To the first portion, was added a few drops of dilute nitric acid and about 1 cm³ of aqueous silver nitrate. (2) (ii) To the second portion of solution **E**, was added about 1 cm³ of lead nitrate solution. (2) (iii) To the third portion of solution **E**, was added about 1 cm³of aqueous sodium hydroxide. The mixture was boiled gently and the gas given off was tested with indicator paper (2)	

d Name the gas given off in test **c**(iii). **(1)**
e Explain the observations in test **b**. **(2)**

(UCLES, IGCSE, Paper 6, October/November 2004)

34 Describe a chemical test to distsinguish between each of the following pairs of substances. An example is given.

oxygen and carbon dioxide

test: glowing splint

result: re-lights in oxygen, no effect with carbon dioxide
a aqueous chlorine and aqueous sodium chloride
   test
   result with chlorine
   result with sodium chloride **(2)**
b aqueous iron(II) chloride and aqueous iron(III) chloride
   test
   result with iron(II) chloride
   result with iron(III) chloride **(2)**
c copper sulfate and copper carbonate
   test
   result with coper sulfate
   result with copper carbonate **(2)**

(UCLES, IGCSE, Paper 6, October/November 2004)

35 Forged Banknote

A fake banknote can be investigated by dissolving the ink off the paper.

You are provided with four different inks from four different criminals. Describe an experiment to show which one of these inks is the same as the ink from the banknote.

You can use a labelled diagram to help you answer the question. **(6)**

(UCLES, IGCSE, Paper 6, October/November 2004)

IGCSE exam questions

1 Ethanoic acid is a colourless liquid at room temperature. It has the typical acid properties and forms compounds called ethanoates.

a A pure sample of ethanoic acid is slowly heated from 0 °C to 150 °C and its temperature is measured every minute. The results are represented on the graph below.

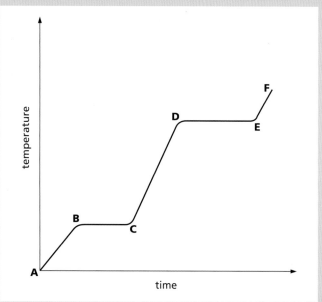

(i) Name the change that occurs in the region D to E. **(1)**

(ii) What would be the difference in the region B to C if an impure sample had been used? **(1)**

(iii) Sketch on the graph how the line would continue if the acid was heated to a higher temperature. **(1)**

(iv) Complete the following table that compares the separation and movement of the molecules in regions C to D with those in E to F.

	C to D	E to F
separation (distance between particles)		
movement of particles	random and slow	
Can particles move apart to fill any volume?		

b Complete the word equations for the reactions of ethanoic acid.

calcium + ethanoic acid → _____ + _____

_____ + ethanoic acid → zinc ethanoate + water **(2)**

c Write the symbol equation for the reaction between ethanoic acid and sodium hydroxide. **(2)**

(UCLES, IGCSE, Paper 3, October/November 2005)

2 The diagrams show four methods of purifying substances.

a Which of these methods, A, B, C or D, is best used for

(i) separating the different colours in a sample of ink?

(ii) separating two liquids with different boiling points?

(iii) separating mud from water?

(iv) making crystals of copper sulfate from copper sulfate solution? **(4)**

b State the name given to the method of separation shown in

(i) diagram A.

(ii) diagram B. **(2)**

c Method A can be modified to separate petroleum into useful fractions. The diagram below shows the different fractions obtained from a fractionating column.

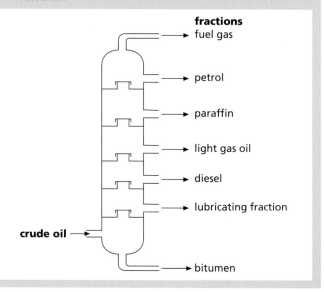

(i) Which of these frctions has the lowest boiling point?

(ii) State **one** use for each of the following fractions.
paraffin
bitumen **(3)**

d Petroleum is a mixture of organic compounds. Which **one** of the following best describes the compounds found in petroleum?

acids   alcohols   carbohydrates   hydrocarbons   (1)

e Before petroleum is fractionated, it is often heated to remove dissolved natural gas. Most of this natural gas is methane, CH_4. Draw a diagram to show how the electrons are arranged in methane.

show hydrogen electrons as   •
show carbon electrons as   × **(2)**

f Methane, ethane and propane belong to a particular homologous series of compounds. State the name of the homologous series to which these three compounds belong. **(1)**
(UCLES, IGCSE, Paper 2, October/November 2003)

3 The diagram shows part of the Period Table.

				He
C	N	O	F	Ne
		S	Cl	Ar
			Br	Kr

a Answer these questions using only the elements shown in the diagram. Write down the symbol for an element which
(i) has five electrons in its outer shell, **(1)**
(ii) has diatomic molecules, **(1)**
(iii) reacts with sodium to form sodium bromide, **(1)**
(iv) is a noble gas, **(1)**
(v) has a giant covalent structure, **(1)**
(vi) has a lower proton number than fluorine, **(1)**
(vii) is the most abundant gas in the air. **(1)**

b Write down a use for each of the following elements.
(i) argon **(1)**
(ii) helium **(1)**
(iii) oxygen **(1)**

c (i) Draw a diagram to show the electronic structure of argon. **(2)**
(ii) Why is argon very unreactive? **(1)**
(UCLES, IGCSE, Paper 2, October/November 2005)

4 The table below gives some information about the elements in Group I of the Period Table.

element	boiling point/°C	density/ g cm^{-3}	radius of atom in the metal/nm	reactivity with water
lithium	1342	0.53	0.157	
sodium	883	0.97	0.191	rapid
potassium	760	0.86	0.235	very rapid
rubidium		1.53	0.250	extremely rapid
caesium	669	1.88		explosive

a How does the density of the Group I elements change down the Group? **(2)**
b Suggest a value for the boiling point of rubidium. **(1)**
c Suggest a value for the radius of a caesium atom. **(1)**
d Use the information in the table to suggest how fast lithium reacts with water compared with the other Group I metals. **(1)**
e State three properties shown by **all** metals. **(3)**
f When sodium reacts with water, hydrogen is given off.
$$2Na(s) + 2H_2O(l) \rightarrow 2NaOH(aq) + H_2(g)$$
(i) State the name of the other product formed in this reaction. **(1)**
(ii) Describe a test, and the result, for hydrogen. **(2)**
g The diagrams below show three types of hydrogen atom.

hydrogen   deuterium   tritium

(i) State the name of the positively charged particle in the nucleus. **(1)**
(ii) What is the name given to atoms with the same number of positive charges in the nucleus but different numbers of neutrons? **(1)**
(iii) State the number of nucleons in a single atom of tritium. **(1)**
(iv) Tritium is a radioactive form of hydrogen. State **one** medical use of radioactivity. **(1)**
(UCLES, IGCSE, Paper 2, October/November 2004)

5 Bromine is an element in Group VII of the Periodic Table.
a State the name given to the Group VII elements. **(1)**
b Bromine has two isotopes. The nucleon (mass) number of bromine-79 is 79 and of bromine-81 is 81.
(i) What is the meaning of the term *isotopes*?
(ii) Complete the table to show the numbers of electrons, neutrons and protons in one atom of bromine-79 and bromine-81. A copy of the Period Table is printed on page 285.

number of	bromine-79	bromine-81
electrons		
neutrons		
protons		

(5)

c Bromine is extracted from seawater by treatment with chlorine. When chlorine is bubbled through a solution of potassium bromide, the solution turns orange-red.
(i) What does this tell you about the reactivity of chlorine compared with bromine?
(ii) Write a word equation for this reaction. **(2)**

d In order to get the maximum yield of bromine from seawater, acid is added during the extraction procedure. The graph shows how the yield of bromine changes with pH.

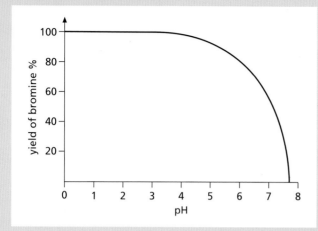

(i) What is the highest pH at which the yield of bromine is 100%?

(ii) The pH scale is used to measure acidity. Some pH values are given below.

pH 3   pH 5   pH 7   pH 9   pH 11

from this list of pH values choose the pH which is most acidic. the pH of a neutral solution. **(3)**

e Bromine water can be used to distinguish between ethane and ethene.

ethane   ethane

Describe what you would observe when bromine water is added to ethene. **(1)**

(UCLES, IGCSE, Paper 2, October/November 2003)

6 Look at the list of five elements below.

argon
bromine
chlorine
iodine
potassium

a Put these five elements in order of increasing proton number. **(1)**

b Put these five elements in order of increasing relative atomic mass. **(1)**

c The orders of proton number and relative atomic mass for these five elements are different. Which **one** of the following is the most likely explanation for this?

The proton number of a particular element may vary.
The presence of neutrons.
The atoms easily gain or lose electrons.
The number of protons must always equal the number of neutrons. **(1)**

d Which of the five elements in the list are in the same group of the Periodic Table? **(1)**

e (i) From the list, choose **one** element which has one electron in its outer shell. **(1)**

(ii) From the list, choose **one** element which has a full outer shell of electrons. **(1)**

f Which **two** of the following statements about argon are correct?
Argon is a noble gas.
Argon reacts readily with potassium.
Argon is used to fill weather balloons.
Argon is used in light bulbs. **(2)**

g Potassium chloride can be made by reacting potassium with chlorine. The bonding in potassium chloride is ionic. What does this information tell you about

(i) the boiling point of potassium chloride, **(1)**

(ii) the electrical conductivity of molten potassium chloride? **(1)**

h Describe the change in the electronic structure of potassium and chlorine atoms when they combine to make potassium chloride.

change in potassium atom
change in chlorine atom **(2)**

(UCLES, IGCSE, Paper 2, May/June 2004)

7 Strontium and sulfur chlorides both have a formula of the type XCl_2 but they have different properties.

property	strontium chloride	sulfur chloride
appearance	white crystalline solid	red liquid
melting point/°C	873	−80
particles present	ions	molecules
electrical conductivity of solid	poor	poor
electricial conductivity of liquid	good	poor

a The formulae of the chlorides are similar because both elements have a valency of 2. Explain why Group II and Group VI elements both have a valency of 2. **(2)**

b Draw a diagram showing the arrangement of the valency electrons in one covalent molecule of sulfur chloride.
Use x to represent an electron from a sulfur atom.
Use o to represent an electron from a chlorine atom. **(3)**

c Explain the difference in electrical conductivity between the following.

(i) solid and liquid strontium chloride **(1)**

(ii) liquid strontium chloride and liquid sulfur chloride **(1)**

(UCLES, IGCSE, Paper 3, October/November 2004)

8 The diagram shows models of various structures.

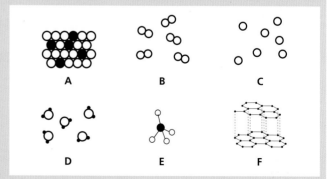

A   B   C

D   E   F

a Which **three** of the structures **A** to **F** represent elements? Give a reason for your answer. **(2)**

b Which one of the structures **A** to **F** represents a gas containing single atoms? **(1)**

c (i) Which one of the structures **A** to **F** represents a gas containing diatomic molecules?
(ii) State the name of a gas which has diatomic molecules. **(2)**

d (i) Which one of the structures **A** to **F** represents graphite?
(ii) State **one** use of graphite. **(2)**

e Structure **D** represents a compound.
(i) State what is meant by the term *compound*.
(ii) Which one of the following substances is structure **E** most likely to represent?
ammonia hydrogen chloride methane water **(2)**

f Hydrogen chloride is a compound.
(i) Draw a diagram to show how the electrons are arranged in a molecule of hydrogen chloride. Show only the outer electrons.
show hydrogen electrons as •
show chlorine electrons as x **(2)**
(ii) State the name of the type of bonding present in hydrogen chloride. **(1)**
(iii) Hydrogen chloride dissolves in water to form an acidic solution (hydrochloric acid). Describe how you would use litmus paper to show that this solution is acidic. **(2)**
(iv) Which one of the following values is most likely to represent the pH of a dilute solution of hydrochloric acid?
pH2   pH7   pH10   pH14 **(1)**
(v) Complete the following equation for the reaction of hydrochloric acid with magnesium.

$Mg(s) + _____ HCl(aq) \rightarrow MgCl_2(aq) + H_2(g)$ **(1)**

(vi) Name the salt formed in this reaction. **(1)**
(UCLES, IGCSE, Paper 2, May/June 2004)

9 It was reported from America that a turbine engine, the size of a button, might replace batteries. The engine would be built from silicon which has suitable properties for this purpose.

a (i) Why are batteries a convenient source of energy? **(1)**
(ii) The engine will run on a small pack of jet fuel. What other chemical is needed to burn this fuel? **(1)**

b Silicon has the same type of macromolecular structure as diamond.
(i) Explain why one atom of either element can form four covalent bonds. **(2)**
(ii) Predict **two** physical properties of silicon. **(2)**
(iii) Name a different element that has a similar structure and properties to silicon. **(1)**

c Silicon is made by the carbon reduction of the macromolecular compound, silicon(IV) oxide.
(i) Balance the equation for the reduction of silicon (IV) oxide.

$SiO_2 + _____ C \rightarrow Si + _____ CO$ **(1)**

(ii) Explain why the silicon(IV) oxide is said to be reduced. **(1)**
(iii) Describe the structure of silicon(IV) oxide. You may use a diagram. **(2)**
(UCLES, IGCSE, Paper 3, May/June 2004)

10a The structure of a typical ionic compound is a regular arrangement of positive and negative ions.

(i) What is the name of this regular arrangement of particles? **(1)**
(ii) Give **two** physical properties of ionic compounds. **(2)**

b Ions are formed by electron loss or gain. The electron distribution of a magnesium atom is 2 + 8 + 2 and of a nitrogen atom is 2 + 5.
(i) Give the formula of the magnesium ion. **(1)**
(ii) Give the formula of the nitride ion. **(1)**
(iii) What is the formula of the ionic compound, magnesium nitride? **(1)**
(iv) In this compound there is an ionic bond. Why are the two ions attracted to each other? **(1)**
(UCLES, IGCSE, Paper 3, October/November 2005)

11 Chemists use the concept of the mole to calculate the amounts of chemicals involved in a reaction.
a Define *mole*. **(1)**
b 3.0 g of magnesium was added to 12.0 g of ethanoic acid.
$Mg + 2CH_3COOH \rightarrow (CH_3COO)_2Mg + H_2$
The mass of one mole of Mg is 24 g.
The mass of one mole of CH_3COOH is 60 g.
(i) Which one, magnesium or ethanoic acid, is in excess? You must show your reasoning. **(3)**
(ii) How many moles of hydrogen were formed? **(1)**
(iii) Calculate the volume of hydrogen formed, measured at r.t.p. **(2)**
c In an experiment, 25.0 cm³ of aqueous sodium hydroxide, 0.4 mol/dm³, was neutralised by 20.0 cm³ of aqueous oxalic acid, $H_2C_2O_4$.
$2NaOH + H_2C_2O_4 \rightarrow Na_2C_2O_4 + 2H_2O$
Calculate the concentration of the oxalic acid in mol/dm³.
(i) Calculate the number of moles of NaOH in 25.0 cm³ of 0.4 mol/dm³ solution. **(1)**
(ii) Use your answer to (i) and the mole ratio in the equation to find out the number of moles of $H_2C_2O_4$ in 20 cm³ of solution. **(1)**
(iii) Calculate the concentration, mol/dm³, of the aqueous oxalic acid. **(2)**
(UCLES, IGCSE, Paper 3, May/June 2004)

12 The diagram shows an electrolysis cell used to extract aluminium.

a What compound of aluminium is used for the electrolyte? **(1)**

b The electrolyte must be molten for the electrolysis to work. Explain why. **(1)**

c (i) State the name of the substance used for the electrodes. **(1)**

(ii) To which electrode do the aluminium ions move during electrolysis? Explain your answer. **(2)**

d Complete the following sentences about the molten electrolyte using words from the list below.

bauxite   chemical   cryolite   decreased

electrical   haematite   increased   light

The melting point of the electrolyte is

_____ by adding

_____. This means that less

_____ energy is needed to melt the

electrolyte. **(3)**

e Aluminium is used in overhead power cables.

pylon

steel cored aluminium cables

aluminium

steel core

The table shows some properties of three metals which could be used for the power cables.

metal	relative electrical conductivity	density/ grams per cm³	price/ £ per kg	relative strength
aluminium	0.4	2.70	18	9
copper	0.7	8.92	15	30
steel	0.1	7.86	2.7	50

(i) Suggest why aluminium is used for overhead power cables rather than copper. **(1)**

(ii) Suggest why steel is not used alone for overhead power cables. **(1)**

(iii) Why is steel used as a core for overhead power cables? **(1)**

(iv) Electrical insulators are used in parts of the pylons which carry the electrical cables. Which one of the following is an electrical insulator?

aluminium   ceramic   graphite   zinc **(1)**

f Aluminium has many uses.

(i) Why is aluminium used for aircraft bodies? **(1)**

(ii) Describe a test, and the result, for aluminium ions. **(3)**

(UCLES, IGCSE, Paper 2, October/November 2005)

13 a Copper has the structure of a typical metal. It has a lattice of positive ions and a 'sea' of mobile electrons. The lattice can accommodate ions of a different metal. Give a **different** use of copper that depends on each of the following.

(i) the ability of the ions in the lattice to move past each other **(1)**

(ii) the presence of mobile electrons **(1)**

(iii) the ability to accommodate ions of a different metal in the lattice **(1)**

b Aqueous copper(II) sulfate solution can be electrolysed using carbon electrodes. The ions present in the solution are as follows:

$Cu^{2+}(aq)$, $SO_4^{2-}(aq)$, $H^+(aq)$, $OH^-(aq)$

(i) Write an ionic equation for the reaction at the negative electrode (cathode). **(1)**

(ii) A colourless gas was given off at the positive electrode (anode) and the solution changes from blue to colourless. Explain these observations. **(2)**

c Aqueous copper(II) sulfate can be electrolysed using copper electrodes. The reaction at the negative electrode is the same but the positive electrode becomes smaller and the solution remains blue.

(i) Write a word equation for the reaction at the positive electrode. **(1)**

(ii) Explain why the colour of the solution does not change. **(2)**

(iii) What is the large scale use of this electrolysis? **(1)**

(UCLES, IGCSE, Paper 3, May/June 2004)

14 Aluminium is extracted from its ore, bauxite. The bauxite is purified to give aluminium oxide. Electrolysis is then used to extract the aluminium from aluminium oxide dissolved in cryolite.

positive electrode (graphite)

+

–

negative electrode (graphite)

steel case

aluminium oxide dissolved in molten cryolite

molten aluminium collecting on the bottom

The melting point of pure aluminium oxide is 2070 °C. The melting point of the mixture of aluminium oxide and cryolite is about 1000 °C.

a Suggest why electrolysis is used to extract aluminium from aluminium oxide rather than reduction using carbon. **(1)**

b How is the electrolyte of aluminium oxide and cryolite kept molten? **(1)**

c What property of graphite makes it suitable for use as electrodes? **(1)**

d State the name given to the negative electrode. **(1)**

e The melting point of steel is about 1500 °C. Suggest **two** reasons why molten aluminium oxide is not used by itself in this electrolysis. **(2)**

f During the electrolysis, hot oxygen is formed at the positive electrodes. Suggest why the positive electrodes have to be replaced frequently. **(2)**

g Aluminium is formed at the negative electrode. Complete the following equation for the reaction at the negative electrode.

$$Al^{3+} + \text{\_\_\_\_\_} \rightarrow Al$$ **(1)**

h Why do aluminium ions move towards the negative electrode? **(1)**

i A sample of bauxite ore had the following composition:

aluminium oxide	120 g
iron(III) oxide	30 g
silica	40 g
titanium(IV) oxide	10 g

Calculate the percentage of aluminium oxide in this sample of bauxite. **(1)**

j Aluminium is a metal in Group III of the Periodic Table. State three physical properties which are typical of most metals. **(3)**

(UCLES, IGCSE, Paper 2, October/November 2003)

15 The electrolysis of a concentrated solution of sodium chloride, provides us with chemicals.

a Sodium chloride has an ionic giant structure. Which **one** of the following is a correct description of a property of sodium chloride.

sodium chloride has a low melting point
sodium chloride conducts electricity when it is solid
sodium chloride has a high boiling point
sodium chloride is insoluble in water **(1)**

b (i) Explain what is meant by the term *electrolysis*. **(1)**

(ii) At which electrode is hydrogen produced during the electrolysis of aqueous sodium chloride? **(1)**

(iii) Name a suitable substance that can be used for the electrodes. **(1)**

c (i) State the name of the particle which is added to a chlorine atom to make a chloride ion. **(1)**

(ii) Describe a test, and the result, for chloride ions. **(2)**

d If chlorine is allowed to mix with sodium hydroxide, sodium chlorate(I), NaOCl is formed. Balance the equation for this reaction.

$Cl_2 + \underline{} NaOH \rightarrow NaCl + NaOCl + H_2O$ **(1)**

e One tonne (1000 kg) of a commercial solution of sodium hydroxide produced by electrolysis contains the following masses of compounds.

Compound	Mass of compound kg/tonne
sodium hydroxide	510
sodium chloride	10
sodiumchlorate(v)	9
water	471
total	1000

(i) How many kilogram sof sodium hydroxide will be present in 5 tonnes of the solution? **(1)**

(ii) All the water from one tonne of impure sodium hydroxide is evaporated. What would the approximate percentage of the remaining impurities be?

0.036%   3.6%   36%   96% **(1)**

f The hydrogen obtained by electrolysis can be used in the manufacture of margarine.

(i) Complete the following sentences about this reaction using words from the list.

catalyst
inhibitor
monomeric
saturated
unsaturated

Hydrogen gas is bubbled through _____ carbon compounds using a nickel _____ which speeds up the reaction. The margarines produced are _____ compounds. **(3)**

(ii) State **one** other use of hydrogen. **(1)**

(UCLES, IGCSE, Paper 2, October/November 2004)

16 Strontium and zinc are both metals with a valency of 2. Strontium is more reactive than zinc. Its chemistry is similar to that of calcium.

a (i) Complete the following table that shows the number of protons, electrons and neutrons in each particle.

Particle	Protons	Electrons	Neutrons
^{88}Sr			
^{90}Sr			
^{65}Zn^{2+}			

(3)

(ii) Explain why ^{88}Sr and ^{90}Sr are isotopes. **(1)**

(iii) Complete the electron distribution of an atom of strontium.

2 + 8 + 18 + _____ + _____ **(1)**

b The major ore of zinc is zinc blende, ZnS.

(i) Describe how zinc is extracted from zinc blende. **(2)**

(ii) Give a use of zinc. **(1)**

c The major ore of strontium is its carbonate, $SrCO_3$. Strontium is extracted by the electrolysis of its molten chloride.

(i) Name the reagent that will react with the carbonate to form the chloride. **(1)**

(ii) The electrolysis of molten strontium chloride produces strontium metal and chlorine. Write ionic equations for the reactions at the electrodes. **(2)**

negative electrode (cathode)
positive electrode (anode)

(iii) One of the products of the electrolysis of concentrated aqueous strontium chloride is chlorine. Name the other two. **(2)**

d Both metals react with water.

(i) Write a word equation for the reaction of zinc and water and state the reaction conditions.

word equation **(1)**
conditions **(2)**

(ii) Write an equation for the reaction of strontium with water and give the reaction condition.

equation **(2)**
condition **(1)**

(UCLES, IGCSE, Paper 3, October/November 2005)

17 a Insoluble compounds are made by precipitation.

(i) Complete the word equation for the preparation of zinc carbonate.

_____ _____ + sodium carbonate
→ zinc carbonate + _____ _____ **(2)**

(ii) Complete the following symbol equation.
$Pb(NO_3)_2 +$ _____ $NaCl$
→ _____ + _____ **(2)**

(iii) Write an ionic equation for the precipitation of the insoluble salt, silver(I) chloride. **(2)**

b 2.0 cm³ portions of aqueous sodium hydroxide were added to 4.0 cm³ of aqueous iron(III) chloride. Both solutions had a concentration of 1.0 mol/dm³. After each addition, the mixture was stirred, centrifuged and the height of the precipitate of iron(III) hydroxide was measured. The results are shown on the following graph.

(i) Complete the ionic equation for the reaction.
$Fe^{3+} +$ _____ $OH^- →$ _____ **(1)**

(ii) On the same grid, sketch the graph that would have been obtained if iron(II) chloride had been used instead of iron(III) chloride? **(2)**

(iii) If aluminium chloride had been used instead of iron(III) chloride, the shape of the graph would be different. How are the shapes of these two graphs different and why? **(2)**

(UCLES, IGCSE, Paper 3, May/June 2004)

18 The diagram shows the salt mines at Bex in Switzerland.

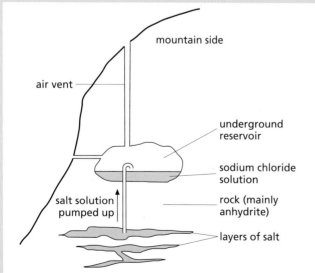

The salt is dissolved by water from underground springs and then pumped up to a reservoir where it is stored as a solution.

a Write the chemical formula for sodium chloride. **(1)**

b Suggest how solid sodium chloride is obtained from the sodium chloride solution. **(1)**

c Sodium chloride has an ionic giant structure. Which one of the following best describes an aqueous solution of sodium chloride?
a mixture of sodium ions and chlorine molecules in water
a mixture of sodium and chlorine atoms in water
a mixture of sodium and chloride ions in water
a mixture of sodium, chloride, oxide and hydrogen ions **(1)**

d Describe a test, and the result, for chloride ions. **(2)**

e The rock surrounding the layers of salt is anhydrite. Pure anhydrite has the chemical formula $CaSO_4$.

(i) State the name of the chemical found in anhydrite. **(1)**

(ii) Calculate the relative formula mass of the chemical in pure anhydrite. **(1)**

(iii) When anhydrite reacts with water, gypsum ($CaSO_4.2H_2O$) is formed. Complete the equation for this reaction.
$CaSO_4 +$ _____ $→ CaSO_4.2H_2O$

(iv) Which one of the following describes this reaction?
**combustion   fermentation   hydration   oxidation
reduction** **(1)**

(v) The chemical in anhydrite can be made by reacting calcium hydroxide with sulfuric acid. Complete the balanced equation for this reaction.

$Ca(OH)_2 +$ _____ $→ CaSO_4 +$ _____ H_2O **(2)**

(vi) The spring water running through the rocks changes anhydrite into gypsum. This reaction is exothermic. Use this information to explain why the temperature of the mine never falls below 17 °C even in cold winters. **(1)**

f The air inside the mine contains 19% oxygen. Which one of the following best describes the oxygen level inside the mine compared with that outside the mine?
the level of oxygen inside the mine is higher
the level of oxygen is the same
the level of oxygen is about a quarter of that of the outside air
the level of oxygen inside the mine is lower **(1)**

(UCLES, IGCSE, Paper 2, October/November 2005)

19 The salt copper(II) sulfate can be prepared by reacting copper(II) oxide with sulfuric acid. Complete the list of instructions for making copper(II) sulfate using **six** of the words below.

blue	**cool**	**dilute**	**filter**
saturated	**sulfate**	**white**	**oxide**

Instructions
1 Add excess copper(II) oxide to _____ sulfuric acid in a beaker and boil it.
2 _____ to remove the unreacted copper(II) oxide.
3 Heat the solution until it is _____.
4 _____ the solution to form _____ coloured crystals of copper(II) _____ **(6)**

(UCLES, IGCSE, Paper 3, October/November 2004)

20 Read the following instructions for the preparation of hydrated nickel(II) sulfate ($NiSO_4.7H_2O$), then answer the questions which follow.
1 Put 25 cm³ of dilute sulfuric acid in a beaker.
2 Heat the sulfuric acid until it is justs boiling then add a small amount of nickel(II) carbonate.
3 When the nickel(II) carbonate has dissolved, stop heating, then add a little more nickel carboante. Continue in this way until nickel(II) carbonate is in excess.
4 Filter the hot mixture into a clean beaker.
5 Make the hydrated nickel(II) sulfate crystals from the nickel(II) sulfate solution.

The equation for the reaction is

$NiCO_3(s) + H_2SO_4(aq) \rightarrow NiSO_4(aq) + CO_2(g) + H_2O(l)$

a What piece of apparatus would you use to measure out 25 cm³ of sulfuric acid? **(1)**
b Why is the nickel(II) carbonate added in excess? **(1)**
c When nickel(II) carbonate is added to sulfuric acid, there is a fizzing. Explain why there is a fizzing. **(1)**
d Draw a diagram to describe step 4. You must label your diagram. **(3)**
e After filtration, which one of the following describes the nickel(II) sulfate in the beaker?
crystals   filtrate   precipitate   water **(1)**
f Explain how you would obtain pure dry crystals of hydrated nickel(II) sulfate from the solution of nickel(II) sulfate. **(2)**
g When hydrated nickel(II) sulfate is heated gently in a test tube, it changes colour from green to white.
(i) Complete the symbol equation for this reaction.

$NiSO_4.7H_2O(s) \rightleftharpoons NiSO_4(s) + \underline{\hspace{2cm}}$ **(1)**
(ii) What does the sign $\rightleftharpoons$ mean? **(1)**
(iii) How can you obtain a sample of green nickel(II) sulfate starting with white nickel(II) sulfate? **(1)**
(UCLES, IGCSE, Paper 2, October/November 2004)

21 a The following method is used to make crystals of hydrated nickel sulfate.
An excess of nickel carbonate, 12.0 g, was added to 40 cm³ of sulfuric acid, 2.0 mol/dm³. The unreacted nickel carbonate was filtered off and the filtrate evaporated to obtain the crystals.

$NiCO_3 + H_2SO_4 \rightarrow NiSO_4 + CO_2 + H_2O$
$NiSO_4 + 7H_2O \rightarrow NiSO_4.7H_2O$

Mass of one mole of $NiSO_4.7H_2O$ = 281 g
Mass of one mole of $NiCO_3$ = 119 g
(i) Calculate the mass of unreacted nickel carbonate.
Number of moles of H_2SO_4 in 40 cm³ of 2.0 mol/dm³ acid = 0.08
Number of moles of $NiCO_3$ reacted = \underline{\hspace{2cm}}
Mass of nickel carbonate reacted = \underline{\hspace{2cm}} g
Mass of unreacted nickel carbonate = \underline{\hspace{2cm}} g**(3)**
(ii) The experiment produced 10.4 g of hydrated nickel sulfate. Calculate the percentage yield.
The maximum number of moles of $NiSO_4.7H_2O$ that could be formed =

\underline{\hspace{6cm}}
The maximum mass of $NiSO_4.7H_2O$ that could be formed = \underline{\hspace{2cm}} g
The percentage yield = \underline{\hspace{4cm}} % **(3)**

b In the above method, a soluble salt was prepared by neutralising an acid with an insoluble base. Other salts have to be made by different methods.
(i) Give a brief description of how the soluble salt, rubidium sulfate could be made from the soluble base, rubidium hydroxide. **(3)**
(ii) Suggest a method of making the insoluble salt, calcium fluoride. **(3)**
(UCLES, IGCSE, Paper 3, October/November 2005)

22 a (i) Write a symbol equation for the action of heat on zinc hydroxide. **(2)**
(ii) Describe what happens when solid **sodium hydroxide** is heated strongly. **(1)**
b What would be **observed** when copper(II) nitrate is heated? **(3)**
c Iron(III) sulfate decomposes when heated. Calculate the mass of iron(III) oxide formed and the volume of sulfur trioxide produced when 10.0 g of iron(III) sulfate was heated.

Mass of one mole of $Fe_2(SO_4)_3$ is 400 g.

$Fe_2(SO_4)_3(s) \rightarrow Fe_2O_3(s) + 3SO_3(g)$
Number of moles of $Fe_s(SO_4)_3$ = \underline{\hspace{2cm}}
Number of moles of Fe_2O_3 formed = \underline{\hspace{2cm}}
Mass of iron(III) oxide formed = \underline{\hspace{2cm}} g
Number of moles of SO_3 produced = \underline{\hspace{2cm}}
Volume of sulfur trioxide at r.t.p. = \underline{\hspace{1cm}} dm³ \underline{\hspace{0.5cm}} **(5)**
(UCLES, IGCSE, Paper 3, October/November 2004)

23 The list shows some oxides.
calcium oxide
magnesium oxide
nitrogen dioxide
sodium oxide
sulfur dioxide
a From this list choose **two** oxides which are basic. Give a reason for your answer. **(2)**
b (i) Which **two** oxides from this list contribute to acid rain? **(2)**
(ii) How do each of these oxides get into the atmosphere? **(2)**
c Calcium oxide is manufactured from calcium carbonate.
(i) Complete the word equation for this reaction.
calcium carbonate $\rightarrow$ calcium oxide + \underline{\hspace{2cm}} **(1)**
(ii) What condition is needed for this reaction to take place? **(1)**
d (i) Explain why calcium oxide and sodium oxide cannot be reduced by heating with carbon. **(1)**
(ii) Copper(II) oxide can be reduced by heating with carbon. Complete the equation for this reaction.

\underline{\hspace{1.5cm}} CuO + C $\rightarrow$ 2Cu + \underline{\hspace{1.5cm}} **(2)**
(iii) What do you understand by the term *reduction*? **(1)**

(UCLES, IGCSE, Paper 2, October/November 2005)

24 One way of making lime from limestone (calcium carbonate) is shown in the diagram.

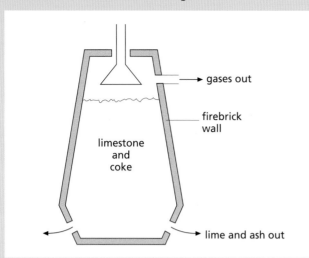

The limestone is mixed with coke and dropped into the linekiln. The coke is burnt and releases heat.

a State **one** use of limestone, other than in making lime. **(1)**

b Coke is mainly carbon. Write a symbol equation for the burning of carbon. **(2)**

c State the name of the type of reaction which releases heat energy. **(1)**

d The heat produced by the burning coke causes thermal decomposition of the limestone. Complete the word equation for the thermal decomposition of calcium carbonate.

calcium carbonate → _____ _____ + _____ _____ **(2)**

e (i) Complete the following equation for the reaction of calcium carbonate with hydrochloric acid.

$CaCO_3$ + _____ HCl → $CaCl_2$ + CO_2 + H_2O
(ii) Describe how you would test for the gas given off in this reaction, and the result. **(3)**

f Quicklime, CaO, is the product of the thermal decomposition of calcium carbonate. When quicklime is heated strongly with coke, calcium carbide is formed.

$CaO + 3C → CaC_2 + CO$
(i) What type of reaction is the conversion of C to CO? Explain your answer.
(ii) When water is added to calcium carbide, CaC_2, acetylene is formed. State a use of acetylene. **(3)**
(UCLES, IGCSE, Paper 2, October/November 2003)

25 The diagram below shows a modern landfill site for the disposal of waste materials.

The waste materials are broken down naturally in several stages.

a In the first stage, micro-organisms (mainly bacteria) break down some of the organic material in the waste to carbon dioxide. What is the name given to the process by which organisms use food to produce carbon dioxide? **(1)**

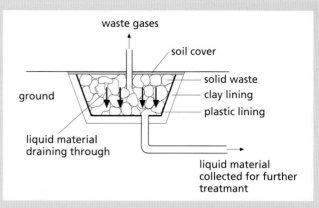

b In the second stage, the micro-organisms break down organic substances to produce ammonia, hydrogen and more carbon dioxide.
(i) Describe a test, and the result, for hydrogen.
(ii) The large volumes of hydrogen produced may be hazardous. Explain why hydrogen may be hazardous when mixed with air.
(iii) Ammonia is a base. Describe a test, and the result, for ammonia. **(5)**

c In the third stage, ethanoic acid is produced. Draw the structure of ethanoic acid showing all atoms and bonds. **(1)**

d In the fourth stage, carbon dioxide reacts with hydrogen to form methane and oxygen.
(i) Complete the equation for this reaction.

CO_2 + _____ → CH_4 + O_2
(ii) State one use of methane.
(iii) Methane is a gas. Which **two** of the following statements about gas molecules are true?
The molecules are far apart.
The molecules are not moving.
The molecules are randomly arranged.
The molecules are arranged in a regular manner. **(4)**

e The list below shows some of the substances which are found in the liquid which drains through the waste.

aluminium
calcium carbonate
iron
lead
magnesium
nickel
sodium sulfate
zinc

From this list choose
(i) a metal used to galvanise iron.
(ii) a transition metal.
(iii) a metal which is in Group IV of the periodic table.
(iv) a substance which will release carbon dioxide when an acid is added.
(v) a metal which is used to make aircraft bodies. **(5)**
(UCLES, IGCSE, Paper 2, October/November 2003)

26 In the following list of ionic equations, the metals are in order of reactivity.

_____ → _____ + _____

Zn → Zn^{2+} + $2e^-$

Sn → Sn^{2+} + $2e^-$

Hg → Hg^{2+} + $2e^-$

Ag → Ag^+ + e^-

reactivity of metals increases ↑

a (i) In the space at the top of the series, write an ionic equation that includes a more reactive metal. **(1)**
(ii) Define *oxidation* in terms of electron transfer. **(1)**
(iii) Explain why the positive ions are likely to be oxidising agents. **(1)**
(iv) Which positive ion(s) can oxidise mercury metal (Hg)? **(1)**

b The following diagram shows a simple cell.

voltmeter
V
tin electrode
zinc electrode
electrolyte dilute sulfuric acid

(i) Predict how the voltage of the cell would change if the tin electrode was replaced with a silver one. **(1)**
(ii) Which electrode would go into the solution as positive ions? Give a reason for your choice. **(1)**
(iii) State how you can predict the direction of the electron flow in cells of this type. **(1)**
(UCLES, IGCSE, Paper 3, October/November 2004)

27 Zinc blende is the common ore of zinc. It is usually found mixed with an ore of lead and traces of silver.
a (i) Describe how zinc blende is changed into zinc oxide. **(2)**
(ii) Write an equation for the reduction of zinc oxide by carbon. **(2)**
(iii) The boiling point of lead is 1740 °C and that of zinc is 907 °C. Explain why, when both oxides are reduced by heating with carbon at 1400 °C, only lead remains in the furnace. **(2)**
b A major use of zinc is to make diecasting alloys. These contain about 4% of aluminium and they are stronger and less malleable than pure zinc.
(i) Give one other large scale use of zinc. **(1)**
(ii) Describe the structure of a typical metal, such as zinc, and explain why it is malleable. **(3)**
(iii) Suggest why the introduction of a different metallic atom into the structure makes the alloy stronger than the pure metal. **(2)**

c A solution of an impure zinc ore contained zinc, lead(II) and silver(I) ions. The addition of zinc dust will displace both lead and silver.
(i) The ionic equation for the displacement of lead is as follows.

change 1

$Zn(s) + Pb^{2+}(aq) → Zn^{2+}(aq) + Pb(s)$

change 2

Which change is reduction? Explain your answer. **(2)**

(ii) Write an ionic equation for the reaction between zinc atoms and silver(I) ions. **(2)**
(UCLES, IGCSE, Paper 3, October/November 2003)

28 Iron is extracted from its ore in a blast furnace using carbon (coke) as a reducing agent and as a source of heat.
a the coke burns in hot air. The equation for this reaction is

$2C(s) + O_2(g) → 2CO(g)$
State the name of the gas produced in this reaction. **(1)**
b Near the top of the blast furnace, the iron(III) oxide in the iron ore gets reduced to iron.

$Fe_2O_3(s) + 3CO(g) → 2Fe(l) + 3CO_2(g)$
Use the equation to explain why the change of iron(III) oxide to iron is a reduction reaction. **(1)**
c In the hottest regions of the furnace, iron(III) oxide is reduced by carbon. Complete the equation for this reaction.

$Fe_2O_3(s) + \underline{\hspace{1cm}} C(s) → \underline{\hspace{1cm}} Fe(l) + 3CO(g)$ **(2)**
d The iron from the blast furnace contains up to 10% by mass of impurities. The main impurities are carbon, silicon and phosphorus. The diagram below shows one method of making steel from iron.

oxygen and powdered basic oxides
slag forming
molten iron from blast furnace

A mixture of oxygen and basic oxides is blown onto the surface of the molten iron.
(i) What is the purpose of blowing oxygen onto the molten iron? **(1)**
(ii) A large amount of energy is released into the process of steelmaking. What name is given to chemical reactions which release energy? **(1)**

(iii) The basic oxides react with the impurities in the iron and form a slag. What information in the diagram suggests that the slag is less dense than the molten iron? **(1)**

(iv) Which one of the following is a basic oxide?
calcium oxide   carbon dioxide   sulfur dioxide water **(1)**

(v) Why is steel rather than iron used for constructing buildings and bridges? **(1)**

e Special steels contain added elements such as vanadium, chromium, cobalt or nickel. These are all transition metals.
State three properties of transition metals which are **not** shown by non-transition metals. **(3)**

f What is the name given to metals which are mixtures of more than one metal? **(1)**

(UCLES, IGCSE, Paper 2, May/June 2004)

29 a Two of the gases in air are nitrogen and oxygen. Name **two** other gases present in unpolluted air. **(2)**

b Two common pollutants present in air are sulfur dioxide and lead compounds. State the source and harmful effect of each.
sulfur dioxide **(3)**
lead compounds **(2)**

c Respiration and photosynthesis are two of the processes that determine the percentage of oxygen and of carbon dioxide in the air.
(i) Name another process that changes the percentages of these two gases in air. **(1)**
(ii) The equation for photosynthesis is given below.

$$6CO_2 + 6H_2O \rightarrow C_6H_{12}O_6 + 6O_2$$

This is an endothermic reaction.
Complete the reaction for respiration.

$$C_6H_{12}O_6 + 6O_2 \rightarrow \underline{\hspace{2cm}} + \underline{\hspace{2cm}}$$

This is an _____ reaction. **(2)**

d The rate of photosynthesis of pond weed can be measured using the following experiment.

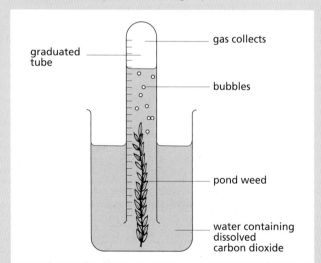

graduated tube — gas collects

bubbles

pond weed

water containing dissolved carbon dioxide

(i) Describe how you could show that the gas collected in this experiment is oxygen. **(1)**
(ii) What measurements are needed to calculate the rate of this reaction? **(2)**
(iii) What would be the effect, and why, of moving the apparatus further away from the light? **(2)**

(UCLES, IGCSE, Paper 3, October/November 2004)

30 The table below shows the composition of the mixture of gases coming from a typical car exhaust.

Gas	% of the gas in the exhaust fumes
carbon dioxide	9
carbon monoxide	5
oxygen	4
hydrogen	2
hydrocarbons	0.2
nitrogen oxides	0.2
sulfur dioxide	less than 0.003
gas **X**	79.6

a State the name of the gas **X**. **(1)**

b The carbon dioxide comes from the burning of hydrocarbons, such as octane, in the petrol.
(i) Complete the word equation for the complete combustion of octane.

octane + _____ → carbon dioxide + _____ **(2)**

(ii) Which **two** chemical elements are present in hydrocarbons? **(1)**
(iii) To which homologous series of hydrocabons does octane belong? **(1)**

c Suggest a reason for the presence of carbon monoxide in the exhaust fumes. **(1)**

d Nitrogen oxides are present in small quantities in the exhaust fumes.
(i) Complete the following equation for the formation of nitrogen dioxide.

$$N_2(g) + \underline{\hspace{1cm}} O_2(g) \rightarrow \underline{\hspace{1cm}} NO_2(g)$$ **(1)**

(ii) State **one** harmful effect of nitrogen dioxide on organisms. **(1)**

e Sulfur dioxide is an atmospheric pollutant which is only found in small amounts in car exhausts.
(i) What is the main source of sulfur dioxide pollution of the atmosphere? **(1)**
(ii) Sulfur dioxide is oxidised in the air to sulfur trioxide. The sulfur trioxide may dissolve in rainwater to form a dilute solution of sulfuric acid, H_2SO_4. State the meaning of the term *oxidation*. **(1)**
(iii) Calculate the relative molecular mass of sulfuric acid. **(1)**
(iv) Sulfuric acid reacts with metals such as iron. Complete the following word equation for the reaction of sulfuric acid with iron.

sulfuric acid + iron → _____ _____ + _____ **(2)**

(v) What effect does acid rain have on buildings made of stone containing calcium carbonate? **(1)**

(UCLES, IGCSE, Paper 2, November 2004)

31 Two of the stages in water purification are filtration and chlorination. The diagram below shows a filter tank.

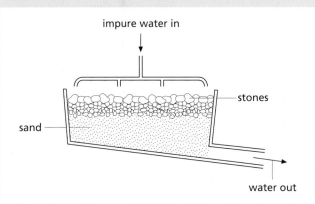

impure water in

stones

sand

water out

a Explain how this filter helps purify the water. **(2)**
b (i) Why is chlorine added during water purification?
(ii) After chlorination, the water is acidic. A small amount of slaked lime is added to the acidic water. Explain why slaked lime is added.
(iii) What is the chemical name for slaked lime?
(iv) State **one** other use of slaked lime. **(4)**
c (i) State the boiling point of pure water. **(2)**
(ii) Describe a chemical test, and the result, for water. **(2)**
(iii) State **one** use of water in the home. **(1)**
d The diagram shows the arrangement of particles in the three different states of water.

A **B** **C**

Which of these diagrams, **A**, **B** or **C**, shows water in a solid state? **(1)**
e Steam reacts with ethene in the presence of a catalyst. Complete the word equation for this reaction.

ethene + steam → _____ **(1)**
f Potassium reacts violently with water. Complete the word equation for this reaction.

potassium + water → _____ _____ + _____ **(2)**
(UCLES, IGCSE, Paper 2, May/June 2004)

32 When lumps of calcium carbonate react with hydrochloric acid, carbon dioxide gas is released.

$CaCO_3(s) + 2HCl(aq) \rightarrow CaCl_2(aq) + CO_2(g) + H_2O(l)$
a Describe a practical method for investigating this reaction, which would enable you to calculate the rate of reaction. **(4)**
b What effect will the following have on the rate of the reaction?
(i) increasing the temperature
(ii) adding water to the acid
(iii) using powdered calcium carbonate instead of lumps **(3)**

c Describe a test, and the result, for calcium ions. **(3)**
d Calcium can be obtained by the electrolysis of molten calcium chloride.
(i) Suggest why calcium must be extracted by electrolysis rather than by reduction with carbon. **(1)**
(ii) Draw the electronic structure of an atom of calcium. **(2)**
(UCLES, IGCSE, Paper 2, May/June 2004)

33 An organic compound decomposes to form nitrogen.

$C_6H_5N_2Cl(aq) \rightarrow C_6H_5Cl(l) + N_2(g)$
a Explain each of the state symbols. **(2)**
b Draw a diagram to show the arrangement of the valency electrons in **one** molecule of nitrogen. **(2)**
c The rate of this reaction can be measured using the following apparatus.

nitrogen gas

solution of organic compound

The results of this experiment are shown on the graph below.

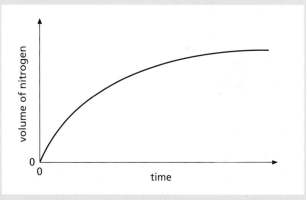

volume of nitrogen

time

(i) How does the rate of this reaction vary with time? **(1)**
(ii) Why does the rate vary? **(2)**
(iii) The reaction is catalysed by copper powder. Sketch the graph for the catalysed reaction on the same grid. **(2)**
(iv) Why is copper powder more effective as a catalyst than a single piece of copper? **(1)**
(UCLES, IGCSE, Paper 3, May/June 2004)

34 Some of the factors that can determine the rate of a reaction are concentration, temperature and light intensity.
a A small piece of calcium carbonate was added to an excess of hydrochloric acid. The time taken for the carbonate to react completely was measured.

$CaCO_3(s) + 2HCl(aq) \rightarrow CaCl_2(aq) + CO_2(g) + H_2O(l)$
The experiment was repeated at the same temperature, using pieces of calcium carbonate of the same size but with acid of a different

concentration. In all the experiments an excess of acid was used.

Concentration of acid/mol dm^{-3}	4	2	2	___
Number of pieces of carbonate	1	1	2	1
Time/s	___	80	___	160

(i) Complete the table (assume the rate is proportional to both the acid concentration and the number of pieces of calcium carbonate). **(3)**

(ii) Explain why the reaction rate would increase if the temperature was increased. **(2)**

(iii) Explain why the rate of this reaction increases if the piece of carbonate is crushed to a powder. **(1)**

(iv) Fine powders mixed with air can explode violently. Name an industrial process where there is a risk of this type of explosion. **(1)**

b Sodium chlorate(I) decomposes to form oxygen and sodium chloride. This is an example of a photochemical reaction. The rate of reaction depends on the intensity of the light.

$$2NaClO(aq) \rightarrow 2NaCl(aq) + O_2(g)$$

(i) Describe how the rate of this reaction could be measured. **(2)**

(ii) How could you show that this reaction is photochemical? **(1)**

c Photosynthesis is another example of a photochemical reaction. Glucose and more complex carbohydrates are made from carbon dioxide and water.

(i) Complete the equation.

$$6CO_2 + 6H_2 \rightarrow C_6H_{12}O_6 + \text{\_\_\_\_\_}$$ **(2)**

(ii) Glucose can be represented as

HO—□—OH

Draw the structure of a more complex carbohydrate that can be formed from glucose by condensation polymerisation.

(UCLES, IGCSE, Paper 3, October/November 2003)

35 Hydrogen peroxide solution, H$_2$O$_2$, decomposes slowly in the absence of a catalyst. Oxygen and water are formed.

$$2H_2O_2(aq) \rightarrow 2H_2O(l) + O_2(g)$$

a Draw a diagram of the apparatus you could use to investigate the speed of this reaction. You must label your diagram. **(3)**

b Catalyst **X** was added to 50 cm^3 of hydrogen peroxide solution at 20 °C and the amout of oxygen given off was recorded over a two minute period. The experiment was repeated with the same amounts of catalyst **Y** and catalyst **Z**. Apart from the type of catalyst, all conditions were kept the same in the three experiments. A graph of the results is shown below.

(i) What is a catalyst? **(1)**

(ii) Which catalyst, **X**, **Y** or **Z**, produced oxygen gas the fastest? Explain your answer. **(2)**

(iii) Why is the final amount of oxygen gas the same in each experiment? **(1)**

(iv) Many transition metals and their oxides are good catalysts. State **two** other properties of transition metals which are not shown by other metals. **(2)**

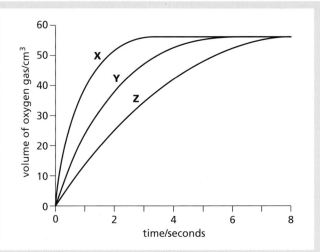

c The experiment with catalyst **Z** was repeated at 40 °C. All other conditions were kept the same. The speed of the reaction increased. Explain why, using ideas about particles. **(2)**

d Some enzymes also catalyse the decomposition of hydrogen peroxide.

(i) State one difference between an enzyme and an inorganic catalyst such as a transition metal. **(1)**

(ii) Enzymes are also responsible for fermentation reactions. Which one of the following equations **A**, **B**, **C** or **D** describes fermentation?

A $C_6H_{12}O_6 + 6O_2 \rightarrow 6CO_2 + 6H_2O$
B $C_2H_4 + H_2O \rightarrow C_2H_5OH$
C $C_6H_{12}O_6 \rightarrow 6C + 6H_2O$
D $C_6H_{12}O_6 \rightarrow 2C_2H_5OH + 2CO_2$

(UCLES, IGCSE, Paper 2, October/November 2005)

36 When fuels are burnt, carbon dioxide and water are formed.

a Complete the equation for the burning of propane.

$$C_3H_8 + \text{\_\_\_\_\_} O_2 \rightarrow 3CO_2 + 4H_2O$$ **(1)**

b Describe a chemical test, and the result, for water. **(2)**

c In which **two** of the following is carbon dioxide produced.
a car driven by a petrol engine
magnesium carbonate reacting with an acid
sodium reacting with water
zinc reacting with hydrochloric acid **(2)**

d The diagram shows a water heater.

If some of the air holes become blocked, a oisonous gas is produced.

(i) State the name of this poisonous gas.

(ii) Explain how this poisonous gas has been formed. **(2)**

e The table below compares the amounts of carbon dioxide and sulfur dioxide formed when 1 kilogram of different fuels are burnt.

Fuel	Mass of carbon dioxide produced/g	Mass of sulfur dioxide produced/g
oil	2900	5.0
gas	2500	0.1
coal	2500	11.0

(i) Which fuel is **least** polluting?

(ii) Which fuel when burnt, contributes most to the formation of acid rain?

(iii) State two harmful effects of acid rain.

(iv) When acid rain falls on the ground, it can react with insoluble aluminium compounds in the soil. A solution of aluminium ions is formed. Describe what you would observe when aqueous sodium hydroxide is added to a solution containing aluminium ions. **(6)**

(UCLES, IGCSE, Paper 2, October/November 2003)

37 The alkenes are a homologous series of unsaturated hydrocarbons.

a The table below gives the names, formulae and boiling points of the first members of the series.

Name	Formula	Boiling point/°C
ethene	C_2H_4	−102
propene	C_3H_6	−48
butene	C_4H_8	−7
pentene	C_5H_{10}	30
hexene		

(i) Complete the table by giving the formula of hexene and by predicting its boiling point. **(2)**

(ii) Deduce the formula of the alkene which has a relative molecular mass of 168. Show your working.

b Describe a test that will distinguish between the two isomers, but-2-ene and cyclobutane. **(3)**

c Alkenes undergo addition reactions.

(i) What class of organic compound is formed when an alkene reacts with water? **(1)**

(ii) Predict the structural formula of the compound formed when hydrogen chloride reacts with but-2-ene. **(1)**

(iii) Draw the structure of the polymer formed from but-2-ene. **(2)**

(UCLES, IGCSE, Paper 3, October/November 2004)

38 Organic substances have many uses.

a Match the substances in the boxes on the left with the descriptions in the boxes on the right. The first one has been done for you. **(3)**

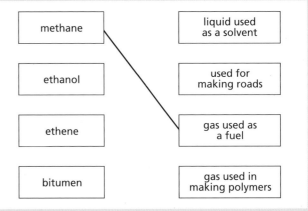

b Which one of the following would be least likely to be obtained from the fractional distillation of petroleum?

bitumen   ethane   ethanol   methane **(1)**

c Some reactions of organic compounds are shown below.

A $n\ CH_2=CH_2 \rightarrow (CH_2 - CH_2)_n$

B $C_3H_8 + 5O_2 \rightarrow 3CO_2 + 4H_2O$

C $C_6H_{12}O_6$ glucose $\rightarrow 2CO_2 + 2C_2H_5OH$

D $C_8H_{18} \rightarrow C_6H_{14} + C_2H_4$

(i) Which **one** of the reactions, **A**, **B**, **C** or **D**, shows fermentation?

(ii) Which **one** of the reactions, **A**, **B**, **C** or **D**, shows polymerization?

(iii) Which **one** of the reactions, **A**, **B**, **C** or **D**, shows combustion?

(iv) Which **one** of the reactions, **A**, **B**, **C** or **D**, shows cracking? **(4)**

d The hydrocarbon C_8H_{18} is an alkane.

(i) What is meant by the term *hydrocarbon*?

(ii) Explain why this hydrocarbon is an alkane. **(2)**

(UCLES, IGCSE, Paper 2, May/June 2004)

39 The structures of some compounds found in plants are shown below.

a Which **two** of these compounds are unsaturated hydrocarbons? **(1)**

b Which **two** of these compounds contain a carboxylic acid functional group? **(1)**

c Write the molecular formula for compound **D**. **(1)**

d Draw the structure of the product formed when compound **A** reacts with bromine. Show all atoms and all bonds. **(1)**

e Strawberry fruits produce compound **A** (ethene) naturally. A scientist left some green strawberry fruits to ripen. The scientist measured the concentration of ethene and carbon dioxide produced by the strawberry fruits over a ten day period. The graph below shows the results.

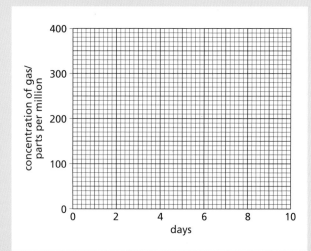

(i) Between which two days does the rate of ethene production increase most rapidly? **(1)**

(ii) What is the name given to the process in which carbon dioxide is produced by living organisms?

acidification combustion neutralization respiration **(1)**

(iii) Carbon dioxide concentration over 350 ppm has an effect on ethene production by the fruits. What effect is this? **(1)**

(iv) Ethene gas spreads throughout the fruit by a random movement of molecules. What is the name given to the random movement of molecules?

aeration diffusion evaporation ionisation **(1)**

(v) Ethene gas promotes the ripening of strawberry fruits. Ripening of strawberries is slowed down by passing a stream of nitrogen over the fruit. Suggest why this slows down the ripening process. **(1)**

(vi) Enzymes are involved in the ripening process. What is an *enzyme*? **(2)**

f Plants make a variety of coloured pigments. A student extracted red colouring from four different plants, **R, S, T** and **U**. The student put a spot of each colouring on a piece of filter paper. The filter paper was dipped into a solvent and left for 30 minutes. The results are shown below.

(i) What is the name given to the process shown in the diagram? **(1)**

(ii) Which plant contained the greatest number of different pigments? **(1)**

(iii) Which two plants contained the same pigments? **(1)**

(UCLES, IGCSE, Paper 2, October/November 2004)

40 The simplest alcohol is methanol.

a It is manufactured by the following reversible reaction.

$$CO(g) + 2H_2(g) \rightleftharpoons CH_3OH(g)$$
$$300\,°C$$
$$30\ atm$$

(i) Reversible reactions can come to equilibrium. Explain the term *equilibrium*. **(1)**

(ii) At 400 °C, the percentage of methanol in the equilibrium mixture is lower than at 300 °C. Suggest an explanation. **(2)**

(iii) Suggest two advantages of using high pressure for this reaction. Give a reason for each advantage. **(5)**

b (i) Complete the equation for the combustion of methanol in an excess of oxygen.

_____ CH_3OH + _____ O_2 → _____ + _____ **(2)**

(ii) Complete the word equation.

methanol + ethanoic acid → _____ + _____ **(2)**

(iii) Methanol can be oxidised to an acid. Name this acid. **(1)**

(UCLES, IGCSE, Paper 3, October/November 2004)

41 Polymers are extensively used in food packaging. Poly(dichloroethene) is used because gases can only diffuse through it very slowly. Polyesters have a high thermal stability and food can be cooked in a polyester bag.

a (i) The structure of poly(dichloroethene) is given below.

Draw the structural formula of the monomer. **(1)**

(ii) Explain why oxygen can diffuse faster through the polymer bag than carbon dioxide can. **(2)**

b (i) A polyester can be formed from the monomers $HO-CH_2CH_2-OH$ and $HOOC-C_6H_4-COOH$. Draw the structure of this polyester. **(2)**

(ii) Name a naturally occurring class of compounds that contains the ester linkage. **(1)**

(iii) Suggest what is meant by the term *thermal stability*. **(1)**

c (i) Describe **two** environmental problems caused by the disposal of plastic (polymer) waste. **(2)**

(ii) The best way of disposing of plastic waste is recycling to form new plastics. What is another advantage of recycling plastics made from petroleum? **(1)**

(UCLES, IGCSE, Paper 3, October/November 2004)

42 In 2002, Swedish scientists found high levels of acrylamide in starchy foods that had been cooked above 120 °C. Acrylamide, which is thought to be a risk to human health, has the following structure.

a (i) It readily polymerises to polyacrylamide. Draw the structure of this polymer. **(2)**

(ii) Starch is formed by polymerisation. It has a structure of the type shown below. Name the monomer. **(1)**

$$-\Box-o-\Box-o-\Box-$$

(iii) What are the differences between these two polymerisation reactions, one forming polyacrylamide and the other starch? **(2)**

b Acrylamide hydrolyses to form acrylic acid and ammonium ions.

(i) Describe the test, and the result, for the ammonium ion. **(2)**

(ii) Given an aqueous solution, concentration $0.1 \ mol/dm^3$, how could you show that acrylic acid is a weak acid. **(2)**

c The structural formula of acrylic acid is shown below. It forms compounds called acrylates.

(i) Acrylic acid reacts with ethanol to form the following compound.

Deduce the name of this compound. What type of organic compound is it? **(2)**

(ii) Acrylic acid is an unsaturated compound. It will react with bromine. Describe the colour change and draw the structural formula of the product of this addition reaction. **(2)**

(UCLES, IGCSE, Paper 3, May/June 2004)

43 The structures of some organic compounds are shown below.

a Name compound **A**. **(1)**

b Which **two** of the compounds **A** to **E** belong to the same homologous series? **(1)**

c (i) Which one of the compounds **A** to **E** has the same functional grop as ethanol? **(1)**

(ii) Draw the structure of ethanol, showing all atoms and bonds. **(2)**

(iii) Describe how ethanol is made in industry from ethene. **(2)**

d (i) Which one of the compounds **A** to **E** is an unsaturated hydrocarbon? **(1)**

(ii) Describe a chemical test, and the result, for an unsaturated hydrocarbon. **(2)**

e Compound **E** is acidic.

(i) State the name of compound **E**. **(1)**

(ii) Describe a test, and the result, to show that compound **E** is acidic. **(2)**

(UCLES, IGCSE, Paper 2, October/November 2005)

44 The alcohols form a homologous series. The first member is methanol and the fourth is butanol.

$$CH_3-OH \qquad CH_3-CH_2-CH_2-CH_2-OH$$
$$\text{methanol} \qquad\qquad\qquad \text{butanol}$$

a (i) Give **two** general characteristics of a homologous series. **(2)**

(ii) Calculate the mass of one mole of the C_8 alcohol. **(2)**

b Give the name and structural formula of the third member of this series. **(2)**

c The structural formula of the fifth member, pentan-1-ol, is drawn below.

$$CH_3-CH_2-CH_2-CH_2-CH_2-OH$$

(i) Draw the structural formula of an isomer of this alcohol. **(1)**

(ii) Predict the names of the product(s) formed when pentan-1-ol

- reacts with an excess of oxygen, **(1)**
- is dehydrated to form an alkene, **(1)**
- is oxidised by acidified potassium dichromate(VI) **(1)**

(UCLES, IGCSE, Paper 3, October/November 2005)

45 Esters occur naturally in plants and animals. They are manufactured from petroleum. Ethyl ethanoate and butyl ethanoate are industrially important as solvents.
 a (i) Explain the term *solvent*. **(1)**
 (ii) Give the formula of ethyl ethanoate. **(1)**
 (iii) Ethyl ethanoate can be made from ethanol and ethanoic acid. Describe how these chemicals can be made.
 ethanol from ethene **(2)**
 ethanoic acid from ethanol **(2)**
 (iv) Name **two** chemicals from which butyl ethanoate can be made. **(1)**
 b The following equation represents the alkaline hydrolysis of a naturally occurring ester.

$C_{17}H_{35}$—CO_2—CH_2 $\qquad\qquad$ CH_2OH
 |
$C_{17}H_{35}$—CO_2—CH + 3NaOH → $3C_{17}H_{35}COONa$ + CHOH
 |
$C_{17}H_{35}$—CO_2—CH_2 $\qquad\qquad$ CH_2OH

 (i) Which substance in the equation is an alcohol? Underline the substance in the equation above. **(1)**
 (ii) What is the major use for compounds of the type $C_{17}H_{35}COONa$? **(1)**
 c A polymer has the structure shown below.

 (i) What type of polymer is this? **(1)**
 (ii) Complete the following to give the structures of the two monomers from which the above polymer could be made **(2)**

 d Esters are frequently used as solvents in chromatography. A natural macromolecule was hydrolysed to give a mixture of amino acids. These could be identified by chromatography.
 (i) What type of macromolecule was hydrolysed? **(1)**
 (ii) What type of linkage was broken by hydrolysis? **(1)**
 (iii) Explain why the chromatogram must be sprayed with a locating agent before the amino acids can be identified. **(1)**
 (iv) Explain how it is possible to identify the amino acids from the chromatogram. **(2)**
 (UCLES, IGCSE, Paper 3, October/November 2003)

46 Fertilisers often contain ammonium nitrate.
 a (i) What effect do fertilisers have on crops? **(1)**
 (ii) Name **one** metal ion which is commonly present in fertilisers. **(1)**
 (iii) Which **one** of the following ions is commonly present in fertilisers? **(1)**
 bromide chloride hydroxide phosphate
 b Describe a test, and the result, for nitrate ions. **(4)**
 c Ammonium nitrate can be made by adding nitric acid to a solution of ammonia.
 (i) What type of reaction is this? **(1)**

 (ii) Complete the symbol equation for this reaction.
 _____ + $HNO_3(aq)$ → $NH_4NO_3(aq)$ **(1)**
 d Which **two** of the following statements about ammonia are true?
 ammonia is insoluble in water
 ammonia turns red litmus blue
 a solution of ammonia in water has a pH of 7
 ammonia has a molecular structure **(2)**
 (UCLES, IGCSE, Paper 2, October/November 2004)

47 Ammonia contains the elements nitrogen and hydrogen. It is manufactured from these elements in the Haber process.

$$N_2(g) + 3H_2(g) \rightleftharpoons 2NH_3(g)$$

The forward reaction is exothermic.
 a (i) Nitrogen is obtained from liquid air by fractional distillation. Why does this technique separate liquid oxygen and nitrogen?
 (ii) Name **two** raw materials from which hydrogen is manufactured. **(3)**
 b The table shows how the percentage of ammonia in the equilibrium mixture varies with pressure at 600 °C.

Percentage ammonia	8	12	15	20
Pressure/atm	200	300	400	500

 (i) Explain why the percentage of ammonia increases as the pressure increases. **(2)**
 (ii) How would the percentage of ammonia change if the measurements had been made at a lower temperature? Explain your answer. **(2)**
 (iii) State **two** of the reaction conditions used in the Haber Process. **(2)**
 c Ammonia is a base.
 (i) Name a particle that an ammonia molecule can accept from an acid.
 (ii) Write an equation for ammonia acting as a base. **(3)**
 d Given aqueous solutions, 0.1 mol/dm³, of sodium hydroxide and ammonia, describe how you could show that ammonia is the weaker base. **(2)**
 e Another compound that contains nitrogen and hydrogen is hydrazine, N_2H_4.
 (i) Draw the structural formula of hydrazine. Hydrogen can form only one bond per atom but nitrogen can form three.
 (ii) Draw a diagram that shows the arrangement of the valency electrons in one molecule of hydrazine. Hydrazine is a covalent compound.
   Use x to represent an electron from a nitrogen atom.
   Use o to represent an electron from a hydrogen atom. **(3)**
 (UCLES, IGCSE, Paper 3, October/November 2003)

48 In 1909, Haber discovered that nitrogen and hydrogen would react to form ammonia. The yield of ammonia was 8%.

$N_2(g) + 3H_2(g) \rightleftharpoons 2NH_3(g)$ the forward reaction is exothermic

Conditions
catalyst platinum
temperature 600 °C
pressure 200 atm

a Describe how hydrogen is obtained for the modern process. **(2)**

b (i) What is the catalyst in the modern process? **(1)**

(ii) Explain why the modern process, which uses a lower temperature, has a higher yield of 15%. **(2)**

c (i) Complete the following table that describes the bond breaking and forming in the reaction beween nitrogen and hydrogen to form ammonia.

Bonds	Energy change/kJ	Exothermic or endothermic
1 mole of N $\equiv$ N broken	+945	
3 moles of H – H broken	+1308	
6 moles of N – H formed	−2328	

(3)

(ii) Explain, using the above data, why the forward reaction is exothermic. **(2)**

(UCLES, IGCSE, Paper 3, October/November 2005)

49 Reversible reactions can come to equilibrium. They have both a forward an a backward reaction.

a When water is added to an acidic solution of bismuth(III) chloride, a white precipitate forms and the mixture slowly goes cloudy.

$$\underset{\text{colourless}}{BiCl_3(aq)} + H_2O(l) \overset{\text{forward}}{\underset{\text{backward}}{\rightleftharpoons}} \underset{\text{white}}{BiOCl(s)} + 2HCl(aq)$$

(i) Explain why the rate of the forward reaction decreases with time. **(2)**

(ii) Why does the rate of the backward reaction increase with time? **(1)**

(iii) After some time why does the appearance of the mixture remain unchanged? **(2)**

(iv) When a few drops of concentrated hydrochloric acid are added to the cloudy mixture, it changes to a colourless solution. Suggest an explanation. **(2)**

b Both of the following reactions are reversible.

reaction 1 $N_2(g) + O_2(g) \rightleftharpoons 2NO(g)$
reaction 2 $2NO(g) + O_2(g) \rightleftharpoons 2NO_2(g)$

(i) Suggest a reason why an increase in pressure does not affect the position of equilibrium for reaction 1. **(1)**

(ii) What effect would an increase in pressure have on the position of equilibrium for reaction 2? Give a reason for your answer. **(2)**

(UCLES, IGCSE, Paper 3, October/November 2005)

50 Sulfur dioxide, SO_2, and sulfur trioxide, SO_3, are the two oxides of sulfur.

a Sulfur dioxide can kill bacteria and has bleaching properties. Give a use of sulfur dioxide that depends on each of these properties.

(i) ability to kill bacteria **(1)**

(ii) bleaching properties **(1)**

b Sulfur trioxide can be made from sulfur dioxide.

(i) Why is this reaction important industrially? **(1)**

(ii) Complete the word equation.

sulfur dioxide + _____ → sulfur trioxide **(1)**

(iii) What are the conditions for this reaction? **(2)**

c Sulfur dioxide is easily oxidised in the presence of water.

$$SO_2 + 2H_2O - 2e^- \rightarrow SO_4^{2-} + 4H^+$$

(i) What colour change would be observed when an excess of aqueous sulfur dioxide is added to an acidic solution of potassium manganate(VII)? **(2)**

(ii) To aqueous sulfur dioxide, acidified barium chloride solution is added. The mixture remains clear. When bromine is added, a thick white precipitate forms. What is the white precipitate? Explain why it forms. **(3)**

d Sulfur dioxide reacts with chlorine in an addition reaction to form sulfuryl chloride.

$$SO_2 + Cl_2 \rightarrow SO_2Cl_2$$

8.0 g of sulfur dioxide was mixed with 14.2 g of chlorine. The mass of one mole of SO_2Cl_2 is 135 g. Calculate the mass of sulfuryl chloride formed by this mixture.

Calculate the number of moles of SO_2 in the mixture = _____

Calculate the number of moles Cl_2 in the mixture = _____

Which reagent was not in excess? _____

How many moles of SO_2Cl_2 were formed = _____

Calculate the mass of sulfuryl chloride formed = _____ g **(5)**

(UCLES, IGCSE, Paper 3, October/November 2003)

51 Sulfur is used to make sulfuric acid. In the UK, the annual production of the acid is about 2.5 million tonnes.

a The reactions in the manufacture of sulfuric acid by the Contact Process are shown below.

Sulfur	$\longrightarrow$	Sulfur dioxide
S	**reaction 1**	SO_2
Sulfur dioxide + oxygen	$\rightleftharpoons$	Sulfur trioxide
$2SO_2 + O_2$	**reaction 2**	$2SO_3$
Sulfur trioxide	$\longrightarrow$	Oleum
SO_3	**reaction 3**	$H_2S_2O_7$
Oleum + water	$\longrightarrow$	Sulfuric acid
$H_2S_2O_7$	**reaction 4**	H_2SO_4

(i) Give a large scale source of the element sulfur. **(1)**

(ii) State another use of sulfur dioxide. **(1)**

(iii) How is sulfur changed into sulfur dioxide? **(1)**

(iv) Name the catalyst used in reaction **2**. **(1)**

(v) Reaction **2** is exothermic. Why is a catalyst, rather than a higher temperature, used to increase the rate of this reversible reaction? **(2)**

(vi) Write a word equation for reaction **3**. **(1)**

(vii) Write a symbol equation for reaction **4**. **(1)**

b About one third of this production of acid is used to make nitrogen and phosphorus-containing fertilisers.

(i) Name the third element that is essential for plant growth and is present in most fertilisers. **(1)**

(ii) Name a nitrogen-containing fertiliser that is manufactured from sulfuric acid. **(1)**

(iii) Rock phosphate (calcium phosphate) is obtained by mining. It reacts with concentrated sulfuric acid to form the fertiliser, superphosphate. Predict the formula of each of these phosphates.

fertiliser	ions	formula
calcium phosphate	Ca^{2+} and PO_4^{3-}	_____
calcium superphosphate	Ca^{2+} and $H_2PO_4^{-}$	_____ **(2)**

(iv) The ionic equation for the reaction between the phosphate ion and sulfuric acid is shown below.

$$PO_4^{3-} + 2H_2SO_4 \rightarrow H_2PO_4^{-} + 2HSO_4^{-}$$

Explain why the phosphate ion is described as acting as a base in this reaction. **(2)**

(UCLES, IGCSE, Paper 3, May/June 2004)

Answers to numerical questions

Chapter 1
Gas laws
3 54.27 cm^3
4 23.81 cm^3
5 14.35 cm^3
Additional questions
8 214.6 °C

Chapter 5
Reacting masses
1
2
Calculating moles
1 a 0.10 mole
   b 0.16 mole
   c 2 moles
2 a 3.20 g
   b 160 g
   c 5.75 g
3 a 0.10 mole
   b 1 mole
   c 10 moles
4 a 162 g
   b 8.50 g
   c 55.83 g
5 a 0.08 mole
   b 10 moles
   c 0.0008 mole
6 a 7.20 dm^3
   b 2.40 dm^3
   c 48 dm^3
7 a 2 mol dm^{-3}
   b 0.20 mol dm^{-3}
8 a 7.98 g
   b 40.40 g
Calculating formulae
1 CaO
2 Empirical formula = CH_3
  Molecular formula = C_2H_6
Moles and chemical equations
1 32 g
2 320 g
3 127 g
4 24 dm^3
5 500 cm^3
6 0.27 mol dm^{-3}
Additional questions
1 a (i) 71 g
     (ii) 72 g
   b (i) 74 g
     (ii) 8.5 g
2 a 300 dm^3
   b 3.60 dm^3
3 a 0.0015 mole
   b 6 moles
4 SiO_2
5 a CHO_2
   b $C_2H_2O_4$

6 d 448 tonnes
7 a VuCl
8 c 0.002 mole
   d 48 cm^3
   e 40 cm^3

Chapter 6
Calculations in electrolysis
2 a 2 moles of electrons
   b 1 mole of electrons
   c 3 moles of electrons
   d 2 moles of electrons
3 a 193 000 coulombs
   b 96 500 coulombs
   c 193 000 coulombs
4 0.2 mole of electrons
Additional questions
4 27 971 s (466.18 min)
5 216 dm^3
6 0.15 mole of electrons

Chapter 7
Crystal hydrates
1 a 36.07%
   b 62.94%
   c 36.29%
Solubility of salts in water
1 a 24.2 g per 100 g of water
   b 18.5 g per 100 g of water
   c 32.4 g per 100 g of water
3 a 56.5 g per 100 g of water
   b 73.8 g per 100 g of water
   c 102.3 g per 100 g of water
4 a 17.3 g (± 0.50 g)
   b 45.8 g (± 0.50 g)
   c 28.5 g (± 0.50 g)
5 6.25 g (± 0.50 g)
Titration
1 0.19 mol dm^{-3}
2 0.18 mol dm^{-3}
Additional questions
4 c 2.33 mol dm^{-3}
6 b (i) 34.0 g per 100 g of water
        41.0 g per 100 g of water
        65.0 g per 100 g of water
        90.0 g per 100 g of water
     (ii) 31.0 g (± 0.50 g)
     (iii) 7 g (± 0.50 g)
8 e (i) 51.22%
     (ii) 0.07   mole of water
        0.01   mole of magnesium sulfate
     (iii) $MgSO_4.7H_2O$

Chapter 8
Additional questions
5 e (iii) 2.40 dm^3
8 a 110 tonnes

Chapter 9

Additional questions

6 c (i) 4.8 tonnes
    (ii) 9.2 tonnes
8 c 1590 tonnes

Chapter 10

Additional questions

1 b (i) 42.9 cm^3
    (ii) 21.45%
5 c 28 720 s (or 478.67 min)
  d (ii) 103 400 tonnes
6 a 184 000 tonnes
  b 3.09 × 10^{12} dm^3
7 a 4.4 × 10^{10} dm^3

Chapter 11

Surface area

2 c 26 cm^3 (± 0.5 cm^3)
  d 1 min 51 s (± 3 s)

Additional questions

2 f 46 s (± 1 s)
  g 43 s (± 1 s)
4 g 12 min
6 c 0.309 g
7 e (i) 6.16 g
    (ii) 3.36 dm^3
    (iii) 13.44 dm^3

Chapter 12

Alkanes

1 a 98.5 °C
  b 126 °C

Other uses of alkanes

2 a 1 mole CH_4 : 1 mole Cl_2
  b 1 mole CH_4 : 4 moles Cl_2

Alkenes

1 a 54 °C

Additional questions

7 a CH
  b C_6H_6

Chapter 13

Chemical energy

1 a −1161 kJ mol^{-1}
  d ethanol 25 kJ
    heptane 49.5 kJ
2 a 364 kJ
  b 3640 kJ
  c 182 kJ
3 a 114 kJ
  b 14.25 kJ
  c 114 kJ

Additional questions

1 b (i) −1461 kJ mol^{-1}
4 b (iii) −114 kJ
  c (iii) 143.75 kJ
5 a 0.50 g
  b 10 080 J
  c 10.08 kJ
  d 20.16 kJ
  e 46 g
  f 927.36 kJ

6 b 100 g
  c 2100 J (2.10 kJ)
  d 0.05 mol
  e 42 000 J (42 kJ)
8 d (i) 110 g
    (ii) 1110 kJ
    (iii) 555 kJ
    (iv) 185 000 kJ

Chapter 14

Additional questions

5 d (i) 363 kJ
       2904 kJ
       90.75 kJ
    (ii) 12 dm^3

Chapter 15

Artificial fertilisers

1 Ammonium nitrate: 35.00%
  Ammonium phosphate: 28.19%
  Ammonium sulfate: 21.21%
  Urea: 46.67%

Additional questions

4 a Sodium nitrate: 16.47%
    Potassium nitrate: 13.86%
  e 39.38 kJ
5 d (i) 32%
    (ii) 19%
  e 25%
6 a 10.25   10.1   10.1   10.15
  b 10.12 cm^3
  c (ii) KOH: 1 mole
        HNO_3: 1 mole
  d (i) 0.0038 mole
    (ii) 0.0038 mole
  e 0.37 mol dm^{-3}

Chapter 16

Additional questions

3 b 7.5 tonnes
  c 15 tonnes
  d 5.6 × 10^6 dm^3
4 b (i) 3%
    (ii) 31%
  d (i) 3 × 10^6 tonnes
    (ii) 550 000 tonnes
    (iii) 900 000 tonnes
6 b 0.01 mole
  c 0.005 mole
  d 0.2 mol dm^{-3}

Data tables

■ *Notes for use in qualitative analysis*

Tests for anions

Anion	Test	Test result
carbonate (CO_3^{2-})	add dilute acid	effervescence, carbon dioxide produced
chloride (Cl^-) [in solution]	acidify with dilute nitric acid, then add aqueous silver nitrate	white ppt.
iodide (I^-) [in solution]	acidify with dilute nitric acid, then add aqueous lead(II) nitrate	yellow ppt.
nitrate (NO_3^-) [in solution]	add aqueous sodium hydroxide, then aluminium foil; warm carefully	ammonia produced
sulfate (SO_4^{2-}) [in solution]	acidify, then add aqueous barium nitrate	white ppt.

Tests for aqueous cations

Cation	Effect of aqueous sodium hydroxide	Effect of aqueous ammonia
aluminium (Al^{3+})	white ppt., soluble in excess giving a colourless solution	white ppt., insoluble in excess
ammonium (NH_4^+)	ammonia produced on warming	-
calcium (Ca^{2+})	white ppt., insoluble in excess	no ppt. or very slight white ppt.
copper (Cu^{2+})	light blue ppt., insoluble in excess	light blue ppt., soluble in excess, giving a dark blue solution
iron(II) (Fe^{2+})	green ppt., insoluble in excess	green ppt., insoluble in excess
iron(III) (Fe^{3+})	red–brown ppt., insoluble in excess	red-brown ppt., insoluble in excess
zinc (Zn^{2+})	white ppt., soluble in excess, giving a colourless solution	white ppt., soluble in excess, giving a colourless solution

Tests for gases

Gas	Test and test result
ammonia (NH_3)	turns damp red litmus paper blue
carbon dioxide (CO_2)	turns limewater milky
chlorine (Cl_2)	bleaches damp litmus paper
hydrogen (H_2)	'pops' with a lighted splint
oxygen (O_2)	relights a glowing splint

Effect of heat on metal nitrates

Metal	Symbol	Effect of heat on nitrate	Equation (some examples)						
potassium	K	form the metal nitrite and oxygen is released	potassium nitrate	$\rightarrow$	potassium nitrite	+	oxygen		
sodium	Na		$2KNO_3(s)$	$\rightarrow$	$2KNO_2(s)$	+	$O_2(g)$		
calcium	Ca	form the oxide and release oxygen	calcium nitrate	$\rightarrow$	calcium oxide	+	nitrogen(IV) oxide	+	oxygen
magnesium	Mg		$2Ca(NO_3)_2(s)$	$\rightarrow$	$2CaO(s)$	+	$4NO_2(g)$	+	$O_2(g)$
aluminium	Al	brown fumes of nitrogen dioxide (nitrogen(IV) oxide) are also produced							
zinc	Zn		zinc nitrate	$\rightarrow$	zinc oxide	+	nitrogen(IV) oxide	+	oxygen
iron	Fe		$2Zn(NO_3)_2(s)$	$\rightarrow$	$2ZnO(s)$	+	$4NO_2(g)$	+	$O_2(g)$
lead	Pb								
copper	Cu		copper(II) nitrate	$\rightarrow$	copper oxide	+	nitrogen(IV) oxide	+	oxygen
			$2Cu(NO_3)_2(s)$	$\rightarrow$	$2CuO(s)$	+	$4NO_2(g)$	+	$O_2(g)$
silver	Ag	decompose to the metal releasing oxygen and nitrogen(IV) oxide	silver nitrate	$\rightarrow$	silver	+	nitrogen(IV) oxide	+	oxygen
platinum	Pt								
gold	Au		$2AgNO_3(s)$	$\rightarrow$	$2Ag(s)$	+	$2NO_2(g)$	+	$O_2(g)$

Effect of heat on metal hydroxides

Metal	Symbol	Effect of heat on hydroxide	Equation (some examples)			
potassium	K	no reaction (i.e. there is no decomposition)				
sodium	Na					
calcium	Ca	decompose to the oxide and release water	calcium hydroxide	$\rightarrow$	calcium oxide	+ water
magnesium	Mg		$Ca(OH)_2(s)$	$\rightarrow$	$CaO(s)$	+ $H_2O(g)$
aluminium	Al					
zinc	Zn		zinc hydroxide	$\rightarrow$	zinc oxide	+ water
iron	Fe		$Zn(OH)_2(s)$	$\rightarrow$	$ZnO(s)$	+ $H_2O(g)$
lead	Pb					
copper	Cu		copper hydroxide	$\rightarrow$	copper oxide	+ water
			$Cu(OH)_2(s)$	$\rightarrow$	$CuO(s)$	+ $H_2O(g)$
silver	Ag	hydroxides are too unstable to exist				
platinum	Pt					
gold	Au					

The periodic table of the elements

The volume of one mole of any gas is 24 dm³ at room temperature and pressure (r.t.p.).

Group																	
I	II											III	IV	V	VI	VII	0

				1 H Hydrogen 1													4 He Helium 2
7 Li Lithium 3	9 Be Beryllium 4											11 B Boron 5	12 C Carbon 6	14 N Nitrogen 7	16 O Oxygen 8	19 F Flourine 9	20 Ne Neon 10
23 Na Sodium 11	24 Mg Magnesium 12											27 Al Aluminium 13	28 Si Silicon 14	31 P Phosphorous 15	32 S Sulfur 16	35.5 Cl Chlorine 17	40 Ar Argon 18
39 K Potassium 19	40 Ca Calcium 20	45 Sc Scandium 21	48 Ti Titanium 22	51 V Vanadium 23	52 Cr Chromium 24	55 Mn Manganese 25	56 Fe Iron 26	59 Co Cobalt 27	59 Ni Nickel 28	64 Cu Copper 29	65 Zn Zinc 30	70 Ga Gallium 31	73 Ge Germanium 32	75 As Arsenic 33	79 Se Selenium 34	80 Br Bromine 35	84 Kr Krypton 36
85 Rb Rubidium 37	88 Si Stronlium 38	89 Y Yitrium 39	91 Zr Zirconium 40	93 Nb Niobium 41	96 Mo Molybendium 42	Tc Technetium 43	101 Ru Ruthenium 44	103 Rh Rhodium 45	106 Pd Palladium 46	108 Ag Silver 47	112 Cd Cadmium 48	115 In Indium 49	119 Sn Tin 50	122 Sb Antimony 51	128 Te Tellurium 52	127 I Iodine 53	131 Xe Xenon 54
133 Cs Caesium 55	137 Ba Banum 56	139 La Lanthanium 57	178 Hf Hafnium 72	181 Ta Tantatium 73	184 W Tungsten 74	186 Re Rhenium 75	190 Os Osmium 76	192 Ir Indium 77	195 Pt Platinum 78	197 Au Gold 79	201 Hg Mercury 80	204 Tl Thallium 81	207 Pb Lead 82	209 Bi Bismuth 83	Po Polonium 84	At Astatine 85	Rn Radon 86
Fr Francium 87	226 Ra Radium 88	227 Ac actinium 89 †															

*58-71 Lanthanoid series
†90-103 Actinoid series

140 Ce Cerium 58	141 Pr Praseodymium 59	144 Nd Neodyium 60	Pm Promethium 61	150 Sm Samarium 62	152 Eu Europium 63	157 Gd Gadolinium 64	159 Tb terbium 65	163 Dy Dysprosium 66	165 Ho Holmium 67	167 Er Erbium 68	169 Tm Thulium 69	173 Yb Ytterbium 70	175 Lu Lutetium 71
232 Th Thorium 90	Pa Protactinium 91	238 U Uranium 92	Np Neptunium 93	Pu Plutonium 94	Am Americium 95	Cm Curium 96	Bk Berkelium 97	Cf Californium 98	Es Einsteinium 99	Fm Fermium 100	Md Mendelevium 101	No Nobelium 102	Lr Lawrecium 103

Key

a	a = relative atomic mass
X	X = atomic symbol
b	b = proton (atomic) number

Acknowledgements

The authors would like to thank Irene, Katharine, Michael and Barbara for their never-ending patience and encouragement throughout the production of this textbook. Also to Lis and Phillipa and the publishing team at Hodder Education.

Examination questions

All IGCSE Chemistry past examination questions are reproduced by permission of the University of Cambridge Local Examinations Syndicate.

p.247 Q1: Paper 5, Q1, May/June 2005; **p.247** Q2: Paper 5, Q2, May/June 2005; **p.248** Q3: Paper 5, Q1, May/June 2004; **p.249** Q4: Paper 5, Q2, May/June 2004; **p.249** Q5: Paper 5, Q1, Oct/Nov 2003; **p.250** Q6: Paper 5, Q2, Oct/Nov 2003; **p.251** Q7: Paper 6, Q1, May/June 2005; **p.251** Q8: Paper 6, Q2, May/June 2005; **p.250** Q9: Paper 6, Q3, May/June 2005; **p.252** Q10: Paper 6, Q4, May/June 2005; **p.252** Q11: Paper 6, Q5, May/June 2005; **p.253** Q12: Paper 6, Q7, May/June 2005; **p.253** Q13: Paper 6, Q8, May/June 2005; **p.253** Q14: Paper 6, Q1, Oct/Nov 2005; **p.254** Q15: Paper 6, Q2, Oct/Nov 2005; **p.254** Q16: Paper 6, Q3, Oct/Nov 2005; **p.254** Q17: Paper 6, Q4, Oct/Nov 2005; **p.255** Q18: Paper 6, Q5, Oct/Nov 2005; **p.255** Q19: Paper 6, Q6, Oct/Nov 2005; **p.255** Q20: Paper 6, Q7, Oct/Nov 2005; **p.256** Q21: Paper 6, Q1, May/June 2004; **p.256** Q22: Paper 6, Q2, May/June 2004; **p.256** Q23: Paper 6, Q3, May/June 2004; **p.256** Q24: Paper 6, Q4, May/June 2004; **p.258** Q25: Paper 6, Q5, May/June 2004; **p.258** Q26: Paper 6, Q6, May/June 2004; **p.259** Q27: Paper 6, Q7, May/June 2004; **p.259** Q28: Paper 6, Q8, May/June 2004; **p.259** Q29: Paper 6, Q1, Oct/Nov 2004; **p.259** Q30: Paper 6, Q2, Oct/Nov 2004; **p.259** Q31: Paper 6, Q1, Oct/Nov 2004; **p.260** Q32: Paper 6, Q4, Oct/Nov 2004; **p.261** Q33: Paper 6, Q5, Oct/Nov 2004; **p.261** Q34: Paper 6, Q6, Oct/Nov 2004; **p.261** Q35: Paper 6, Q7, Oct/Nov 2004; **p.262** Q1: Paper 3, Q2, Oct/Nov 2005; **p.262** Q2: Paper 2, Q1, Oct/Nov 2003; **p.263** Q3: Paper 2, Q1, Oct/Nov 2005; **p.263** Q5: Paper 2, Q1, Oct/Nov 2004; **p.263** Q5: Paper 2, Q4, Oct/Nov 2005; **p.264** Q6: Paper 2, Q5, May/June 2004; **p.264** Q7: Paper 3, Q5, Oct/Nov 2004; **p.264** Q8: Paper 2, Q1, May/June 2004; **p.265** Q9: Paper 3, Q1, May/June 2004; **p.265** Q10: Paper 3, Q1, Oct/Nov 2005; **p.265** Q11: Paper 3, Q7, May/June 2004; **p.265** Q12: Paper 2, Q6, Oct/Nov 2005; **p.266** Q13: Paper 3, Q5, May/June 2004; **p.266** Q14: Paper 2, Q6, Oct/Nov 2003; **p.267** Q15: Paper 2, Q6, Oct/Nov 2004; **p.267** Q16: Paper 3, Q5, Oct/Nov 2005; **p.267** Q17: Paper 3, Q4, May/June 2004; **p.268** Q18: Paper 2, Q2, Oct/Nov 2005; **p.268** Q19: Paper 3, Q2, Oct/Nov 2004; **p.269** Q20: Paper 2, Q3, Oct/Nov 2004; **p.269** Q21: Paper 3, Q6, Oct/Nov 2005; **p.269** Q22: Paper 3, Q7, Oct/Nov 2004; **p.269** Q23: Paper 2, Q4, Oct/Nov 2005; **p.270** Q24: Paper 2, Q3, Oct/Nov 2003; **p.270** Q25: Paper 2, Q2, Oct/Nov 2003; **p.271** Q26: Paper 3, Q4, Oct/Nov 2004; **p.271** Q27: Paper 3, Q3, Oct/Nov 2003; **p.271** Q28: Paper 2, Q6, May/June 2004; **p.272** Q29: Paper 3, Q1, Oct/Nov 2004; **p.272** Q30: Paper 2, Q4, Oct/Nov 2004; **p.273** Q31: Paper 2, Q2, May/June 2004; **p.273** Q32: Paper 2, Q3, May/June 2004; **p.273** Q33: Paper 3, Q3, May/June 2004; **p.273** Q34: Paper 3, Q2, Oct/Nov 2003; **p.274** Q35: Paper 2, Q3, Oct/Nov 2005; **p.274** Q36: Paper 2, Q5, Oct/Nov 2003; **p.275** Q37: Paper 3, Q8, Oct/Nov 2004; **p.275** Q38: Paper 2, Q4, May/June 2004; **p.275** Q39: Paper 2, Q2, Oct/Nov 2004; **p.276** Q40: Paper 3, Q3, Oct/Nov 2004; **p.276** Q41: Paper 3, Q6, Oct/Nov 2004; **p.277** Q42: Paper 3, Q6, May/June 2004; **p.277** Q43: Paper 2, Q5, Oct/Nov 2005; **p.277** Q44: Paper 3, Q4, Oct/Nov 2005; **p.278** Q45: Paper 3, Q4, Oct/Nov 2003; **p.278** Q46: Paper 2, Q5, Oct/Nov 2004; **p.278** Q47: Paper 3, Q1, Oct/Nov 2003; **p.278** Q48: Paper 3, Q7, Oct/Nov 2005; **p.279** Q49: Paper 3, Q3, Oct/Nov 2005; **p.279** Q50: Paper 3, Q5, Oct/Nov 2003; **p.279** Q51: Paper 2, Q2, May/June 2004.

Source acknowledgements

pp. 56, 57, 58, 59, 176, 179, 180, 198, 199, 200, 204, 206
The molecular models shown were made using the Molymod® system available from Molymod® Molecular Models, Spiring Enterprises Limited, Billingshurst, West Sussex RH14 9NF England.

Photo credits

Cover © Andrew Brookes/Corbis; **p.1** *l* © helo-pilote – Fotolia.com, *tr* fstop2/Alamy, *br* © Adam Booth – Fotolia.com; **p.2** *l* Robert Harding Photo Library, *tr* Andrew Lambert Photography/Science Photo Library, *br* GeoScience Features Picture Library/Dr B. Booth; **p.3** *t* Science Photo Library/James King-Holmes, *b* Science Photo Library/Andrew Syred; **p.4** *t* Science Photo Library/Michael W Davidson, *b* © ONOKY – Photononstop/Alamy; **p.5** James McCauley/Rex Features; **p.6** *all* Andrew Lambert Photography/Science Photo Library; **p.7** *both* Last Resort Picture Library/Dick Makin; **p.8** *l* Mary Evans Picture Library, *r* © DJ – Fotolia.com; **p.11** *l* Science Photo Library, *r* Mary Evans Picture Library; **p.12** *t* Robert Harding/Getty Images, *c* © icholakov – Fotolia.com, *b* © Mihaela Bof – Fotolia.com; **p.13** *t* Science Photo Library/Ron Sutherland, *c* © Christie's Images/The Bridgeman Art Library, *b* BL Images Ltd/Alamy; **p.14** Andrew Lambert Photography/Science Photo Library;

p.15 Andrew Lambert Photography/Science Photo Library; **p.16** Science Photo Library/Simon Fraser/Searle Pharmaceuticals; **p.17** *l* Science Photo Library, *tr* © Sean Gladwell – Fotolia.com, *cr & br* Andrew Lambert Photography/Science Photo Library; **p.19** *c & r* Andrew Lambert Photography/Science Photo Library; **p.20** *tl & bl* Klaus Guldbrandsen/Science Photo Library, *tr* Andrew Lambert Photography/Science Photo Library, *br* Ricardo Funari/BrazilPhotos/Alamy; **p.21** *l* Andrew Lambert Photography/Science Photo Library, *tr* Sipa Press/Rex Features, *br* Still Pictures/Al Grillo; **p.22** *both* Andrew Lambert Photography/Science Photo Library; **p.23** *tl* Rex Features, *bl* BOC Gases, *tr* Science Photo Library/Alex Bartel, *br* Andrew Lambert Photography/Science Photo Library; **p.24** *l* Andrew Lambert Photography/Science Photo Library, *tr* Science Photo Library/JC Revy, *br* Jenny Matthews/Alamy; **p.25** *tl* Science Photo Library/Geoff Tompkinson, *bl & all r* Last Resort Picture Library/Dick Makin; **p.26** *tl* © Roger Scruton, *bl* UpperCut Images/Alamy, *tr* PurestockX, *br* Andrew Lambert Photography/Science Photo Library; **p.27** *l* Science Photo Library/Dr Jeremy Burgess, *tr* Science Photo Library/Eye of Science, *br* The Laser Centre, Banbury; **p.28** *l* Science Photo Library/Manfred Kage, *r* Biophoto Associates; **p.32** Science Photo Library/Mehau Kulyk; **p.37** Science Photo Library; **p.39** *tl* © Windsor – Fotolia.com, *bl* © jon11 – Fotolia.com, *tr* © Debby Moxon, *cr* © Mihaela Bof – Fotolia.com; **p.39** *br* Science Photo Library/Mehau Kulyk; **p.40** Andrew Lambert Photography/Science Photo Library; **p.41** *tl & bl* Andrew Lambert Photography/Science Photo Library, *tr* Science Photo Library/Charles D Winters; **p.42** Andrew Lambert Photography/Science Photo Library; **p.43** *l* Andrew Lambert Photography/Science Photo Library, *r* © Mohd Haka Khambali – Fotolia.com; **p.44** *tl & bl* Andrew Lambert Photography/Science Photo Library, *r* Martyn F Chillmaid; **p.45** *tl, bl & tr* Last Resort Picture Library/Dick Makin, *br* Art Directors & Trip Photo Library/Helene Rogers; **p.46** *both* Science Photo Library; **p.47** *tl* Last Resort Picture Library/Dick Makin, *bl* Art Directors & Trip Photo Library/Helene Rogers, *tr* Koichi Kamoshida /Getty Images, *cr & br* Last Resort Picture Library/Dick Makin; **p.48** *t* Andrew Lambert Photography/Science Photo Library, *b* Last Resort Picture Library/Dick Makin; **p.51** *l* © Germany Images David Crossland/Alamy, *tr* Andrew Lambert Photography/Science Photo Library, *br* Last Resort Picture Library/Dick Makin; **p.54** Science Photo Library/Science Source; **p.56** Andrew Lambert Photography/Science Photo Library; **p.57** Andrew Lambert Photography/Science Photo Library; **p.58** *both* Andrew Lambert Photography/Science Photo Library; **p.59** Andrew Lambert Photography/Science Photo Library; **p.60** *t* Science Photo Library/E R Degginger, *b* © Manfred Vollmer/Das Fotoarchiv./Still Pictures; **p.61** Science Photo Library/Philippe Plailly; **p.62** *l* Robert Harding Photo Library, *tr* Science Photo Library/Ricjard Megna/Fundamental, *crt* Tobias Titz /Fstop/Getty Images, *crb* Science Photo Library/Sheila Terry, *br* Science Photo Library/Manfred Kage; **p.63** *br* RIBA Library Photographs Collection; **p.64** *l* Last Resort Picture Library/Dick Makin, *r* © Orsovolante – Fotolia.com; **p.69** *l* Science Photo Library/Rosenfeld Images Ltd, *tr & br* Andrew Lambert Photography/Science Photo Library; **p.70** Andrew Lambert Photography/Science Photo Library; **p.75** Alan Murray Tilcon Ltd; **p.79** *t* © Alexander Maksimenko – Fotolia.com, *c* Last Resort Picture Library/Dick Makin, *b* Robert Harding Photo Library; **p.81** © Howard Davies/Corbis; **p.82** Rex Features; **p.83** *tl* Larry MacDougal/Rex Features, *br* Rex Features, *r* © Nicola Stratford/ iStockphoto.com; **p.84** © O.Alamany & E.Vicens/Corbis; **p.85** *l* Kind permission of ICI Chemicals & Polymers, *r* Andrew Lambert Photography/Science Photo Library; **p.88** Andrew Lambert Photography/Science Photo Library; **p.89** Clynt Garnham/Alamy; **p.90** Science Photo Library/Adam Hart Davis; **p.91** *both* Andrew Lambert Photography/Science Photo Library; **p.92** Mary Evans Picture Library; **p.96** *both* Last Resort Picture Library/Dick Makin; **p.97** *tl & tr* Andrew Lambert Photography/Science Photo Library, *bl &br* Science Photo Library/Jean Loup Charmet; **p.98** *l* Science Photo Library/Sheila Terry, *r* Andrew Lambert Photography/Science Photo Library; **p.99** © Clyde H. Smith/Still Pictures; **p.101** *tl & bl* Last Resort Picture Library/Dick Makin, *r* Andrew Lambert Photography/Science Photo Library; **p.102** *both* Andrew Lambert Photography/Science Photo Library; **p.103** *all* Andrew Lambert Photography/Science Photo Library; **p.104** *tl* PurestockX, *bl* © rgbdigital.co.uk – Fotolia.com, *r* Andrew Lambert Photography/Science Photo Library; **p.105** *both* Andrew Lambert Photography/Science Photo Library; **p.106** *both* Andrew Lambert Photography/Science Photo Library; **p.113** Ronald Evans/Alamy; **p.114** *tl* Last Resort Picture Library/Dick Makin, *bl* Science Photo Library/Martin Land, *r* © yang yu – Fotolia.com; **p.115** *l* Agripicture Images/Alamy, *tr* Rex Features, *br* © gaelj – Fotolia.com; **p.116** Alan Murray, Tilcon Ltd; **p.117** *all* Andrew Lambert Photography/Science Photo Library; **p.118** *l* Science Photo Library/Roberto de Gugliemo, *c* GeoScience Features Picture Library, *r* Science Photo Library/Maximillian Stock Ltd; **p.119** *tl* PurestockX, *cl* © EyeMark – Fotolia.com, *bl* Andrew Lambert Photography/Science Photo Library, *cr* Christopher Nash/Alamy, *br* Science Photo Library/Sheila Terry; **p.120** *l* © lefebvre_jonathan – Fotolia.com, *r* Andrew Lambert Photography/Science Photo Library; **p.121**

Andrew Lambert Photography/Science Photo Library; **p.123** Andrew Lambert Photography/Science Photo Library; **p.124** Andrew Lambert Photography/ Science Photo Library; **p.128** *l* Andrew Lambert Photography/Science Photo Library, *tr* Digital Vision/Getty Images, *br* Robert Harding Picture Library Ltd/Alamy; **p.129** Andrew Lambert Photography/Science Photo Library; **p.130** *t* © govicinity – Fotolia.com, *c* © Owen Franken/Corbis, *b* Antony Nettle/Alamy; **p.131** Andrew Lambert Photography/Science Photo Library; **p.132** *all* Andrew Lambert Photography/Science Photo Library; **p.134** *all* GeoScience Features Picture Library; **p.135** Corus; **p.136** *t* Science Photo Library/Dr Jeremy Burgess, *bl* Science Photo Library/Martin Bond, *br* Science Photo Library; **p.137** *t* Kuttig – People/Alamy, *b* The Travel Library/Rex Features; **p.138** *t* Science Photo Library/Hank Morgan, *b* Digital Vision/Getty Images; **p.139** *tl* Art Directors & Trip Photo Library/Helene Rogers, *cl* Cameron Spencer/Getty Images, *bl, tr, cr & br* Last Resort Picture Library/Dick Makin; **p.140** *l* Alan Wiseman/Alamy, *r* Leslie Garland Picture Library/Alamy; **p.141** *l* © Phoenix – Fotolia.com, *r* Rex Features; **p.142** Science Photo Library/ Maximillian Stock Ltd; **p.143** *l* Robert Harding Photo Library, *r* Science Photo Library/Manfred Kage; **p.147** *l* Purestock/Alamy, *tr* Design Pics Inc./Alamy, *br* Science Photo Library/O I Miller; **p.148** *t* © Beboy – Fotolia.com, *b* Carolyn Clarke/Alamy; **p.150** Andrew Lambert Photography/Science Photo Library; **p.151** *l* BOC Gases, *r* Science Photo Library/Penny Tweedie; **p.152** *l & tr* Last Resort Picture Library/Dick Makin, *cr* Science Photo Library/Alexander Tsiaras, *br* Science Photo Library/Vaughan Fleming; **p.153** *l* Alan Puzey /Stone/Getty Images, *r* Art Directors & Trip Photo Library/Ken Maclaren; **p.154** *t* © Shariff Che'Lah – Fotolia.com, *b* Last Resort Picture Library/Dick Makin; **p.156** *t* David McNew/Getty Images, *b* Digital Vision/Getty Images; **p.158** *l* Peter Treanor/Alamy, *r* Photodisc; **p.162** *tl* © OutdoorPhotos – Fotolia.com, *bl* Cephas Picture Library/Alamy, *tr* © Arnaud Santini – Fotolia.com, *br* Motoring Picture Library/Alamy; **p.163** *all* Andrew Lambert Photography/Science Photo Library; **p.164** *all* Andrew Lambert Photography/Science Photo Library; **p.167** *t* © VisualHongKong/Alamy; **p.168** *bl* Science Photo Library/Alfred Pasieka, *t* Last Resort Picture Library/Dick Makin, *br* © Wellcome Trust; **p.169** *l* Science Photo Library/Rosenfeld Images Ltd, *r* Maria L Antonelli/Rex Features; **p.173** *tl* © Charles Taylor – Fotolia.com, *bl* © Matt K – Fotolia.com, *tr* © Dmitry Sitin – Fotolia.com, *cr* © iofoto – Fotolia.com, *br* © Rene Drouyer – Fotolia.com; **p.174** *tl* Andrew Lambert Photography/Science Photo Library, *c, bl, cr & br* Last Resort Picture Library/Dick Makin, *tr* Science Photo Library; **p.175** Science Photo Library/Stevie Grand;

p.176 *all* Andrew Lambert Photography/Science Photo Library; **p.177** *t* © nikkytok – Fotolia.com, *b* Andrew Lambert Photography/Science Photo Library; **p.178** *both* Last Resort Picture Library/Dick Makin; **p.179** *all* Andrew Lambert Photography/Science Photo Library; **p.180** Andrew Lambert Photography/Science Photo Library; **p.181** *all* Andrew Lambert Photography/Science Photo Library; **p.185** *t* GeoScience Features Picture Library, *b* © Kevin Burke/Corbis; **p.186** *c* Science Photo Library/Richard Folwell, *r* Science Photo Library/Martin Bond; **p.187** Science Photo Library/Martin Bond; **p.188** *l* Science Photo Library/Novosti, *tr* Science Photo Library/John Mead, *br* Images&Stories/Alamy; **p.189** Andrew Lambert Photography/Science Photo Library; **p.190** *tl* Science Photo Library/Philippe Plailly, *bl* Danita Delimont/Alamy, *t* Science Photo Library/David Hall, *b* Rex Features; **p.193** Andrew Lambert Photography/ Science Photo Library; **p.194** Science Photo Library/NASA; **p.198** *both* Andrew Lambert Photography/Science Photo Library; **p.199** *t* Andrew Lambert Photography/Science Photo Library, *b* Martyn F Chillmaid. Thanks to Molymod.com for providing the model.; **p.200** *all* Andrew Lambert Photography/ Science Photo Library; **p.201** *t* Science Photo Library/James Holmes/Zedcor, *c* © Marc Dietrich – Fotolia.com, *b* Photodisc/Getty Images; **p.202** *t* Sally Morgan/Ecoscene, *b* Andrew Lambert Photography/ Science Photo Library; **p.203** Andrew Lambert Photography/Science Photo Library; **p.204** *tl, tcl, bcl & bl* Martyn F Chillmaid. Thanks to Molymod.com for providing the model. *r* Andrew Lambert Photography/Science Photo Library; **p.205** *tl* Science Photo Library/Biophoto Associates, *bc* Science Photo Library, *r* Andrew Lambert Photography/Science Photo Library; **p.206** *all* Martyn F Chillmaid. Thanks to Molymod.com for providing the model. **p.207** *t* Paul Cooper/Rex Features, *b* Last Resort Picture Library/Dick Makin; **p.208** *t* © Maria Brzostowska – Fotolia.com, *b* Last Resort Picture Library/Dick Makin; **p.210** *all* Andrew Lambert Photography/Science Photo Library; **p.211** *t* Science Photo Library/Richardson, *b* Science Photo Library/James King Holmes; **p.212** Rex Features; **p.221** Andrew Lambert Photography/Science Photo Library; **p.222** Science Photo Library/Malcom Fielding, Johnson Matthey plc; **p.223** Andrew Lambert Photography/Science Photo Library; **p.227** *t* GeoScience Features Picture Library, *b* Andrew Lambert Photography/Science Photo Library; **p.228** *t* Still Pictures, *b* © Bojan Pavlukovic – Fotolia.com; **p.229** *c & b* Andrew Lambert Photography/Science Photo Library, *t* ICI Chemicals & Polymers; **p.231** *all* Andrew Lambert Photography/Science Photo Library; **p.236** *tl* © Efinfoto – Fotolia.com, *bl & r* Martyn F Chillmaid; **p.237** Martyn F Chillmaid.

Index

Note: page numbers in *italics* indicate references to suggestions for practical work.

absolute temperature 8, 9
abuse of drugs 212, 213
acetic acid (ethanoic acid) 99, 204, 206–7, *244*
acid rain 99, 158, 228
acids 96–8, 110, *241*
   neutralisation 100–1, 108–9
   reaction with alkalis 102–3
   reaction with carbonates 102
   reaction with insoluble bases 103
   reaction with metals 101–2, 129, *240*
   relative strengths 99
acid salts 104, 110
acid soil, neutralisation 115
activation energy 165, 170, 192
   effect of catalysts 166
active sites 168
addiction 212
addition polymerisation 199–200, 213
addition reactions 180
aerobic decay 185, 195
aerobic respiration 122
air
   fractional distillation 151, 159
   percentage of oxygen 150
   reaction with metals 129, 130
alanine 209
alcohols 203, 213
alkali metals 40–2, *239*
alkaline earth metals 42, *239*
alkalis 96, 99, 110, *241*
   reaction with acids 102–3
   see also bases
alkanes 176–9, 182, *244*
   catalytic cracking 180
   see also methane
alkenes 179–81, 182
allotropes 60, 66
   of carbon 60–3
alloys 141, 143, 144
   steel 141–2
aluminium 12, 81, 129, 133
   reaction with halogens *239*
   recycling 138
   uses 83
aluminium oxide 133
   electrolysis (Hall–Héroult process) 81–2
aluminium salts, identification 87, 283
Alzheimer's disease 83
amino acids 209, 213, *245*
ammonia 57, 58, 99, 220–1, *245*
   Haber process 219
   in production of nitric acid 222

test for 283
ammonia solution 102
ammonium chloride 100
   reaction with calcium hydroxide 220
ammonium ion, identification 220
ammonium nitrate 100, 218, 223
ammonium salts, identification 87, 283
amperes 92
amphoteric substances 98, *242*
   hydroxides 133, 144
amylase 203
anaerobic decay 185, 186, 195
anaerobic respiration 205
anhydrite 119
anions 79
   tests for 283
anode 79, 94
anodising 83
aqueous cations, identification 87
argon 46, 149, 151, 152
   arrangement of electrons 36
Arrhenius, Svante 98
artificial fertilisers 155, 218, 223, 224
ascorbic acid (vitamin C) 207
aspirin 207, 212
Aston, Francis 35
atheroma 205
atmosphere 147–9, *243*
   composition 149–50
atmospheric pollution 157–8, 159
atmospheric pressure 4, 9
atomic absorption spectroscopy 16
atomic mass unit (amu) 32, 49
atomic number (proton number, Z) 33
atoms 11, 12, 13, 29, 32
   arrangement of electrons 36–7
   proton and nucleon numbers 33
   structure 32–3
Avogadro's constant 69
Avogadro's Law 72, 77

baking 205
baking soda 104
balanced equations 16
barium 42
barium chloride, reaction with sodium sulfate 103
bases 97–8, 102, 110
   reaction with nitric acid 222
   relative strengths 99
basic oxygen process 142
bauxite 81, 134
beryllium 42
biodegradable plastics 202, 213

biological washing powders 168
biomass and biogas 190
biopolymers 203, 213
biotechnology 205–6, 213
Biuret test 210, 213
blast furnaces 115, 135–6, 144
blood, centrifuging 20
Bohr, Niels 32
boiling points 4, 5, 9, 12
   in assessment of purity 25, *238*
   of atmospheric gases 151
   effect of pressure *238*
bond energy 191–2, 195
bonding
   covalent 56
   ionic 51–3
   metallic 65
bone, composite nature 28
Boyle, Robert 8, 11, 12
Boyle's Law 8, 9
Bragg, Sir William and Sir Lawrence 53
brass 143
brine, electrolysis 86–7
bromides, identification 104
bromine 43–4, 45, 46, 153, *239–40*
   extraction and uses 153
   test for unsaturated compounds 181, 182
Brønsted, Johannes 98
Brønsted–Lowry theory of acids and bases 98
bronze 143
Brownian motion 7, *238*
buckminsterfullerene 63
'bucky balls' 63
Bunsen, Robert 133
butane 176, 177
butanoic acid 206
butene 179

caesium 40
caesium chloride, ionic structure 54
calcium 13, 33, 42, 129
calcium carbonate 100
   calcite 113, 114
   reaction with hydrochloric acid 118, 163–4
   see also limestone
calcium chloride 100
   ionic bonding 53
calcium hydrogencarbonate 118, 119, 120
calcium hydroxide (slaked lime) 102, 116–17

reaction with ammonium chloride 220
calcium oxide (lime) 75, 116
calcium salts, identification 87, 283
calcium sulfate 100
carbohydrates 203, 213, *244*
carbon 12, 13, 33
  allotropes 60–3
  isotopes 34
carbonated drinks 123
carbonates 101, 118, 125, *241*
  identification 105, 110, 283
  reaction with acids 102
  reaction with nitric acid 222
carbon cycle 121, 122
carbon dioxide 59, 121–2, 125, 149, *242*
  laboratory preparation 123
  limewater test 117, 283
  properties 124
  sublimation 5
  uses 123
carbonic acid 124
carboxylic acids 206–8, 213
Carothers, Wallace 200
cast (pig) iron 136
catalase 168
catalysts 165, 166, 170, 222, 230
  enzymes 168–9, 203, 206
catalytic converters 157, 167, 170, 217
catalytic cracking 180, 182
cathode 79, 94
cations 79
  tests for 283
cement 115
centrifuging (centrifugation) 19–20, 29
ceramics 64, 66
CFCs (chlorofluorocarbons) 178–9, 182
Chadwick, James 32
chain reactions 178, 182
  nuclear fission 187, 195
chalcopyrite 134
chalk 113, 114
Charles, Jaques Alexandre César 7
Charles' Law 7–8, 9
cheese making 169
chemical cells 193, 195, *244*
chemical changes 15, 29
chemical energy 191–2, *244*
chemical equilibrium 219, 224
chemical formulae 15–16, 29
  calculation of 73–4
  of ionic compounds 55
Chernobyl power station 188
chlor-alkali industry 86–7
chlorides 101
  identification 104, 110, 283
chlorine 13, 43–4, 45, 46, *239–40*
  isotopes 34

reaction with iodine *245*
reaction with methane 178
test for 283
uses 86
chlorine molecule 56
chloroform (trichloromethane) 178
chlorophyll 122
cholesterol 205
chromatography 23–4, 29
  in assessment of purity 25, *239*
chrome alum crystals 2
chromium 47
citric acid 208
coal 185, 186, 187
cobalt(II) chloride 105
colloids 25–7, 29
colours, transition elements 48
combining power *see* valency
combustion 191–2, 195
  enthalpy of 192, *244*
competition reactions 131–2, 144, *242*
complete combustion 191
composite materials 27–8, 29
compounds 15–16, 29
  calculating moles 70–1, 77
  covalent 57–60
  detection and identification 16–17
  difference from mixtures 17–18, *239*
  ionic 54–5
compressibility 2
concentrated nitric acid 222
concentrated sulfuric acid 231–2, *246*
concentration, effect on reaction rates 164–5
concrete 115
condensation 4, 5, 9
condensation polymerisation 200–1, 209, 213
conductivity 13
conservation of mass, law of 75
constant composition, law of 15, 29
Contact process 166, 229–30, 233
contraction of solids 2
cooling curves 5
copper 12, 47, 129
  corrosion 141
  extraction from ore 137
  purification 88–90
  reaction with nitric acid 222, 223
copper(II) carbonate
  reaction with nitric acid 102
  thermal decomposition 118
copper(II) nitrate, reaction with zinc 132
copper(II) oxide 102
  reaction with sulfuric acid 103, 230
copper(II) sulfate 105–6
  crystal hydrate 105–6
  preparation 103

copper(II) sulfate solution, electrolysis 88
copper nitrate, preparation 102
copper pyrites 134, 137
copper salts, identification 87, 283
corrosion 140–1, 144
  rusting 138–40
coulombs 92, 94
covalent bonding 56, 66
covalent compounds 57–9, *240*
  properties 60
covalent structures 59–60
cross-linking 201, 213
crude oil 173–4, *244*
  fractional distillation 23
cryolite 81, 83
crystal hydrates 105–6, 110, *241*
crystallisation 29, *239*
  fractional 25
  in separation of mixtures 20
crystals 2–3, 53–4
cupro-nickel 143

Dalton, John 11, 32
Davy, Sir Humphry 84, 86, 98, 133
Dead Sea 153
decanting 19, 29
dehydrating agents 231
deliquescence 106
delocalised electrons 61, 65
desalination plants 21
detergents, soapless 209, 214, 232, 233
Devarda's alloy 105
1,6–diaminohexane 200
diamond 60, 61, 62, *240*
diaphragm cell 87
diatomic molecules 14, 29, 56
dibasic acids 230, 233
dichloromethane 178
diffusion 6, 9, *238*
dilute nitric acid 222
dilute sulfuric acid 230–1
dipeptides 209
disaccharides 203
displacement reactions 44, 49, 132, *239*, *242*
dissolving 18, 192
distillation 21, 29, *239*, *244*
  fractional 22–3, 151, 175, *238*, *244*
  of hard water 121
DNA (deoxyribonucleic acid) 211, 213
Döbereiner, Johann 37
dodecane, catalytic cracking 180
dolomite 118
double covalent bonds 59
double decomposition 103, 110
Down's cell 84
drug abuse 212, 213
drugs (pharmaceuticals) 211–12, 213

testing for purity 25
dry ice 123
ductility 13, 65
duralumin 143
dusts, explosion hazard 164
dynamic equilibrium 219

efflorescence 106, 110
Einstein, Albert 32
electrical energy 193–4, *244*
electrodes 79, 94
electrolysis 79, 94, *240–1*
    of aluminium oxide (Hall–Héroult
        process) 81–3
    calculations 92–3
    chlor-alkali industry 86–7
    copper purification 88–90
    of copper(II) sulfate solution 88
    guidelines 90
    of lead(II) bromide 80
    sodium extraction 84–5
    of water 85–6
electrolytes 79, 94, *240*
electron density maps 54, 56
electronic balances 236
electronic structure (electron
        configuration) 37, 49
    of group 0 (noble gases) 46
    of group 1 (alkali metals) 41–2
    of group 7 (halogens) 44
    relationship to periodic table 40
electrons 32–3, 49
    arrangement in atoms 36–7
electron shells 32, 36–7
electrophoresis 24
electroplating 90–1, 94, *241*
electrostatic force of attraction 32, 49
electrovalent (ionic) bonding 51–3
elements 11, 12, 14, 29, *239*
    calculating moles 70, 77
    detection and identification 16–17
    in sea water 153–4
    symbols 13
    see also periodic table
empirical formulae 74, 77
emulsifiers 26–7
emulsions 25, 26–7, 29, *239*
endothermic reactions 192, 195, *244*
end-point of titration 103
energy level diagrams 191, 192
energy levels, electrons 32, 36–7, 49
energy sources
    biomass and biogas 190
    fossil fuels 185–7
    geothermal energy 188
    hydroelectric power 188
    hydrogen 190
    nuclear power 187–8
    renewable and non-renewable 187
    solar energy 189

tidal power 189
    wave power 189
    wind power 189
enthalpy (*H*) 192, 195, *244*
enthalpy changes (*H*) 192, 195
    enthalpy of combustion 192
    enthalpy of fusion and vaporisation
        193
    enthalpy of neutralisation 192
'E-numbers', of emulsifiers 27
enzymes 168–9, 170, 203, 206
equations 16
    and moles 75–6
equilibrium processes 219, 230, *245*
ester links 201
esters 207, 213, *244*
ethane 176
ethanoates 101
ethanoic acid (acetic acid) 99, 204,
        206–7, *244*
ethanol 204, *244*
    fractional distillation 22
    as fuel for cars 190
    production from ethene 180–1
ethene 179
evaporation 4, 9
    separation of mixtures 20
exothermic reactions 18, 29, 106, 191,
        195, *244*
expansion of solids 2

faraday 92, 94
Faraday, Michael 92
fermentation 205
fertilisers 155, 218, 223, 224
fibrous proteins 209
filtrates 19, 29
filtration 19
fire extinguishers 123
flame colours 43
flocculation 27, 29
flue gas desulfurisation (FGD) units
        158
fluorine 33, 43, 46
    comparison with hydrogen 48
foams 25, 26, 27, 29, *239*
formulae *see* chemical formulae
fossil fuels 185–7, 195
'fountain flask experiment' 220, 221
fractional crystallisation 25
fractional distillation 22–3, *238, 239,
        244*
    of air 151, 159
    oil refining 175
fractions of crude oil 175
francium 40
Frasch process 227, 233
free radicals 177, 182
freezing 4, 5
froth flotation 136, 137

fuel cells 194, *244*
fuels 186
    see also energy sources
functional groups 203, 213
fusion, enthalpy of 193, 195

galena 134
galvanising 140
gas, natural 185–6, 187
    see also methane
gases 2, 3
    calculating moles 71–2, 77
    chemical equations 76
    compressibility *237*
    diffusion 6
    separation from air 23
    tests for 283
gas laws 7–9, *238–9*
gels 25, 26, 27, 29, *239*
genetic engineering 206
geothermal energy 188
giant ionic structure 53–4, 66
giant molecular (macromolecular)
        structures 60, 66
glasses 64, 66, 116, *240*
glass-reinforced plastic (GRP) 27
globular proteins 209
glucose 203, *244*
glyceryl stearate, saponification 208
glycine 209
gold 12, 128, 129, 134, 137, 141
graphite 60–1, 62, *240*
graphitic compounds 61
greenhouse effect 122, 125, 179
groups of periodic table 38, 39, 49
    group 0 (noble gases) 46
    group 1 (alkali metals) 40–2
    group 2 (alkaline earth metals) 42
    group 7 (halogens) 43–6
Grove, Sir William 194
gypsum 119

Haber process 165, 166, 218–19
haematite 134, 135
Hall, Charles Martin 81
Hall–H roult cell 81–2, 94
halogenoalkanes 177–8, 182
halogens 43–6, *239–40*
hardness of water 119–21, 125, *242*
hard steel 142
heating curves 5
heat of reaction 192
helium 33, 46, 149, 151, 152
Héroult, Paul 81
hexane 176
hexanedioic acid 200
HIV 212, 213
Hofmann voltameter 85
homologous series 176
hydrate crystals 105–6

hydrocarbons 174, 182
  alkanes 176–9
  alkenes 179–81
hydrochloric acid 43, 99
  reaction with calcium carbonate 118, 163–4
  reaction with metals 129
  reaction with sodium hydrogencarbonate 118
  reaction with sodium hydroxide 100, 103, 108–9
  reaction with sodium thiosulfate 164–5
hydroelectric power (HEP) 82, 188
hydrogen 12, 33
  arrangement of electrons 36
  isotopes 34
  position in periodic table 48
  production from steam re-forming 218
  test for 283
  uses 86
hydrogenation 180
hydrogencarbonates 104, 118, 241–2
hydrogen fuel cells 194
hydrogen halides 43
hydrogen ion donors and acceptors 98
hydrogen molecule 56
hydrogen peroxide, decomposition 9, 166, 168, 243
hydrogensulfates 104
hydrolysis 203, 213
hydrophilic heads, soaps 208–9
hydrophobic chains, soaps 208
hydroxides
  amphoteric 133, 242
  identification of metal ions 132–3
hygroscopic substances 106, 110

ibuprofen 212
immiscible liquids 21, 29
  separation 22
incendiary bombs 131
incomplete combustion 191
indicators 96–7, 110
inert electrodes 79
infrared spectroscopy 16–17
insoluble bases 102
  reaction with acids 103
insoluble salts, preparation 103
insoluble solids 18, 29
instrumental methods of analysis 16–17, 29
intermolecular bonds 59, 66
intimate mixing 6
intramolecular bonds 59, 66
iodides, identification 104–5, 283
iodine 43–4, 45, 46
  reaction with chlorine 245
  sublimation 23

ion exchange 121
ionic (electrovalent) bonding 51–3, 66
ionic compounds 240
  formulae 55
  properties 54
ionic equations 100, 110
ionic structures 53–4
ions 34, 49
  in electrolysis 79
iron 12, 33, 47, 115, 128, 129
  extraction from ores 135–6
  rusting 138–40, 243
  use as catalyst 166
iron(II) sulfate 100, 101, 105
iron(III) chloride, reaction with sodium hydroxide 132
iron(III) oxide 102
'iron problem', water treatment 157
iron salts, identification 87, 283
isomerase 169
isomers 177, 182
isotopes 34–5, 49

kaolinite 64
kiersite 119
kinetic theory of matter 2, 9
Kirchhof, Gustav 133
krypton 46, 149, 151, 152

lattices 53, 66
Lavoisier, Antoine 75, 97
lead 12, 13, 129
lead(II) bromide, electrolysis 80
lead(II) oxide 102
'lean burn' engines 157
Le Chatelier, Henri 219
leguminous plants 217
light, effect on chemical reactions 166
lime (calcium oxide) 75, 116, 125
limestone 113, 125
  direct uses 114–15
  indirect uses 116–17
limewater 117, 124
liquid crystals 4
liquid/liquid mixtures, separation 21–3
liquids 2, 3
  diffusion 6
lithium 40, 42, 239
  arrangement of electrons 36
  comparison with hydrogen 48
litmus 96
locating agents 24
Lowry, Thomas 98
LPG (liquid petroleum gas) engines 157

macromolecular structures 60, 66
magnalium 143
magnesium 12, 33, 42, 129, 130
  extraction and uses 154

reaction with hydrochloric acid 129
reaction with nitric acid 101–2, 222
magnesium chloride 55
magnesium hydrogencarbonate 118, 119
magnesium nitrate, preparation 101–2
magnesium oxide 73–4, 102
  ionic bonding 52–3
magnesium sulfate 100, 105, 106
magnetic separation 17, 23
magnetite 135
malachite 118
malleability 13, 65
maltose 203
manganese(IV) oxide, use as catalyst 166
manganese steel 142
marble 113, 114
margarines 180
mass
  law of conservation 75
  measurement of 236
mass number (A, nucleon number) 33
mass spectrometers 35, 49
matter 1, 9
  kinetic theory 2
  see also states of matter
MAZIT metals 101
measurements 236–7
melting points 4, 5, 9, 12
  in assessment of purity 25, 238
membrane cell 87, 94
Mendeleev, Dmitri 37–8
mercury cell 87
mesosphere 149, 159
metal compounds, action of heat 243
metal ions
  flame colours 43
  identification using aqueous ammonia 221
  insoluble hydroxides 132–3, 144
metallic bonding 65, 66
metalloids 38, 39, 49
metals 12–13, 29, 38, 39, 128, 240, 242
  alloys 141
  corrosion 138–41
  discovery of 133
  extraction from ores 134–7, 144
  properties 65
  reaction with air/oxygen 130
  reactions with acid 129
  reaction with water 130
  recycling 138
methane 57, 176, 177, 178
  combustion 191–2
methanoic acid 206
methyldopa 212
methyl orange 96
methyl red 96
mica 81

mild steel 142
miscible liquids 21, 29
    separation 22–3
mixtures 17–18, 29
    colloids 25–7
    composite materials 27–8
    difference from compounds 239
    separation 18, 239
        liquid/liquid mixtures 21–3
        solid/liquid mixtures 18–21
        solid/solid/mixtures 23–4
molar solution 72
molar volumes 240
molecular formulae 74, 77
molecules 14, 29
moles 69, 77
    and chemical equations 75–6
    of compounds 70–1
    of electrons 92, 94
    of elements 70
    of gases 71–2
    of solutions 72
monatomic molecules 14, 30
monel 47
monomers 199, 213
monosaccharides 203, 213
Montreal Protocol 179
mortar 117
Moseley, Henry 32, 38
motor vehicles
    atmospheric pollution 157–8
    catalytic converters 167

national grid 187
natural gas 185–6, 187
    see also methane
neon 33, 46, 149, 151, 152
neutralisation 100–1, 108–9, 110
    enthalpy of 192
neutrons 32–3, 49
Newlands, John 37
nickel 12, 47
    use as catalyst 166, 218
ninhydrin 210
nitrates 101, 217–18, 222
    identification 105, 283
    water pollution 155
nitric acid 222–3, 245
    reaction with copper(II) carbonate
        102
    reaction with magnesium 101–2
nitrogen 12, 33, 149, 151, 217–18,
    245
    uses 152
nitrogen cycle 217–18, 224
nitrogen fixation 217, 224
nitrogen oxides 157
noble gases 46, 149, 151
    uses 152
non-electrolytes 79

non-metals 12–13, 30, 38, 39
non-renewable energy sources 187
normal salts 100, 110
nuclear fission 187, 195
nuclear power 187–8
nucleon number (A, mass number) 33,
    49
nucleus 32
nylon 200–1
'nylon rope trick' 244

oceans, resources 153–4
Octaves, Law of 37
Oersted, Hans Christian 81
oil formation 185–6
oil refining 174–5, 182
'oil rig' mnemonic 82
oleum 230
optimum temperature 219, 224, 230
ores 133, 144
    extraction of metals 134–7
organic chemistry 174, 182
Ostwald, Wilhelm 222
oxidation 15, 30
    in electrolysis 82, 94
oxidation states 55
oxides 15, 242
oxidising agents 223
oxygen 12, 33, 149
    arrangement of electrons 36
    isotopes 34
    percentage in air 150
    reaction with metals 129, 130
    test for 283
    uses 151
ozone 148–9, 159
ozone layer holes 178–9

paper chromatography 23–4
paracetamol 212
particles
    Brownian motion 7
    kinetic theory 2
particulates, atmospheric pollution
    158, 159
Peak District 113
penicillin 212
pentane 176
peptide bonds 209, 210
percentage yield 77
periodic table 37–9, 49, 284
    group 0 (noble gases) 46
    group 1 (alkali metals) 40–2
    group 2 (alkaline earth metals) 42
    group 7 (halogens) 43–6
    hydrogen 48
    relationship to electronic structure
        40
    transition elements 47–8
periods of periodic table 38, 39, 49

permanent water hardness 119, 120–
    1, 125
Perspex 200
petrol 175
pewter 143
pH 97, 110
pharmaceuticals (drugs) 211–12, 214
    testing for purity 25
phenolphthalein 96
photodegradable plastics 202, 214
photographic film 166
photosynthesis 122, 125, 148, 166, 192
physical changes, changes of state 4
pig (cast) iron 136
planets, composition of atmospheres
    150
plastics 60, 244
    addition polymerisation 199–200
    condensation polymers 200–1
    disposal 202
    electroplating 91
    poly(ethene) 198–9
    thermosoftening and thermosetting
        201, 214
platinum 129, 141
    use as catalyst 166
pollution 159
    atmospheric 157–8
    from aluminium plants 83
    of water 155–6, 223
polyamides 200
polyester 201
poly(ethene) (polythene) 198–9, 200
polymerisation 199, 214
polymers 199, 214, 244
    biopolymers 203
    see also plastics
poly(propene) 200
polysaccharides 203, 214
polystyrene 200
potassium 33, 40, 41, 42, 129, 130, 239
    arrangement of electrons 36
potassium dichromate(VI) 204
potassium hydroxide 102
potassium nitrate 100
powders, reaction rates 164
power stations 186–7
    atmospheric pollution 158
practical assessment 237
pressure 7
    effect on boiling points 238
    effect on equilibrium processes 219,
        230
    effect on reaction rates 165
    gas laws 8–9
primary atmosphere 147, 159
products 69
propane 176, 177
propanoic acid 206
propene 179

proteases 169
proteins 209–10, 214, 217
proton number (Z, atomic number) 33
protons 32–3, 49
PTFE (poly(tetrafluoroethene)) 199–200
purity of substances 25, 238
PVC (polyvinyl chloride) 199, 200
Pyrex 64

quartz 60

radioactive waste 188
radioactivity 34, 49
radioisotopes 34, 49
radium 42
radon 46
Ramsay, Sir William 46
Rayleigh, Lord John William Strutt 46
reactants 69
reacting masses 68–9
reaction rates 162, 170, 243
    effect of catalysts 166–9
    effect of concentration 164–5
    effect of light 166
    effect of pressure 165
    effect of surface area 163–4
    effect of temperature 165
reactivity of halogens 44
reactivity of metals 128–30
reactivity series 129, 144
    competition reactions 131–2
recycling 138, 142, 144, 202
redox reactions 15, 30, 124, 132, 237, 240, 241, 242, 243, 244
    Thermit reaction 131
reduction 15, 30
    in electrolysis 80, 82, 94
relative atomic mass (A_r) 35, 49, 68, 70
relative formula mass (RFM) 68, 77
relative molecular mass (M_r) 70
renewable energy 187
residues 19, 30
respiration 149, 205
reversible reactions 99, 106, 218, 224, 245
R_f values, chromatography 24, 30
rock salt 134
rubidium 40, 133
rusting 138, 144, 243
    prevention 138–40
Rutherford, Ernest 32, 38

sacrificial protection of steel 140
safety 237
salt
    separation from sea water 20, 21, 154
    see also sodium chloride
Salters' 'Duck' wave machine 189

salt hydrates 105–6, 110
salts 100–1, 104, 241
    identification 104–5
    insoluble, preparation 103
    solubility 107
    soluble, preparation 101–3
saponification 208
saturated molecules 176
saturated solutions 20, 30
scuba diving 8
sea water, extraction of elements 153–4, 159
secondary atmosphere 148, 159
separating funnels 22
separation of mixtures 239
    liquid/liquid mixtures 21–3
    solid/liquid mixtures 18–21
    solid/solid mixtures 23–4
sewage treatment 157
silver 12, 129, 134, 137
silver bromide 100, 101, 166
simple molecular structures 59, 66
slag 115, 135, 136
slaked lime (calcium hydroxide) 102, 116–17
    reaction with ammonium chloride 220
soapless detergents 209, 214, 232, 233
soaps 208–9, 214, 245
    effect of hard water 120
soda glass 64, 116
sodium 13, 33, 40, 41, 128, 129, 130, 239
    arrangement of electrons 36
    extraction by electrolysis 84–5, 134
    uses 85
sodium alkyl benzene sulfonates 209
sodium carbonate 100, 105, 118
    Solvay process 115
sodium chlorate(I), decomposition 243
sodium chloride 100
    electrolysis 84–5, 86–7
    extraction from sea water 20, 21, 154
    formula 55
    ionic bonding 51–2
    ionic structure 54
    preparation 103, 108–9
    solubility 107
sodium hydrogencarbonate 118
sodium hydrogensulfate 105
sodium hydroxide 99, 102
    identification of aqueous cations 87
    identification of metal ions 132–3
    reaction with hydrochloric acid 100, 103, 108–9
    reaction with sulfuric acid 230, 231
    uses 86
sodium ions 41
sodium stearate 100, 120, 208

sodium sulfate 105
    reaction with barium chloride 103
sodium thiosulfate, reaction with hydrochloric acid 164–5
solar energy 189
solder 143
solid/liquid mixtures, separation 18–21
solids 2, 3
    solubility 18, 30
solid/solid mixtures, separation 23–4
sols 25, 26, 30, 239
solubility 107, 110
    of salts 241
solubility curves 107, 110
soluble bases 102
soluble salts, preparation 101–3
soluble solids 18, 30
solutes 18
solutions 18, 27, 30
    calculating moles 72, 77
    chemical equations 76
Solvay process 115
solvent extraction 24
solvents 18
SØrenson, SØren 97
sparingly soluble substances 103
spectator ions 100
spectroscopy 16–17
spot tests 104–5
stainless steel 142
stalactites and stalagmites 120
starch 203, 244
states of matter 1–3, 9, 237
    changes of state 4, 5, 237–8
        enthalpy changes 193
    liquid crystals 4
steam re-forming 218
steel 115
    different types 142
    production 141–2
    recycling 142
    rust prevention 138–40
steroids 205
stopwatches 236
stratosphere 149, 159
strong acids 99, 110
strong alkalis 99, 110
strontium 42
    isotopes 34
structural isomerism 177
sublimation 5, 9, 12, 238
    iodine 23
substitution reactions 178
sucrose, reaction with concentrated sulfuric acid 231
sulfates 101, 230, 231, 233
    identification 104, 110, 283
sulfur 12, 33, 227–8, 245
sulfur dioxide 158, 228–9
sulfuric acid 229–32, 245–6

reaction with copper(II) oxide 103
supercooled liquids 64
surface area, effect on reaction rates
    163–4
suspensions 19, 66
symbols for elements 13, 14

tapping off, blast furnaces 135
temperature
    during changes of state 4, 5
    effect on equilibrium processes 219,
        230
    effect on reaction rates 165
    in gas laws 7–9
    measurement of 236
temporary water hardness 12, 119,
    120
terylene 201
theoretical yield 77
thermal decomposition 116, 125, 192,
    241
Thermit reaction 131, 242
thermometers 236
thermosoftening and thermosetting
    plastics 201, 214
thermosphere 149
Thomson, Joseph 32
tidal power 189
time, measurement of 236
tin(II) fluoride 100
titration 102–3, 108–9, 110, 241
transition elements 38, 39, 47–8, 240
trends 40

triads of elements 37
troposphere 149, 159
tungsten steel 142

universal indicator 97
unsaturated fats 180
unsaturated molecules 179
    testing for 181, 182
uranium-235, fission 187
uranium, isotopes 34

valency (combining power) 55, 66
Valium 212
vanadium(V) oxide, use as catalyst
    166, 230
van der Waals' bonds (forces) 59, 61
vaporisation, enthalpy of 193, 195
verdigris 141
vitamin C (ascorbic acid) 207
volume
    in gas laws 7–9
    measurement of 237
von Liebig, Justus 98
vulcanisation of rubber 228

washing soda 120
water 15, 57–8, 119, 243
    electrolysis 85–6
    hardness 119–21, 125, 242
    reaction with alkali metals 40, 41
    reaction with alkaline earth metals
        42
    reaction with metals 129, 130

water of crystallisation 105, 110
    calculation 106
water cycle 155, 159
water pollution 155–6, 159, 223
water treatment 27, 156
    'iron problem' 156
wave power 189
weak acids 99, 110
weak alkalis 99, 110
weak electrolytes 79
wind power 189
Winkler, Clemens 38
Wöhler, Friedrich 81, 133
wood, composite nature 28

xenon 46, 149, 151, 152
X-ray diffraction (crystallography) 3,
    53–4, 66

yeast 205
yoghurt production 169

zinc 12, 33, 47, 129
    extraction from ore 136
    reaction with copper(II) nitrate
        132
    reaction with sulfuric acid 230–1
zinc blende 134
zinc hydroxide 133
zinc oxide 133
zinc salts, identification 87, 283